I0043619

Archibald Henry Swinton

Insect Variety - its Propagation and Distribution

Treating of the odours, dances, colours, and music in all grasshoppers, Cicadae, and

moths

Archibald Henry Swinton

Insect Variety - its Propagation and Distribution
Treating of the odours, dances, colours, and music in all grasshoppers, Cicadae, and moths

ISBN/EAN: 9783337088187

Printed in Europe, USA, Canada, Australia, Japan

Cover: Foto ©berggeist007 / pixelio.de

More available books at **www.hansebooks.com**

SEASONAL DIMORPHISM IN BUTTERFLIES.
(Araschnia Prorsa).

Fig. 1. Levana, spring brood.   Fig. 2. Prorsa, summer brood.

INSECT VARIETY:

ITS PROPAGATION AND DISTRIBUTION.

TREATING OF THE ODOURS, DANCES, COLOURS, AND MUSIC IN ALL GRASSHOPPERS,
CICADÆ, AND MOTHS; BEETLES, LEAP-INSECTS, BEES, AND BUTTERFLIES;
BUGS, FLIES, AND EPHEMERÆ; AND EXHIBITING THE BEARING
OF THE SCIENCE OF ENTOMOLOGY ON GEOLOGY.

BY

A. H. SWINTON,

MEMBER OF THE ENTOMOLOGICAL SOCIETY OF LONDON.

"Les passions sont les vents, qui font aller notre vaisseau, et la raison est le pilote
que le conduit. Le vaisseau n'irait pas sans les vents, et se perdrait sans le pilote."
—ESPRIT DES ESPRITS.

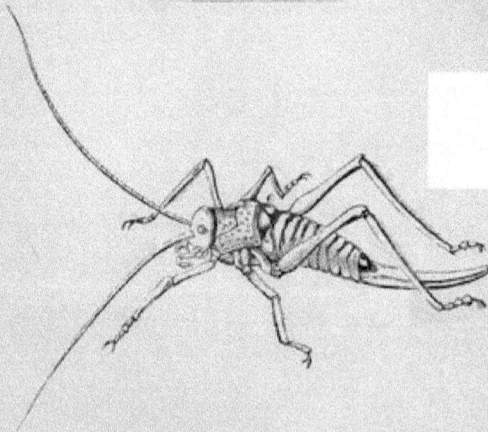

CASSELL, PETTER, GALPIN & CO.:
LONDON, PARIS & NEW YORK.

CONTENTS.

INTRODUCTION 1

CHAPTER I. — Metaphysical Incentives to Reproduction and Distribution —The Germs of the Passions, Fear, Rivalry, Love, and Maternal Care, shown to exist in Insecta : and their Expression in Contractions and Secretions—Corresponding Senses of Touch and Smell ... 15

APPENDIX. — Tabular View of the Secretions of Insects 70

CHAPTER II. — Display and Dances Material Agents in Reproduction and Distribution—Corresponding Sense of Sight ... 76

APPENDIX. — A List of Luminous Insects 101

CHAPTER III.—Instrumental Music considered as a Material Agent in Reproduction and Distribution 102
    Stridulation of the Neuroptera 105
    Stridulation of the Hymenoptera ... ... ib.
    Stridulation of the Hemiptera 108
    Stridulation of the Lepidoptera 112
    Stridulation of the Coleoptera 127

CHAPTER IV.—Instrumental Music considered as a Material Agent in Reproduction and Distribution 149
    Stridulation of the Orthoptera ... ... ib.
    Of the Locustina (Leaf-crickets) 153
    Of the Acridiidæ (Locusts and Grasshoppers) 164
    Of the Gryllidæ (Crickets) 182
    Of the Gressoria 190
    Of the Cursoria 191

APPENDIX. — Tabular View of Genera containing Insects that Stridulate 194

CHAPTER V. — Wing Beating and Vocal Music considered as a Material Agent in Reproduction and Distribution ... 207
    Music of Flies, Bees, and Cicadæ, &c. ... ib.

CHAPTER VI.—The Organ of Hearing in Insecta 230
    Organ of Hearing in the Cicadidæ 231
    Organ of Hearing in the Acridiidæ 234
    Organ of Hearing in the Locustidæ and Gryllidæ 239
    Organ of Hearing in the Lepidoptera 243
    Organ of Hearing in the Coleoptera 249

CHAPTER VII.—The foregoing Phenomena supplemented by Migration, which induces Variation and Natural Selection.. ... 253

APPENDIX. — A Table of Remarkable Cycles of Insect Multiplication and Migration 318

A 2

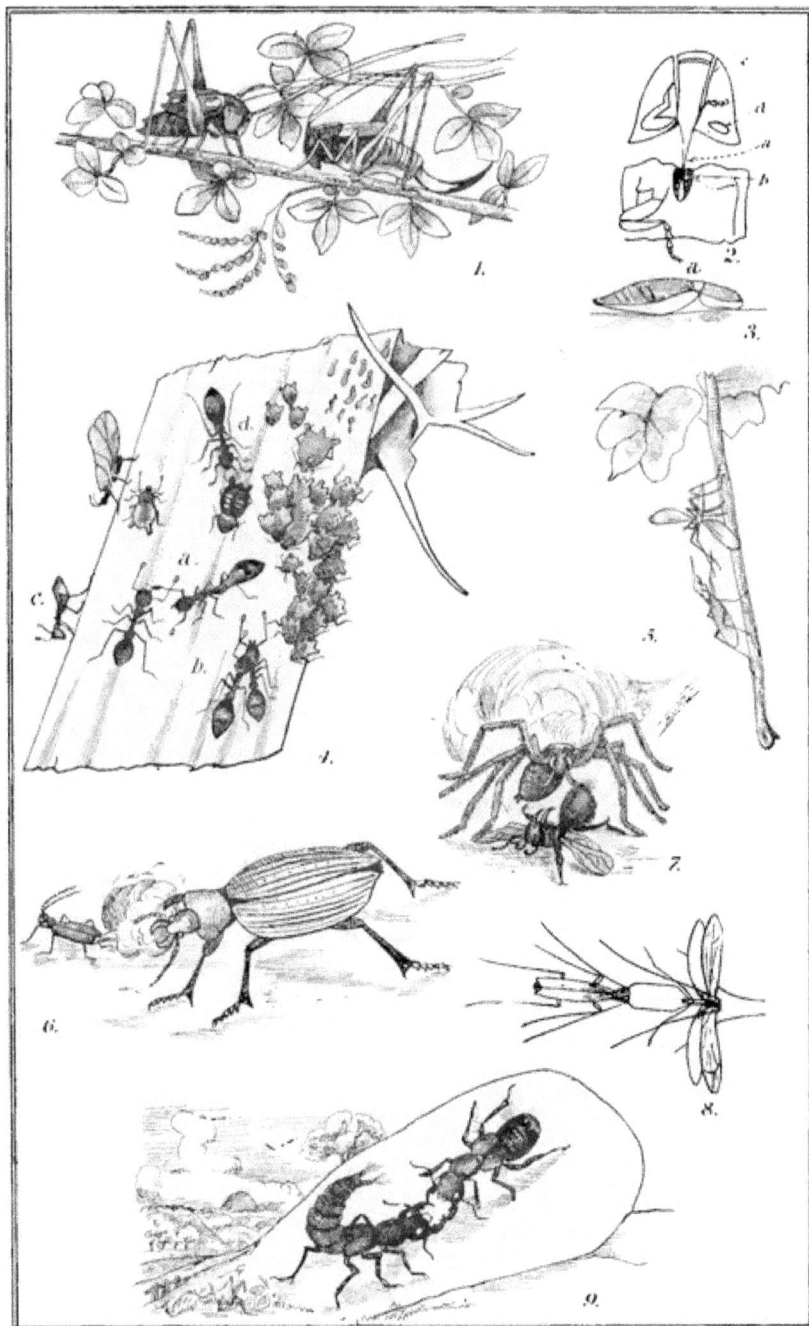

1. 2. 3. 4. 5. 6. 7. 8. 9.

EXPLANATION OF PLATES.

—◆◇—

FRONTISPIECE.

SEASONAL VARIETIES OF A BUTTERFLY (ARASCHNIA PRORSA).

Fig. 1. The spring variety, *Levana*, of the butterfly.
Fig. 2. The summer variety, *Prorsa*.

PLATE I.

THE EMOTIONS THAT SWAY THE INSECT WORLD, AND THEIR MEANS OF COMMUNICATION ILLUSTRATED.

Fig. 1. *Decticus Brachyptera*, a male Leaf-cricket, encountering his female, and greeting her with a serenade.

Fig. 2. Under surface of the anterior extremity of the body of a coleopteron of the *Elateridæ* (magnified), showing the grooves, *c, d,* that receive the antennæ and legs; and also the spine, *a,* a posterior prolongation of the prothorax, which on being rapidly jerked into a groove, *b,* beneath the metathorax, by reason of ts elasticity, propels these click-beetles into the air in the fashion of a somersault.

Fig. 3. Shows the same beetle fallen on its back, preparing to right itself by a spring.

Fig. 4. A group of Ants, *Myrmica ruginodis,* as seen through a glass on a flower-head of the *Cnicus arvensis,* or common thistle: eagerly engaged in the sun, sucking sap exuding from the teat-like ducts of the aphides or blight. *a,* Two communicating by touch of the antennæ; *b,* others quarrelling and attacking with their jaws; *c,* one stridulating or emitting music by a vertical movement of the abdomen; *d,* one sucking the sap exuding from the aphis.

Fig. 5. The courtship of two flies (*Empis ignota*) on a crow's-foot, exemplifying communication by the tarsi of the feet.

Fig. 6. Explains the manner in which certain small beetles of the tribe *Geodephaga* and genera *Brachinus* are asserted to arrest the onset of a larger predacious beetle, by exploding a volatile secretion from the anus.

Fig. 7. A Bumble Bee defending itself from the attacks of a House Spider.

Fig. 8. Shows the way in which the male *Trichoptera* choose their partners when dancing over the surface of the water.

Fig. 9. Two Rove-beetles (*Goërius*) as they stand on the defensive with open jaws, protruding two tubular ducts from the anal extremity of the reversed abdomen, and diffusing a pungent vinegary secretion.

PLATE II.

Fig. 1. A, Internal anatomy of a common Green Grasshopper, *Stenobothrus viridulus*. *t, t*, are the main tubes of the trachea, which pass down the line of spiracles or external respiratory openings at either side of the body ; *t', t', t'*, central air bladders with which they are in communication ; *m, m'*, membrana tympana, or drum of the organ of audition ; F, an opening suitable to act as an Eustachian tube ; n is a more highly magnified view of the membrana tympana, *m*, showing the horny pieces on its disc, the inwardly attached vesicle containing a watery fluid, and the acoustic nerve, *n*, issuing from beneath it ; E, as before.

Fig. 2. A, Shows one of the double rimae or slits in the tibiae of the forelegs of *Locusta viridissima*, the great Green-leaf Cricket ; B, shows, *m*, a membranum tympanum of the auditory apparatus, exposed by the removal of the operculum to this cavity ; *c*, E, is a tube of the trachea that penetrates the forelegs of this insect, and swells out into a vesicle opposite the tympana, *m* ; *n* is the acoustic nerve attaching itself to the vesicle in a ganglion. (These figures are reduced from Fischer's " Orthoptera Europea.") D, shows the position of E and *a*, *in situ*.

Fig. 3. *Acridium migratoria*, the Migratory Locust. *a*, the entrance of the auditory cavity ; the blur, *a', b'*, indicates the movement of the hind leg in stridulation ; *x* and *y* are the veins on which the lima, or file, is mostly placed.

Fig. 4. A, Lima of *Acheta domesticus*, the House-cricket ; *B*, lima of *Gryllus campestris*, the Field-cricket (from Landois) ; c, is the file of *Locusta viridissima*. (All highly magnified.)

Fig. 5. Anal fields of the wing-covers of *Locusta viridissima*. A, under surface of the anal field of right elytron ; *sm*, mirror ; *s*, rounded edge, over which the lima or file, *l*, on the under surface of the anal field of the left elytron, B, passes during the music.

Fig. 6. Femoral joint of the hind or leaping legs of the *Stenobothrus viridulus*, the common Green Meadow Grasshopper, showing the lima or file on the inner side, *l*, employed by the insect in stridulating or singing. (Highly magnified.)

Fig. 7. The lima in *Stenobothrus pratorum* more highly magnified (from Landois).

Fig. 8. Right wing-cover of *Gryllus campestris*, the Field-cricket. Under side, showing the vein carrying the lima or file, *l*, with which the insect sings ; *s*, a veinlet raised at the upper side of the elytra, over which the lima passes when the wing-covers, or elytra, are rubbed together in the action of stridulation ; *sm*, a little glassy space thought to be the undeveloped counterpart of the mirror of the Leaf-crickets. (Magnified.)

PLATE III.

Fig. 1. View of the ventral surface of *Geotrupes stercorarius*, the common Watchman Beetle. *l*, Lima on the coxal joint of the hind legs ; *s*, faint groove,

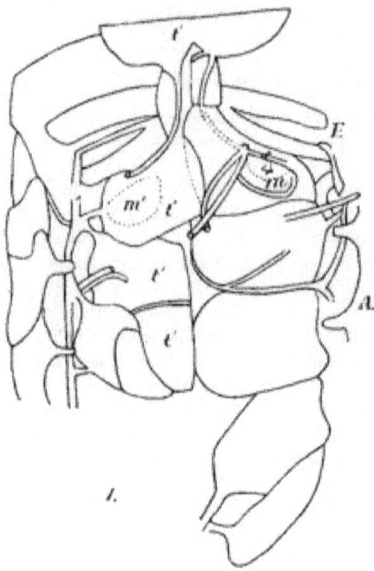
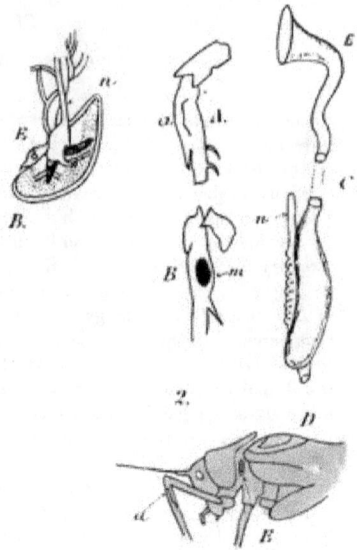
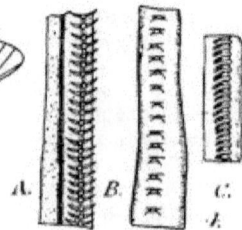
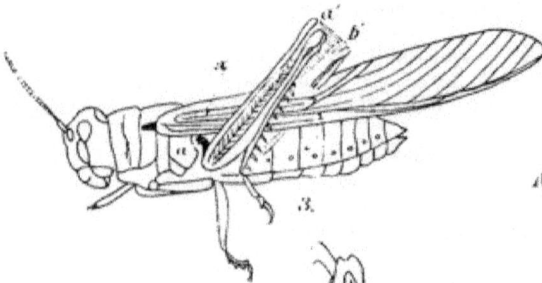
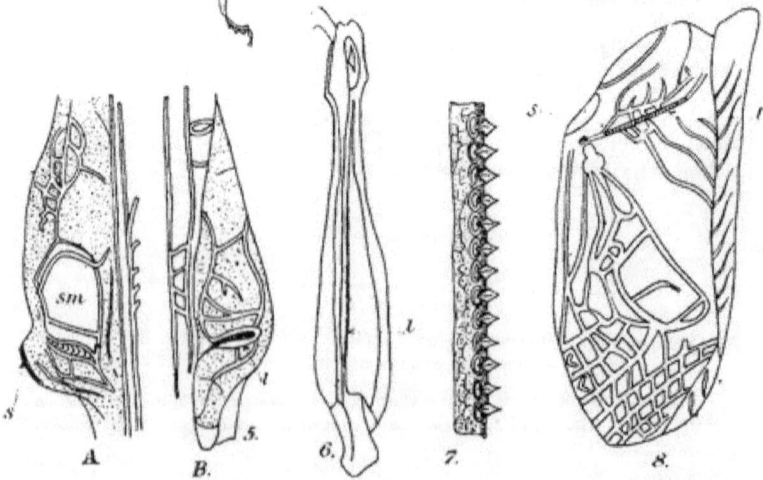

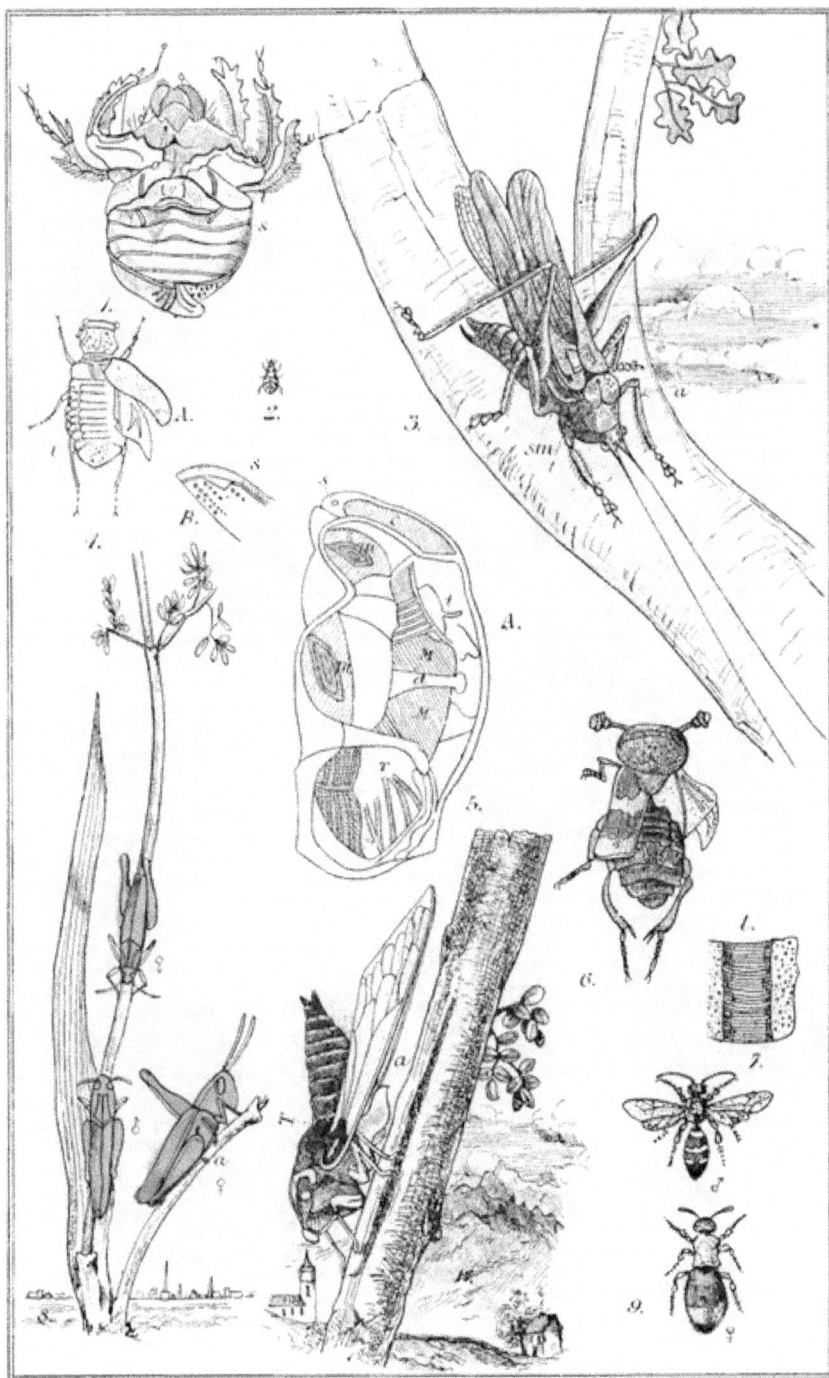

indicating the direction in which *l* moves on the abdomen when the beetle creaks. (Magnified.)

Fig. 2. *Acalles misellus*, a minute stridulating weevil. (Natural size.)

Fig. 3. *Locusta viridissima*, the Great Green Leaf-cricket. *sm*, Mirror of the organ of stridulation; *t*, mouth of the prothoracic spiracle; *a*, slit or rima on the tibiæ.

Fig. 4. A, Magnified view of *Elaphrus riparius*, right elytron and wing removed, left elytron and wing expanded; *l*, musical lima. B, Under surface of right elytron, showing a protuberance, *s*, that acts as clasp to *l*.

Fig. 5. A, Abdomen of a male cicada dissevered from the thorax and magnified. The species (*Cicada Plebeja*, Oliv.) being cryptotympanous, or having its drum-covers concealed by a lap of the dermis, this skin lap, *e*, has been cut away from the left drum, so as to expose the ribbed membrane, *r*. To the point indicated, *p*, the tendon *t* of the motor muscle M is attached, and by this means the membrane is drawn inward during the music, the sound resulting from its vibration on each rebound. *d*, Shows the diaphragm separating the large internal air cavity; *s*, the opening of the first abdominal spiracle; *m*, mirrors of the cicada, showing an iridescent spot of various colours centrally, to which a little styliform thickening proceeds from their margin. The internal aspect of this part, which has the essentials of an insect ear, is shown on Plate VI., Fig. 5. B, *Cicada Plebeja* singing on an acacia spray at the foot of the Superga Hills, near Turin. Motion of the abdomen indicated by a blur. Sketch from nature, with the Alps in the background.

Fig. 6. A magnified representation of *Necrophorus vespitillo*, a common Gravedigger Beetle, with the left elytron removed. *l*, The lima; *s*, the extremity of the elytron that acts as clasp.

Fig. 7. *l*, A magnified view of the lima of *Necrophorus vespitillo* (after Landois).

Fig. 8. Illustration of the biology of the grasshoppers. ♂, Males stridulating in rivalry; ♀, female attracted by their music; *a*, shows the lunate opening of the organ of audition of Müller and Siebold. (Sketched in the fields at Maida Vale.)

Fig. 9. *Mutilla Europea*, a musical Solitary Ant. ♂, Winged male; ♀, apterous female.

PLATE IV.

ORGANS OF STRIDULATION AND AUDITION OF THE LEPIDOPTERA, ETC.

Fig. 1. Organs of instrumental music of *Vanessa Io*, the Peacock Butterfly *l*, The submedian or anal vein of the forewing, or that nearest its inner margin. Its filed aspect at the under surface of the wing when submitted to a strong magnifying power is indicated. *s*, Costal vein of the hind wing, over which the vein *l* plays its lima when the insect rubs its wings together in stridulation; it is prominent on the upper surface of the wing. *sm*, Shows a raised pucker at the base of the hind wing devoid of scales, probably serviceable in impressing the vibrations caused by the friction of the veins on the ambient air. Fig. 12 shows how the wings overlap at the basal point of friction, and illustrates the contrast in the dark under and brilliant upper surface of the wings of this butterfly.

Fig. 2. *Dicrorampha sequana*, a small moth of the blunt-winged group of the Tortricina. (Natural size.)

Fig. 3. A, *Halias prasinana*, the common Silverlines Moth; n, a callosity running along the under side of the inner edge of the forewing at its base before the elbow, that locks the unexpanded forewings to a side piece of the metathorax, *l*. The moth is musical, and stridulates on the wing.

Fig. 4. Shows the filing on the inner surface of the lower joint of the labial palpi, *l*, and the appearance of the adjacent surface of the proboscis, *s* (magnified); by the mutual friction of which *Acherontia atropos*, the Death's-head Moth, produces its squeak.

Fig. 5. *Bombus lapponicus*, a bee common to Europe and North America.

Fig. 6. Organ of audition of *Catocala Nupta*, the Red Underwing Moth. *a*, Entrance of the ear, closed inwardly by a tender iridiscent membrane, *m*, representing the membranum tympanum. Within the tympanum or drum is a second acoustic chamber, *b*, in which is observed an inflated membrane, attached to the tympanum by two arms, one of which takes its rise from a horny piece on the margin; from the other end of this vesicle proceeds the acoustic nerve, *n*. There is an interior chamber, *c*, parted from *b* by an oval iridiscent membrane, and a third most interior, *d*. A tube answering to the Eustachian communicates air to the cells, *c*, *d*, *b*, successively, counteracting the atmospheric pressure on the tender tympanum, *m*. In Fig. 8 of Plate V. are seen the external openings of the ear and Eustachian tube, E, *in situ*. (Highly magnified.)

Fig. 7. *Sericomya borealis*, a Hover-fly.

Fig. 8. The Death's-head Moth, *Acherontia Atropos*. *l*, Lima; *s*, clasp.

Fig. 9. Another preparation of the lepidopterous ear, like the preceding, highly magnified; introduced to show its *accessories* in *Xylophasia polyodon*, the common Dark Arches Moth, and in most of the *Noctuina*; *a*, *c*, *d*, *m*, as before; *c*, *e*, two large air-bladders communicating with the trachea, into which the exterior boundary, *b*, of the auditory apparatus projects; *m*, *m'*, are two iridiscent membranes in the common partition of the chambers, *c*; and separating the chambers or cells, *b*, *c* (Fig. 6); *w*, is a little projecting piece, protecting the meatus externus, or entry of the ear; *v*, a movable membranous valve; E, shows the course of the counterpart of the Eustachian tube; *f*, fan of the male moth; *r*, its containing fold.

Fig. 10. The Lily Beetle, *Crioceris merdigera*. A, The coleopteron; n, the same magnified to show its musical apparatus; *l*, abdominal lima; *s*, elytral clasp.

Fig. 11. Shows the singular vesicle beneath the hind femora in *Chelonia pudica*, a musical moth. *l*, Striæ and raised impressions on its membrane; *s*, femur.

Fig. 12. *Vanessa Io*, the Peacock Butterfly.

Fig. 13. *Coranus subapterus*, covered with dust. *s*, Rostrum and its striated groove.

PLATE V.

ORGANS OF VOCAL MUSIC IN DIPTERA AND HYMENOPTERA, ANTICS OF A DIPTERON, SEXUAL SECRETIONS OF LEPIDOPTERA, ETC.

Fig. 1. Love and rivalry, traits in the biology of *Dolichopus nobilitatis*, à little fly.

Fig. 2. A, Stridulating organ of *Clytus arietis*, a Longhorn Beetle; *m*, *m*,

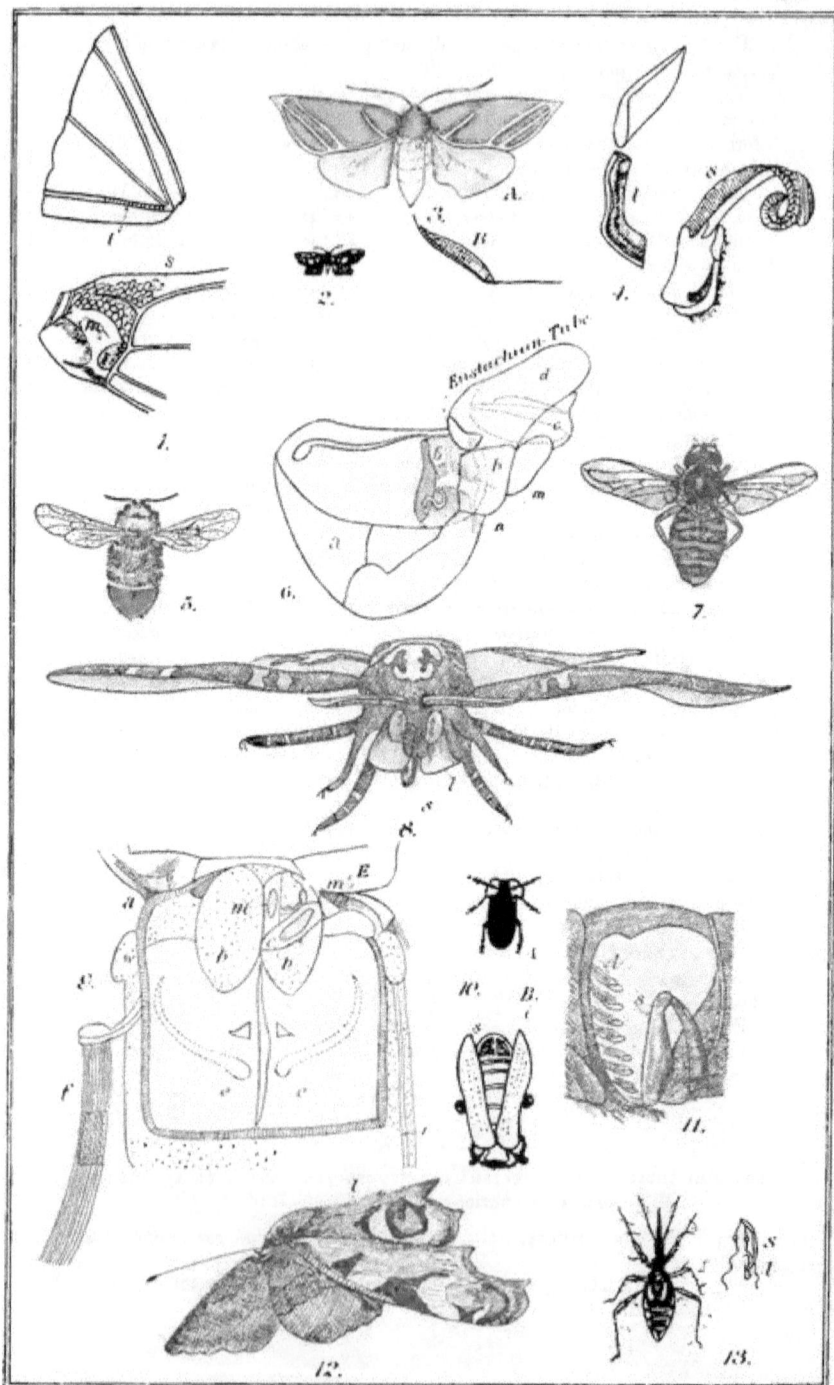

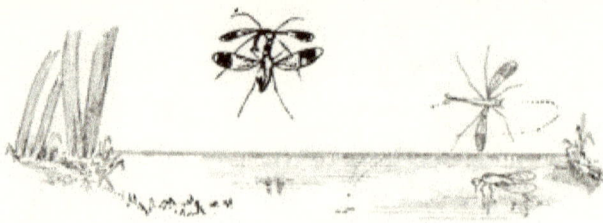

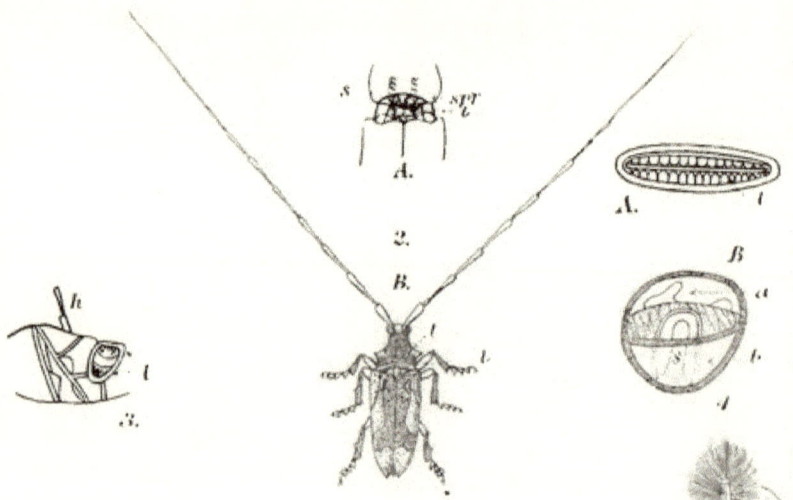

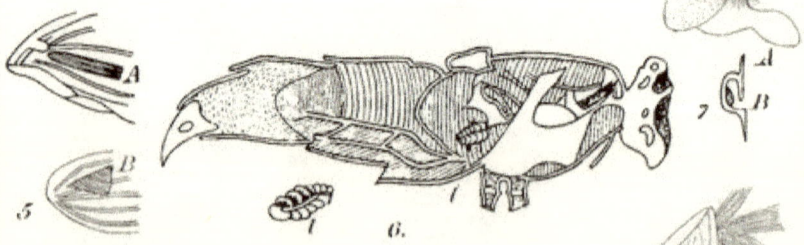

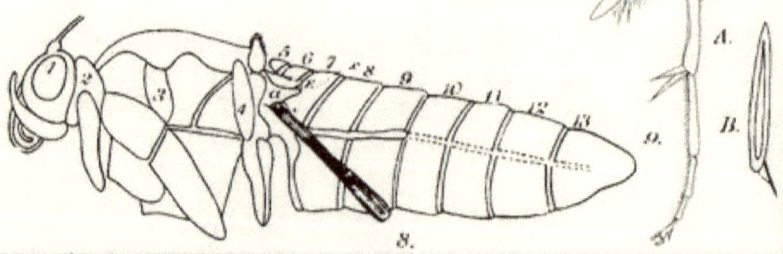

muscles moving the prothorax ; *l*, lima on metathorax ; *s*, clasp ; *spr*, prothoracic spiracle. B, *Astynomus ædilis*, the Woodman Beetle ; *t*, prothorax ; *l*, lima.

Fig. 3. Metathoracic spiracle of *Musca vomitoria*, the Blue-bottle Fly. *l*, Laminæ on the operculæ of the spiracle ; *h*, haltera (internal aspect).

Fig. 4. A, Metathoracic spiracle of *Melolontha vulgaris*, the common Cockchafer ; *l*, undulated lips or operculæ. B, Metathoracic spiracle of the Humble Bees (*Bombus*), internal structure ; *s*, external opening of the spiracle ; *b*, cup-shaped cavity ; *a*, movable internal opercula. (After Landois.)

Fig. 5. A, Fan of the male *Cidaria prunata* ; B, fan of the male *Cidaria populata*. In these moths, belonging to the Geometrina, the fans are inserted at the base of the fore wings.

Fig. 6. Internal view of the integument of *Heliophilus pendulus*, a Hover-fly. *l*, Laminæ at the opening of the mesothoracic spiracle. These are again shown beneath, as seen under a somewhat greater magnifying power.

Fig. 7. A, Fan of the male Death's-head Moth, *Acherontia Atropos*, expanded ; B, the same lying in its fold.

Fig. 8. Exterior anatomy of *Mamestra Brassicæ*, the Cabbage Moth. 1, The head ; 2 to 4, the segments of the thorax ; 5 to 13, those of the abdomen ; *a*, entrance or adit of the auditory organ ; E, that of the Eustachian tube ; *x*, fan of the male, showing the portion stained by the secretion these organs diffuse darker, as likewise their point of insertion.

Fig. 9. Fan of the male *Catacala Nupta*, or common Red Underwing Moth. inserted at the upper extremity of the tibia. A, Fan expanded ; B, the groove into which it shuts.

PLATE VI.

ORGANS OF AUDITION IN ORTHOPTERA AND OF STRIDULATION IN COLEOPTERA AND HYMENOPTERA, DRUM OF THE CICADIDÆ, ORGAN OF SMELL IN BEES AND GNATS, ORGANS OF CIRCULATION. (VARIOUSLY MAGNIFIED.)

Fig. 1. The right ear of the grasshopper, *Stetheophyma grossum*, from without. *a*, The opening in the horny ring ; *b*, *b'*, composite horny piece seen through the membrane of the drum ; *c*, the triangular horny piece seen through the membrane of the drum. (From a paper by Von Siebold.)

Fig. 2. The Müllerian ganglion of *Truxalis nasuta*, L., from China. Isolated and prepared with potash, and tinted with carmine. (From a paper by Herr Vitis Graber.) · *m*, Rod-shaped corpuscles : *l*, hexagonal cells.

Fig. 3. Part of the antennæ of a Hive Bee, *Apis mellifica*, and of a Gnat. *Ctenophora bimaculata*, showing the nerve, *n*, ramifying to the pores, *p*. (From Dr. Hick's paper in the "Transactions of the Linnæan Society.")

Fig. 4. Drum of a male cicada. *s*, Metathoracic spiracle ; *p*, point of insertion of motor tendon.

Fig. 5. Mirror of a male cicada from within. *n*, Supposed acoustic nerve.

Fig. 6. Undeveloped organ of stridulation in a female Leaf-cricket, *Decticus verrucivorus*.

Fig. 7. Organ of stridulation in the Ant, *Myrmica ruginoides*. *p*, Pedicle : *a*, abdomen ; *l*, striæ.

Fig. 8. Organ of stridulation in a Solitary Ant, *Mutilla*. *p*, Pedicle ; *a*, second joint of abdomen ; *l*, lima or file.

Fig. 9. Interior view of the structure at the base of the abdomen in the

·

Crickets, *Gryllidæ*, from within.   τ, Membrane ; *m*, muscle attached ; *n*, nerve serving it with ramifications.   (From Herr Vitis Graber.)

Fig. 10. View of the membrane, τ, from without ; after Dr. Landois.

Fig. 11. System of circulation in the abdomen of a Privet Hawk Moth, *Sphinx Ligustri.* n, Dorsal canal ; c, ventral canal ; A, vessel in which they unite ; *a*, afferents ; s, intestines ; *t, t, t,* tracheæ ramifying around the lower vessel.

Fig. 12. Organ of stridulation in the Chafer, *Serica brunnea.* *l*, Lima.

PLATE VII.

THE NERVOUS SYSTEM IN THE VARIOUS ORDERS OF INSECT.

Fig. 1. Nervous system in *Cicada orni;* after Léon Dufour.   n, Auditory nerve.

Fig. 2. Nervous system in the Grasshopper, *Epacromia Thalassina.* n, Auditory nerve.

Fig. 3. Nervous system in the Red Underwing Moth, *Catacala Nupta.* n, Auditory nerve.

Fig. 4. Nervous system in a Dung Beetle, *Geotrupes stercorarius.* n, Auditory nerve.

Fig. 5. Nervous system in Drinker Moth Caterpillar, *Odonestis potatoria.* Dotted Ganglions.   ·

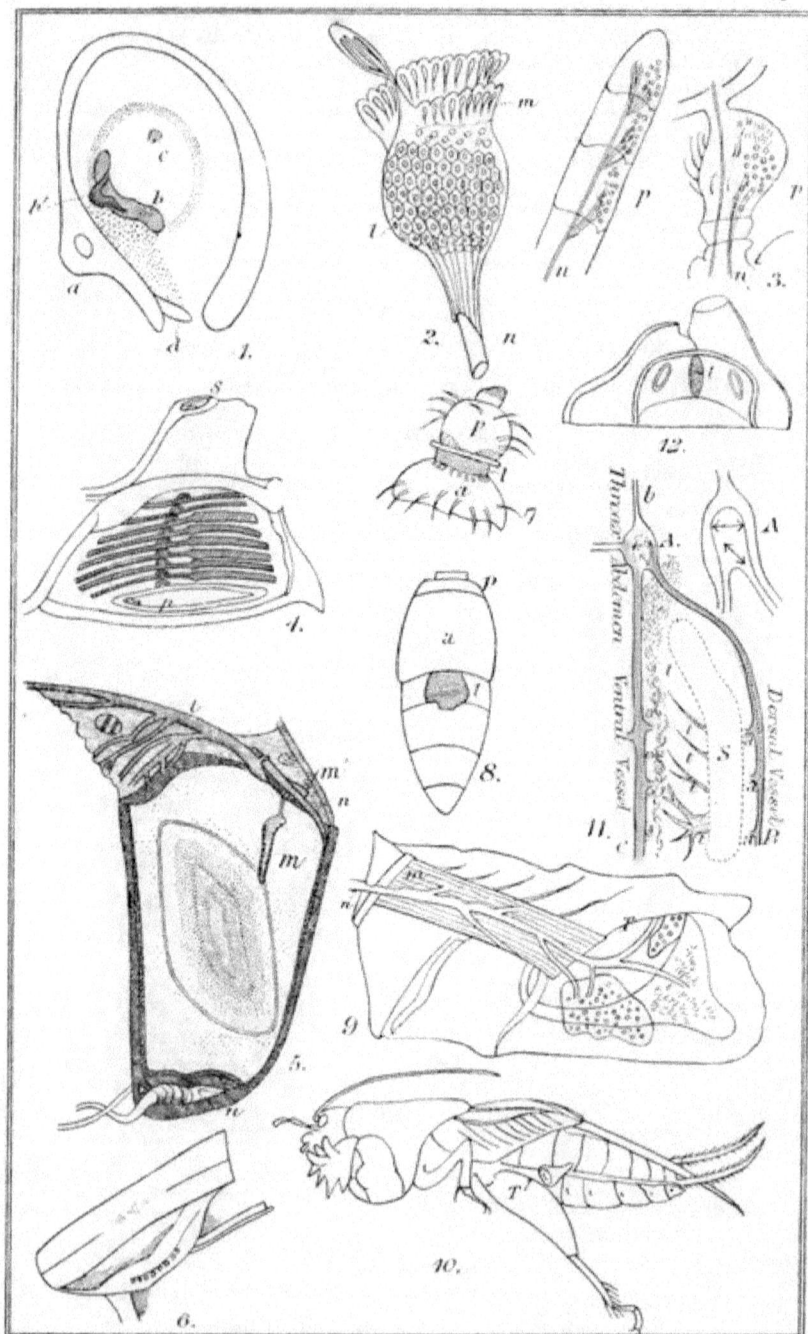

1. *2.* *3.* *12.* *4.* *7.* *8.* *11.* *5.* *9.* *6.* *10.*

Thorax Abdomen Ventral Vessel Dorsal Vessel

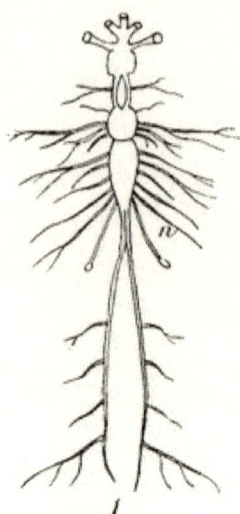

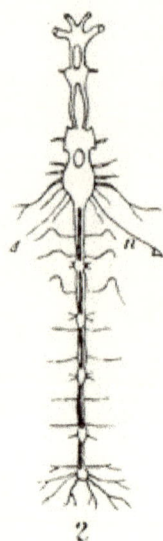

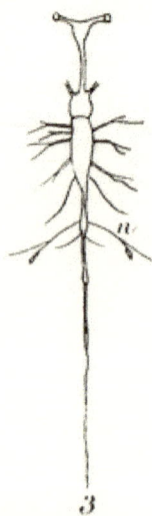

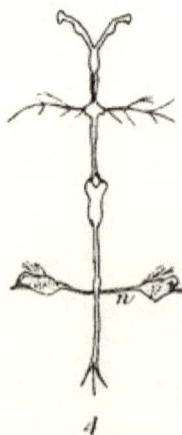

THE CAUSES WHICH PROPAGATE AND DISTRIBUTE INSECT VARIETY.

INTRODUCTION.

THOUGH by education and conviction no egotist, I am free to confess that were the first personal pronoun uniformly eliminated from books much loss would ensue as regards our general literature, and some of the traditions of Natural History would be especially violated. Should the gentle reader share that opinion, his indulgence will embolden me to introduce by way of preface a slight chronological account of the present work, so far as to trace some of the causes which led to its being ever undertaken, and pursued under not, I think, invariably favourable circumstances.

I had indeed, I may say, given much attention to the popular, but in these days often maligned, pursuit of Butterflies and Moths, and amassed a considerable collection of our familiar English sorts, before the topics here treated of intruded themselves upon my notice. My first interest in insects dates back to my school days at Southampton, and the editorial labours of Mr. Stainton, whose numerous periodicals, the forerunners of a new era of thought, duly introduced me to the mysteries of the butterfly net, with its accompaniments of caterpillar rearing and chrysalis digging; sallow beatings in the spring and patient watchings at sugared tree-trunks, ivy flowers, and street lamps at autumn, with other expedients employed to obtain those delicate scale wings termed Lepidoptera—a pursuit much fostered by the glow of charming colours, an inborn love of sport, and perfect rage for novelties. The prosaic and technical sweeping net, and nice laborious arts of bark-searching

B

and carcase-baiting for beetles were less esteemed ; and pond-dragging for water-dwellers was alone resorted to, because the longed-for puddles were out of bounds and necessitated the breaking through of an obnoxious blackthorn hedge. I, at this time, first heard tell of the oak-haunting Purple Emperor and scarce Black Paphia of the New Forest, and visited Rufus' Stone and the woods and ruins of Netley Abbey. The Norman hunting-ground, already robbed of the beautiful high deer to discourage poaching, but still dank, dark, and lonely in its recess, resounding to the echoing tap of the climbing wood-pecker, and thrilling through all its valleys to the soft voice of the distant cuckoo : the " Abbey of the purple abbots, tranquil yet lively," covered with the enchanted rubbish-heaps of ages, as yet undesecrated by the griding rail and mechanical spade of the navvies—when can such happy picnics occur again ?

Subsequent residence in a remote country house in the neigh-bourhood afforded me opportunities of ransacking the early gossamer lawns, quiet lanes, and shady copse clearings of Hampshire ; and now sometimes the dull, wet mornings and vacant fireside evenings were alleviated by the task of identify-ing the obscurer kinds of moths I captured. On the 13th of April, 1867, a year the snow lay long, I was congratulated by Mr. Stainton on the addition of the minute and gnat-like *Solenobia conspurcatella* to his cabinet. The subject of the communication ran thus :—

" DEAR SIR,—I am favoured with your letter of the 2nd inst., inquiring respecting a little moth you have taken near the Southampton Water last month.

" Your capture is a Solenobia, new to this country ! I have Belgian specimens of it, and apprehend it has never yet been described : so will write at once to Mons. Fologne on the subject. I rather fancy I saw some specimens at Fontainebleau, which had been captured there the end of last month. Annexed you have all the information I at present possess respecting it."

" Extract from a letter from Mons. Fologne, of Brussels :—

" ' *March 15th*, 1851.

" ' I have bred a species of *Talæporia* differing from *Inconspicuella*. It is a brilliant grey, spotted with brown, the fringes neatly chequered with brown. I found the cases containing pupæ three weeks ago, whereas the last-named species is still now in the larva state. They were under the bark of a dead tree.' "

"To this I replied that I imagined from the sketch the insect was Bruand's *Lapidicella*. On the 28th of March, however, Mons. Fologne wrote as follows :—

"'I send you three little boxes, one of which contains the insect I spoke of, which I find tolerably common on the wing in the locality where I had found the cases. If you want specimens for your collection I can give you some. All the specimens are very similar to that which I send you, and the species does not appear to vary much.'"

The result of this correspondence was that the little gregarious international, whose existence Mons. Fologne and the writer were the first to recognise, appeared figured in the *Entomologists' Annual* for the next year, under its present designation. More recently, as late as August, 1874, I received a note from the Rev. Mr. Marshall, containing matter, as far as I know, never made public, I having informed him of the discovery by my sister of the knotty gall of a little Bee beside the same richly-wooded river banks. Herr Müller had previously published his paper, and kindled quite a passing rage for galls and their artificers in this country :—

"DEAR SIR,—I am much gratified by the galls you have been kind enough to send. Until they hatch, it would be rash to pronounce upon them. But it seems most likely that they are the galls of *Xestophanes potentillæ*, Vill.; the *Cynips brevicornis* of Curtis, a pretty species, distinguished by its red abdomen. I have several times identified it from the descriptions, not from the gall, which I never before saw. Others have not been so fortunate as myself, and I believe no one else has yet recognised the species. I hope I shall be able to bring it forward now more distinctly, its habits being likely to be known. There is, however, much uncertainty in breeding from galls. They frequently produce nothing, or parasites come forth instead of the real insect."

In the year 1857 my pen first faltered on the first lines of a naturalist's journal, to record, like White of Selborne, the happy idleness and visionary moments of a wonderland of sunny bramble and coppice; a hurried jotting down of caterpillars bred and chrysalids discovered, with due tale of the moths that stole at summer's shadowy close to my lantern, and butterflies hunted down over the thicket-grown dingle and upland furze. This artless scrap had long been forgotten; but one day it having fallen out of a cobwebbed corner covered with the precious *débris* of time, I was astonished on re-perusal to find in so careless a document, breathing the fragrance of flowers and country air, so true a chronicle of the waxing and waning of surrounding life. The opening page informed me that the Peacock Butterfly first displayed its blue inlaid eyes among the garden plots on the 25th of February of that year; and that the woolly-bear caterpillars of the Tiger Moth were observed hurrying over their fleshy dock-leaves before the 7th of March. Easter Monday, falling on the 13th of April, was the epoch of finding a violet-ringed Oak Eggar caterpillar. On the 22nd the grey moth *Lithorhiza* was at the crevice of its accustomed paling, and the second week in May brought out on the landscape the beautiful Orange Tip and the two commoner Cabbage Whites. This springtide is then on the next sheet, unwittingly thrown into contrast with the ensuing, which proved much more backward; for the Brimstone Butterfly was not awake and flitting along his sequestered lanes before the 17th of March, when the large Tortoiseshells were already sunning high up on the rustling elms at the porch, nor was the first annual brood of smaller Tortoiseshells capering along the warm river embankments before the 21st of June. On the day following a heavy-flighted six-spot Burnet Moth was put up in the flowering grass with yellow-wing spots instead of red—a certain mark to the adept of previous uncongenial conditions; and then as by magic the Black-veined Butterflies were everywhere, in the river marshes, at the wood side, and at the coach gate—a species I never met with in the locality before or after. The year 1865 will be remembered by entomologists as an *annus mirabilis* of Death's Heads, Humming Bird Hawks, and Gamma Moths; and the exceeding abundance of the latter Bacchanalian revellers allowed me to remark their singular habit of passing the mornings on the common lands and meadows, where they sunk into inactivity at noontide. But no sooner came on the lapse of soft and invigorating twilight than they hastened in shoals to the scent distilling from the petunia beds, where stragglers lingered probing the flowers until early dawn.

Before this property—whose venerable volumes, drilled by the silvery worm and redolent of the days of Johnson and of Addison, still breathe from the shelves around me—had with its yachting, driving, and shooting, passed to other and happier possessors, the tide of land enclosure insensibly setting in around the purlieus was rendered painfully evident in my summer rambles, by a farmer, a very Sir Oliphant, enclosing a rude slice of nice damp common land, long frequented by that fast disappearing butterfly the Fairy Artemis, and ruthlessly turning its ragged resort of furze and brown fern to turnips and potatoes.

Verily it has seemed, looking back on the past, as if our insect hunters of the future were doomed to a sport composed of an influx of Colorado Beetles, White Butterflies, Hessian Flies, and woody Oak-galls; and that the flowery wood clearings and purple heaths of our forefathers, with their basking and fluttering fauna, were all to be heartlessly swept away in the present era of steam and telegraphy. It is quite certain that our butterflies, especially the Orange Fritillaries, are fast disappearing. In the days of Moses Harris' book, and in the good time before, the Marsh Fritillary flew near Kingsbury and adorned the Wormwood Scrubs; now it is even scarce at a distance from London. Mr. J. P. Barrett, when preparing a county list for the South London Entomological Society, writes me tenderly concerning it and other of the fauna of Surrey :—

"DEAR SIR,—In addition to the species enumerated in your list of Diurni, I am anxious to know whether others may not occur in your district, especially the Marsh Fritillary and the Brown and White Letter Hair Streaks. The first occurs at Haslemere in Surrey, and Sandwich in Kent, and at a few intermediate localities sparingly. I have been told it is to be met with near Leith Hill, and shall esteem it a favour if you will kindly make inquiries of any entomologist resident in the neighbourhood of Blackheath, near Chilworth, whom you meet in your excursions, respecting it.

"The Hair Streaks are frequently overlooked in the imago state. I have no record whatever of the occurrence of Betulæ in Surrey except my own. For several years past I have beaten the larvæ off blackthorn near Epsom, on the clay; and I therefore see no reason why it should not also occur to the north of Guildford, in the London basin. *W. album* occurs near Croydon on wych elm, and Stephens records it at Ripley in the imago state, but he was ignorant of the food plant of these butterflies. Should you have an opportunity of beating blackthorn and wych elm about the beginning of June, I shall be very curious to know the result."

But to turn from vain regrets to inquire why the world around us is so insensibly changing. Lately I revisited the scenes of my childhood—where, to me, all was long lapped in the realm of mediæval enchantment, and tree and lane

whispered audibly legends of spirit land—mainly to discover
why such dependent things are so proverbially transient.
Wending once more over the hummocky heather land stretching
out along the sail-studded Southampton Water, here and there
traversed by a tinkling brooklet, incessantly eating deep and
deeper through an irony stratum of angular chalk, flint, and
water-rounded pebble into a thick layer of yellow and green sand
beneath. A silent turn in the road brought again to my view
the winding creek, bordered with its well-known white mansions
mirrored glassy and grey. Over the cockly mud flats and fucus
wash, made in hurriedly the clear spring flood with a silver
ripple until it kissed the shingles at my feet; and then there
arose a voice of the tutelary river nymph, " Behold your fluvio-
marine limestones of the future!" The graceful yachts lay
sleeping mid-stream, moored on a magic shadow; and nought
was stirring save the drowsy village echoes and cry of the
gliding seamew from the shadowy banks. Away and beyond the
silvery gauze of fresh bromine and iodine undulated the upheaved
cloud-cliffs of Vectis, "The deep sea formation continued the
wavelets of the secondary." The salt plant clung to the dripping
river dyke, where the Dragon-fly was mating, the red with the
olive. And now as the morning fled its prime, a fleeting cats-
paw escaping the russet tangle hung with coral red, sat momen-
tarily in the bulging jib of the creeping smack, and caused its
keel to career sleepily along the waving water-grass; or now,
the indolent ducklings again and again put forth from the red
brick angle, to undulate on the cool reflections. Chronologically
it was autumn, a definable period in unlimited time.

Yet maybe we naturalists and insectmongers are too prone
to thus philosophise over our delectable spots and localities;
though it is never well to ignore wholly the written transactions
plainly inscribed around us. A great forest will, if ransacked,
prove to contain more compressed information than our trim
shrubberies and hanging rosaries; a neighbouring heath or marsh
land tells more tales than rich Levantine orchards and cattle-
cropped parks. And therefore, not improbably, the young entomologist expending his time and means in storing boxes and
cabinets with local specimens, and his shelves with technical
pamphlets, on account of the comparative exhaustion of our

insular fauna, and the extreme unlikelihood of the discovery of any new species, will sooner or later insensibly seek to extend his views to the remarkable races and sub-races of insects that occur elsewhere; or, if he have leisure, employ himself in biographies of the little entities around, or seek to understand their place in the earth's history.*   To many economic entomology is one great inducement to this result; the vernal Canker-worm in the sweet-scented rose and "Pug" in the snowy apple petals; the moths secretly destroying our embroidered hangings, and perforating wind-fallen apples and dessert-figs; the Cheese-hoppers, the Weevils in the granaries and Wire-worms in the turnip crop, the Letter-writers scoring over the lank and sickly plantation clumps, one and all of such ministers of destruction and unbidden retainers are sufficiently obnoxious to rid a proud lord of the Salique soil of some of his inborn dignity and uninquisitive disposition.   Yet this science rather should be considered an application of entomology to humanity than the pure intellectual cult itself, as many would have us to believe.

My occasional summers spent among those dark romantic lochs that indent the deep depressions in the picturesque clay-slate mountains of the Western Highlands of Scotland, first brought me face to face with the great problem of the influence exerted by climate over our fauna, when the gloomy glen and heathery hill disclosed the existence of species unknown in the genial South.   The dark Scotch Argus Butterfly fluttering in the shady bushes, the globular papery nest of the Tree Wasp hung at the rushing burnside—no less than the sooty aspect assumed alike by the Garden Moth and Braeside Butterfly, with their late and uncertain appearance after the June rains—failed not to exert a special charm for a youthful eye nurtured in the soft South.   Among the Alpine flowers that nestle in the wild crags of Loch Rannoch and Braemar this peculiar fauna has its centre, probably from the conditions of climate being more congenial.   Then while climbing lonely declivities, where the sheep's bell and heatherbloom assert their silent reign, the undomesticated condition of certain Clothes Moths had a passing interest; for these, I questioned not, were a luxurious race,

* Address of Mr. H. W. Bates, President of the Entomological Society, 1879.

whose effeminate habits are coeval with the wheel and loom, and who formerly found the fleecy backs of the mountain flocks suffice their wants.   But how, again, could Clothes Moths have existed before sheep were?   Was there ever such an age?

These and such like speculations must from time to time spontaneously arise in the mind of the collector, especially if he happen to have carried a geological hammer and bottle of acid into the wilds to unclose the record of the rocks; and, as so happens, a life aspect is all the latter science wants when dealing with its successive births of extinct beings; for these must have lived and been subject to existing law—they are found entombed in various stages of growth, entombed as male and female, and besides are in no wise incomplete, deformed, or of monstrous growth, as certain poets have fabled.   And this is true of the oldest insects discovered in our coal mines, as of the earliest king crabs.

In 1871 the sounds produced by insects and their import began to occupy attention, and about the same time the " Descent of Man " was in the zenith of its reputation.   My notice nevertheless was independently drawn to this subject from a query in " The Annual " by Dr. Knaggs, as to whether the Silver Lines Moth was capable of uttering a cry.   I soon after picked up copies of Kirby and Spence and Rennie's " Miscellanies " at an old bookstall, and finding myself within a short walk of the libraries and natural history collections of the metropolis, was not only enabled to dissect the insects I found in the fields and eligible building plots around Maida Vale, but likewise to systematically read up the investigations of others in English and foreign works.   The organs of hearing in moths I stumbled on by the merest chance during the course of the summer of 1871, and for a long time I fancied these structures had some connection with the sound made by insects during flight; under which impression some amicable notes passed between myself, Dr. Knaggs, and Mr. M'Lachlan, which eventually led to certain communications finding publicity in the *Entomologists' Monthly Magazine*, for which I have to tender acknowledgments to the editors, especially to Mr. Douglas.

But the mighty heart of London towards the turn of the leaf grows lethargic, and in autumn there are few engaged in business or study but contrive to snatch a week or so of country air to soothe their jaded spirits before the crown of fog fairly descends on the mart of nations. And thus it has come to pass that the seaboard of England presents many a quiet and historic resort known to the entomologist, where in a happy hour some strange species has been met with, either the importation of commerce or the natural and hereditary migrant over the winds and tides. In many obscure shipyards Long-horned Beetles have been found crawling, transported with timber from out the virgin woods of North and South America; a dozen such were known and described by Stephens. And on the verge of the Dover cliffs arrive from time to time snowy flocks of White Butterflies, Queens of Spain, Camberwell Beauties, and flights of southern Sphinx Moths, winging periodically from their banks of oleander to the cold North. So at Penzance or Pembroke, the points of impact of the Gulf Stream, foreign bees and butterflies reach *terra firma* in their rarer visits from the warmer scenery of the west. This constant influx of life has established and sustained a 'longshore fauna of alien insects at localities on the coast line, where, lapped in the luxury of fresh breezes and salt foam, they constitute no little inducement to a tarry by the seaside.

The latter days of cheap return tickets opened out to myself the lore of many a down and thicket, until finally one autumn found me at Deal, the favourite hunting-ground of the late Mr. Frederick Smith, a very paradise of thymy chalk hills and sand dunes shagged with buckthorn, sweetly melodious and busy with those vagrant bees, whose admirer has now, alas! passed away. During my stay in the port of Cæsar I was rejoiced by capturing both sexes of our rarest Thorn Moths at the evening street lights; and here, when wearied out with the sun's heat, I used to sit and gaze over a molten water at the far-off French cliffs looming and vanishing cloud-like on the horizon. Nor had many years flown before the packet-boat, whose ten o'clock smoke I had so often tracked into the blue distance, was itself in requisition, the summer strait was crossed, and I found myself *vis-à-vis*, reclining in the cornfields of

Blanc Nez, lulled by the sleepy, hazy dirl of the large Leaf
Crickets, amid a rare profusion of bluets and poppies. Scour-
ing in my search for insects the skirts of that vasty quiescent
plain, the projected prolongation of the Deal flats stretching
out northward from the Calais bluffs away to the pine-clad
Baltic, I soon began to recognise the relict of that secret
habitat which in olden time replenished our south-eastern fauna,
and conferred on it its charming alluvial trait.   Once a dreary
but faunistically rich fenland, where baronial castles lay girt
about with whispering reeds and clamorous marsh fowl, the
forgotten home, as I doubt not, of our large Copper Butterflies,
where Swallow Tails still fly—this low-lying district may be now
known by a black peaty deposit, shelving beneath the Channel
waves and running into reefs; but shoreward, arid and covered
with sand drift, incessantly pounded from transported chalk
flints and sea shells by the tides tumbling their breakers along
the shore.

Here, around sluice and ditch, still struggles on a remnant of
a once rich coleopterous fauna of lacustrine aspect, ill at ease
amid the changed and changing surroundings.   Many of these
species of beetle are common to Calais and our own Fen district,
others are rare or never found on this side of the Channel.   Of
the latter, when strolling along the sand dunes, I have myself
picked up two, perhaps the largest and most conspicuous,
Carabus coriaceus and the Jardinier (*Auratus*); and at Sand-
gate, the small Chafer, *Rhizotrogus rufescens*, Lt., widely distributed on the Continent, becomes quite common in summer;
and yet this is a species as far as I can learn that has *never
reached*, or at least never propagated on, British soil.   Nearer
Lille occurs the Purple-edged Copper, seldom or never seen in
this country; and this butterfly, with the Queen of Spain
Fritillary, and a Blue-winged Locust, confers a thoroughly
Gallic aspect on the spot.   The local moths of our south-
eastern shores also might be all with a little trouble hunted up
by an enterprising collector in their colonies and fastnesses
around Calais and Boulogne. *

Now it so happened, among other things, in the year

* For information concerning the Highland Insect Fauna, see "Fauna
Perthensis," by Dr. F. Buchanan White, and Review, by R. C. R. Jordan, M.D.,

of Our Lord 1877, the singing, or more properly the
music of the Cicadæ, had begun to engage my attention,
and as I increased my store of knowledge, the popular
accounts of the organisation of these people of the pines, hack-
neyed since the days of the great French chemist, appeared
more and more unsatisfactory. That it is the males alone that
vie in these lively airs, which when the Sun-god ruled the theatre
and circus attained a celebrity Jenny Lind or Patti might have
envied, there can be little doubt, as a naturalist would say in a
general way. We have the fact, indeed, handed down from the
time of old Zenachus, who used to chide his wife with

" Happy the Cicadæ live,
    Since they all have voiceless wives."

And when Dr. Bennett was investigating the natural history of
the antipodes, the aborigines used to inform him concerning
these females, "Old woman galang, galang; no got; no make a
noise." And what she had not got were two little parchment
drums on either side of the body under the wings, that exist
alone in the males, and which though noticed by mediæval
Italian physicists, were left to the originator of the thermometric
scale to specialise. But then Réaumur in describing these little
drums drew his views as to their action from dead and dried
specimens, and his notions were not at all borne out in their
entirety by subsequent experience. So that in this age of
railroad it occurred to me one might easily enjoy what to the
French savant was ever an unrealised day-dream, and personally
hear this music, the delight of chiefs and sages. I should
mention I had previously trod in vain the renowned stretches of
whitethorn, interspersed with bracken, in the New Forest, in
hopes of finding a musical straggler of our native Cicada; and
an alternative scheme to plod the ploughed fields around the
Fontainebleau forest, seemed, if sportsmanlike, to promise but a
poor bag.

Many and various are the emotions evoked by a holiday

Ent. Annual, 1873, p. 70. The Cumberland and Welsh mountains have been
indicated as forming an elevated line of migration of these species southward.
Regarding the Lepidoptera of an ancient raised beach in Norfolk, see paper by
C. G. Barrett, Ent. Mon. Mag., Feb., 1871, &c.

in Italy.   To float on the sea, to angle at Venice of a summer's
evening, and watch the sun decline behind the misty campaniles
and palaces of the Doges ; a morning's reverie with philosophic
Germans at Castellamare or Sorrento, watching Vesuvius eternally
steam a summery cloud with hectic glow over the cindery heaps
of Pompeii, and sheer beneath a rippling wash of starch-blue
water ; to sit on Ana Capri, and strain the eye toward Cape
Misenum and Baja, till the galleys of Æneas loom in the sea
mist, or the phantom ship of the apostle veers on the low
volcanic hill behind Pozzuoli ; to gaze from the Capitoline over
queenly Rome and the warm and flowery Campagna, until with
shouts the long triumph winds upward, and the marble forms
around start and glow with life ; to turn to watch the urban
veils and handsome officials, enhanced with a certain mediæval
lustre acquired from classic books, paintings, pedigree, and not
a little set off by those traditional tales of the large knife and
rapier, in which a Mrs. Radcliffe and Lord Byron used to glory,—
such is Italy to the many, Italy in some respects, may we hope,
as it was.

Another aspect has Italy for the dilettanti in her works of
art. Many see nothing beyond the middle ages ; and Raffael,
Rienzi, Dante, or Petrarca are the spirits that animate every
grove and fountain. To forego blue sky and spring flowers for
a silent zephyry seat where Madonnas float on the light arras,
sombre martyrs sink death-pale, and little children dance in ring ;
to walk echoing aisles of mouldering mosaic, while music
sweeter than the nightingale's awakes the holy trance in colours
of the camera obscura or woof of Iris ; to sit on marble steps
amid ivory tracery, where princes once walked and rustling silks
swept by, but where now the townsfolk hold gala and pass
simple jokes ; to trace historic streams where war has left a
sanguine streak, and Guelphs and Ghibellines drew a fratricidal
sword to storm the gate and charge the foe,—this, with the
addition of ices and lemonade, makes a comprehensive Italy to
another class of philosophers.

But every country acquires a mundane dress with every-day
occupations, and in Italy, despite the *dolce far niente* of the
cafés, the pursuit of insects has asserted its rights as a study or
relaxation ; so that the Società Entomologica, founded, I under-

stand by an Englishman, has come under distinguished patronage. Italy conjointly has dropped off some of that disagreeable prestige for malaria and brigands, rendering the investigation of her fauna the more possible: not that either for a moment is absent or suppressed in the southern portion of the peninsula, although their spheres are circumscribed. But with all these advantages and disadvantages, the poetic fields and classic ground of Italy probably do not evince for our fly-catcher the charm they have manifested in the case of the artist and antiquarian. Old in civilisation, the virgin coronet of deer-forest and brake has long fallen from her brow. Here and there, it is true, as south of Pescara, his heart will beat at the little stations with their oak copses and sea cliff; making twilight with rank and lush tangles of cisti, vetches, and tall boraginous spikes; inviting to rambles such as may be found in remote nooks in the fairy south of our own island. But otherwise Italia is emphatically an exotic garden, a land of vine or olive, or a waste of waving grain; and among such prodigality the weary collector may sit down, wipe his brow, and sigh for the thread of Dædalus. Even the more open plain of Lombardy, with its methodical rows of willow, over which a sea rippled in Tertiary times, is far less tempting than the leafy gorges of the Tyrol and Rhone valley. Indeed, all the more attractive butterflies seem to hold the Swiss pastures and Sicilian high ground, where the glacial epoch, it is said, left them. The bare-backed Apennines and fertile Maritime Alps are equally ignored by Italian savants; a few varieties of our northern favourites perchance grace Florentine gardens; and around Pisa, according to Rossi, you may just capture enough Diurni to convince yourself you are abroad.

Yet, if Southern Europe be comparatively poor in regard to its insect fauna when compared with the virgin tropical bush, those few of the bolder beauties it possesses, bathed in Levantine air, have a two-fold charm of borderland, whether we draw the line for place or time. Not only have we here the summer Cicadæ drumming among the boughs, but many other objects of classic interest. The Golden Wasp (*Scolia hortorum*) yet lazily wheels around the tufted fountain. The sacred Scarabæi, in the ravines, yet roll their miniature globes, as the Egyptians imagined, to procreate. The ants store up flower-seed against

winter's dearth.   There also are found the Ant Lions and Trap-
door Spiders.   The Italian Fire Flies nightly flash along the
rivers; florid species of butterflies, *Charaxes* and *Danais*, flutter
in localities.   These sights and sounds emphatically signalise
the Mediterranean basin, and, like the warm Afric sirocco,
breathe an alien character over the Continent's southern hem.
A visit to the palatial museum of Marseilles will allow us
further to enlarge our scope, and learn how Europe in hoar
antiquity was even more a part of the globe as regards productions, clothed even as late as Tertiary time with fan-leaved palms
that seem to have escaped the trim gardens of the Riviera,
among which flitted Brown Butterflies, Moths, Tipulæ, and Bees,
with little-snouted Weevil Beetles, that fraternise strangely with
the latter-day legions of pink destroying Chafers and Humming-
bird Moths.

Even in this minute department of Nature the hammer of
the Great Workman was ringing, and previous to the hour when
the Greek keels beached on the shore they deemed so beautiful,
Nature had been moulded, so to speak, and decked in her attractive robes.   And were we to historically classify the works of
man as indicative of progress, we should likewise find these
running through their cycles obedient to exterior law.   Arts
and commerce pass from one race to another, and wax and wane.
Assyria, Greece, Rome had their day.   Races arose from the
East, West, North, and South, and swept over these gay cities
like a deluge, mingling and commingling.   Spread the light
lateen on the kissing land-breeze, and waft round those misty
headlands renowned in battle and song.   The sea, no longer red
with Punic blood-stains, changes colour like a dolphin in the
eventide, and the hum of busy cities lingers on the ear.   'Tis
Aphrodite that leads her chariot over the wave, or the Nereides
that wail among shaftless columns and marish damp?   No ; now
all the very creed is changed, and the strains of Ave Maria
arise as the prow rushes through the foam.

But it is not migration pure and simple that has writ on our
southern *fauna* and *flora* Afric's brand and India's swarthy hue.
We have some evidence of a time when the Mediterranean was
not, and Italy formed a promontory of a southern land.   The
Ovidian lyre speaks of a vague tradition or sailor's tale then

rife, which imagined an age when Sicily was one with Italy,
and no sea as yet roared between Scylla and Charybdis. And
some score of years back, Dr. Faulkener made the discovery of
the skeletons of elephants, a gigantic tortoise, and immense
mouse in Malta, an island only seventeen miles long and
seven broad, not so large as our own Isle of Wight; while
in Sicily, bones of the hippopotamus have been exhumed
similar to those of the denizens of the Upper Nile. So that
these lands, the argument goes, must have been originally
connected.

Italy, then, I concluded, was the country where the Muse
president over the present work especially dwelt recluse, and to
Italy I must wend; so packing up my MS., and entering the
train at Calais, while the skirts of winter yet lay on the landscape, a ride over the beds of the old cretaceous ocean brought
me to Paris ere the now famous electric lights were kindled
before the Opera House. Here I waited some days to meet a
compagne de voyage, and then away down the endless river-
valleys of the Allier and Rhone, gazing wearily at the poplars
hung with mistletoe-balls, and endeavouring to form some idea
of local temperatures from the soft foliage kindling on their
boughs, till Orange, that gave a scion to the British Throne, at
length reared high her Roman arch, and bade enter the garden
of the South. Avignon, whilom abode of the Popes and
Petrarca, shows a matter-of-fact little station with gigs in
waiting, and the sound of the Durance is heard pouring its
chaplet of roses on the ripples. And then the sea! the sea!
Marseilles—with her fountains and avenues of grey Oriental
plane, where voices blend with the music of the boughs, birds,
and bells, and every window is barred against riot and revelry—
has passed like a dream. The little hot gullies of Nice and
Cannes, with the French ironclads practising in the offing;
Genoa and its ships, much as Columbus left it; Pisa, with the
celebrated tower that hangs, as indeed, as far as statics are concerned, do some of the campaniles at Venice; Florence, with
its pictures and low-trellised vineyards; Milton's Vallombrosa
—a round, bare hill apparently;—Rome, with omnibuses, and
the Tiber swerving a yellow flood resembling a gravelly gutter
after a thunder-shower,—all are passed, and at length we reach

Parthenope at the auspicious advent of the month of Flora!
One can descend from the whirligig and breathe.

After having visited the Lucrine Lake and Baja, where
Painted Lady Butterflies were swarming at roadside nettles and
around the dilapidated walls of the Temple of Venus, I ransacked
the sepia green chestnut covert above Castellamare with no
great result, and then proceeded by steamer to Capri, where the
sunny hill-side, looking towards Naples, proved not unproductive
in insects in measure of northern type. I now began to inquire
after the object of my quest. Having learnt in English classics
that Cicadæ are the heralds and harbingers of the spring, I was
not a little disconcerted on learning none appeared in Italy
before June. One gentleman suggested I should post off
to Algeria, as the Cicadæ at Naples were little things, and another
that I should accompany him to the Greek islands. However,
having previously provided myself with a circular ticket, Alta
Italia I felt should be my destiny.

On the fifth of June, one summery morning, when descending
the gardens of the Palazzo Giusti, where the blue Burnet Moths
were already floating in the shadows, and whence I had been
surveying the spires of Verona, traversed by the swift-footed
Adige, I espied the first nymph of Hæmatodes crossing my path,
besmeared with the soil from whence it had just risen. After
this first auspicious encounter in the land of Virgil and city of
Catullus, no further intimation had I of my quest until I was
settled for a season among the villas at the banks of the Po.
Strolling along the acacia thickets that hang over the dark
margin—it was one Church festival towards mid-June, when a
storm-cloud on passing had hung out the long phantasmagoria
of Alpine peaks, now about to fade again into cloudland—suddenly my soliloquies were interrupted by the loquacious bull-
frogs in the slime. Sitting down mechanically on the partially
dried grass, while the trees around drew in their shadow, I
listlessly watched the clouds form in the clear sky around the
frosty pinnacle of the Grand Paradiso; and then mustering
insensibly, stretch out their gauzy veil over the couchant mass
of Monte Bianco and the distant Cenis, till one by one the
grand old cordon of barrier giants, together with the ruddy
lower spurs of the Albigensian valleys, were replaced by a

curtain of opal grey sky, from which stood out sharply the long willow rows fringing the opposite bank, the neighbouring farm, and herded cow.

"Ella si sta pur com' aspr' Alpe a l'aura."

As the heat increased, the man who was deepening the torrent bed with an antique ladle, having successively divested himself of his coat, shirt, and inexpressibles, now pushed in to moor his punt at the sedgy brink; and eventually, fairly overcome with the fierce noontide, lay down to ponder while the river fretted on ironically over its stones, whirling away to dreamland. Suddenly I fancied I heard a frog quacking in a bush at hand. And then again came the sound *Pip! pip!* Turning over the brittle boughs, which have an unpleasant fashion of breaking off short, and leaving you to the mercy of anything; from being taken into custody, to the shotted blunderbuss of an infuriated peasant, I turned up a drowsy Cicada on a damp spray, who was attuning his lyre to the stray glints that crept in among the dense soft foliage.

But can this be the Cicada of one's school days? I exclaimed. It is nothing like a "Grasshopper," as elegant writers such as Pope and Dryden maintain; nor does it seem as if it would "hop," as Wordsworth and Goethe would make out. No, it is not a "Tree Hopper." Cowley said it "danced." No, I don't think it dances. And it is not a Cricket, as another wiseacre, a German, has it! Nor a Leaf-cricket with a curly tail, as La Fontaine illustrates it! It used to turn its eyes and wink at St. Franciscus; but alas! its optics have become immovable. Well, here is my pocket Virgil and the explanation. "These insects differ essentially from our Grasshoppers; being found in warm climates alone, they have not, indeed, any English name. Their habit, noticed in the text of sitting on trees, would alone make a distinction. (Hem!) In form they are more round and short than our Grasshopper; they make a much louder noise, which begins when the sun grows hot and continues till it sets. Their wings have silvery streaks, and are marked with brown; the inner pair of twice the length of the outer—(hem!)—and more variegated." Well, but these are Lord Byron's " People

c

of the pine, making their summer lives one ceaseless song." They are not a bit like the Cicada before me, pure and simple. I must describe it for our northern literati. Well, it carried itself, I think I may say, with somewhat the air of a gigantic bee, but in form it closely resembled the little froth insect of quickset hedges, to which it is near akin. In colour it was black, elegantly lined with blood-red on the body and wing-veins, or if Latin should be preferred, *Nigra abdominis incisuris alarumque nervis sanguinis.* Any way it was a Cicada, sometime known as *hæmatodes,* whose generic name is undecided. *Cicada hæmatodes,* the Blood Cicada, satisfied Linnæus; Fabricius baptised it *Tettigonia;* and lately it has been proposed to surname it *Melampsalta,* and christen it *Cicadella.* But this is getting as bad as the poets.

Having likewise read of the suavity of the Cicadæ in France, and of a certain physician at Aix who found, Eurydice-like, they came at a whistle to sit on his walking-stick, thinking no harm, I placed the Cicada on my hand in the sun to see whether he would "sing." But alas for preconceived notions! the thing took to its wings, and left me in chagrin. Returning to the same spot a few days afterwards, I spied another sounding lustily on a poplar sapling just out of reach, where he was probing ecstatically the tepid sap. I approached and violently shook the tree to dislodge him; but this was useless; the creature beat his wings as if it were the rising west wind that rocked him, and screaked the louder. At this juncture a Piedmontese with a peeled fishing-rod passed, whom I invited in *la bella lingua* to take a "swipe." The angler politely complied, but with an acclamation "Securo!" for the bavard was already crowing with his compeers on the tree-top, *Pip! pip! pee . . .!*

It is part of the economy of the Cicadæ that the matron is provided with a sharp ovipositor wherewith she pierces dry twigs, and thrusts her eggs well into the pith to hatch in process of time. On leaving their sylvan nurseries, the young burrow in the ground, where they subsist on the roots of plants. Here they become so plump that they were sought for by the ancients as food, and termed Mother Cicadæ, *Tettigometra.* Living for about a year thus entombed, they crawl to light, and clambering up the plants and trees, their skin first hardens and then bursts, allowing the escape of the winged sylph. This metamorphosis did not escape the keen, though uninformed, eye of antiquity, and consequently takes shape as a pretty mythological trait. Aurora, it is said, had a failing for two young men, Cephalus and Tithonus, and, as the legend goes, carried them to heaven ; but having forgotten to ask perpetual youth for Tithonus, he became old and decrepid, and, like an infant, was rocked in a cradle. Hereupon, weary of life and immortality in such a form, he asked but power to die. The goddess of roseate dawn said this was not possible, but, exerting her magic, she transformed him to a Cicada, which moults when it is old and grows young again.

Proceeding to ramble in the mowing-grass at the end of June, I found the brown paper-like cases whence the Cicadæ had emerged, not unfrequently clinging on to rank mints and grass stalks ; and sometimes I came upon a newly-fledged male sounding his double pipes low down in a bush, or even sitting ignominiously on the ground, *Pip! piping!* like an angry infant on the point of losing its equilibrium. Some of these I caught, and when grasped by the hand they made a guttural noise like a young bird, strongly beating their wings and raising their abdomens. Then walking out about 10 a.m., where the sun fell on the foliage, and the quick-eyed green lizard frisked over the tree trunks and walls, I invariably heard the squallers at the accustomed spots; and by dint of careful scrutiny the minstrel could be brought in view. Their matutinal music then resembled most innumerable watches set on the trees and let run down one after another.

It was now midsummer by the banks of the Po, and the brisk movement of Italian life had given way to languor in the heavy and fragrant evening air. Groups of gaily-clothed shopgirls walk past my window flirting their fans; across the river wings to and fro a flat-bottomed barquetta, to bear Amaryllis and Phyllis to romp and lounge where the shadows descend on the grass and flowers, and at late twilight to return and arrange their hair by the watery mirror. Bathers plunge in the flood, are borne down by the current, and struggle back to the alders. And now all is gloom ; the village lads chat in groups and fireflies flash, where before was so much subdued animation.

c 2

But sights and sounds in Italy live in retrospect. Over these enamelled meads and through these woodland paths strayed Virgil in his early years, when pursuing his studies at Cremona and Milan; and at Mantua lay his farm, where the "Eclogues" were commenced to commemorate the munificence of Augustus in restoring his patrimony after the contest at Philippi, quietly made over to some old veteran. There, on yon once sheep-clad hills, he may have sat pouring over Theocritus. Here the flowers and bushes of his childish romps yet spring,—tiger lilies in the dusk thickets; buttercups and floating white chalices on the water; poppies and blue-bottles hanging in the corn, bound with the most fugitive of convolvuli. Here wave the wall-flowers or white violets on the ruins, and everywhere is profusion of fragrant mints and dank potherbs, suitable to fill a panniera and twine the garland for young and transient beauty. And there, too, the old-world trellised vineyards yet remain, where the arbustæ, or propping trees, are annually rent with the shrill and querulous piping of Hæmatodes, with as little doubt the Cicada Virgil deems so harsh.

> " Young Corydon, th' unhappy shepherd swain,
> The fair Alexis lov'd, but lov'd in vain:
> And underneath the beechen shade, alone,
> Thus to the woods and mountains made his moan :—
> Is this, unkind Alexis, my reward?
> And must I die unpitied and unheard?
> Now the green lizard in the grove is laid,
> The sheep enjoy the coolness of the shade,
> And Thestylis wild thyme and garlic beats
> For harvest hinds o'erspent with toil and heats;
> While in the scorching sun I trace in vain
> Thy flying footsteps o'er the burning plain.
> The creaking locusts with my voice conspire;
> They fried with heat, and I with fierce desire!"

The sheep howbeit are now all stall-fed, and no longer rove the meads in Northern Italy; and Dryden, no less inapposite in translating, has made locusts out of Virgil's Cicadæ, and, proceeding with the same artificial diction, turns a love of romping into fierce desire. So, again, in the following quotation from the "Georgics," he, as a change, transforms the Cicadæ into Grasshoppers to suit the metre :—

" But when the western winds with vital power
Call forth the tender grass and budding flower,
Then, at the last, produce in open air
Both flocks; and send them to their summer fate.
Before the sun while Hesperus appears,
First let them sip from herbs the pearly tears
Of morning dews, and after break their fast
On green-sward ground—a cool and grateful taste.
But when the day's fourth hour has drawn the dews,
And the sun's sultry heat their thirst renews ;
When creaking grasshoppers on shrubs complain,
Then lead them to their wat'ring troughs again."

The end of the Blood Cicada was dire and classic. About the commencement of July, there appeared as if by magic certain greyish insectivorous birds with a harsh and guttural note, among the sunny vines and woody knolls where the Cicadæ had established their coteries; and these, sitting on the low brambles, sometimes two together, knavishly whistled a tune until an unwary chanticleer was inveigled to respond, and so betray his hiding. The obnoxious intruders then flew at him, and brought him to the ground in their beak and claws screaking most piteously, *Whee ! whee !* But sometimes they missed their mark, and the ill-used one took wing; whereupon a chase would ensue high over wood and dale, the Cicada exclaiming most lustily the while. The very ear was held in involuntary suspense to catch the final snap of the beak and death-scream of the quarry. If one remembers right, the warblers of Plato's retirement are accused of this delinquency in the " Anthologia " :—

" Attic maid, with honey fed,
    Bear'st thou to thy callow brood
Yonder locust from the mead,
    Destined their delicious food ?

" Ye have kindred voices clear,
    Ye alike unfold the wing ;
Migrate hither, sojourn here,
    Both attendant of the spring.

" Ah ! for pity drop the prize ;
    Let it not with truth be said
That a songster gasps and dies
    That a songster may be fed."

Yet if the nightingale be truant, no less has our poet Cowper tarnished his bays. Not only does he conventionally term the Cicada

locust, but he seems to have convinced himself they were synony-
mous, translating τὸν ξεῖνον the guest in both cases " migrate
hither ; " and the summer one, τὸν θερινόν, he renders " attendant on
the spring." Rather should it be : " He the guest, she the guest ;
he the summer one, she the summer one ? " The Cicadæ, exerting their capacity of song in thick boughs, were considered
enamoured of the summer. July in Italy ! The leaves and hedgerows crusted over with thick dust ; the diligence horses sweltering
under loads of worn-out travellers, wooing with sunshade and
fan the slightest breath of air, and as they rumble along turning
up flocks of the no longer scarce Swallow-tail Butterflies from off
the village gutter ; the poplars prematurely brown, and melancholy
songs of the peasantry echoing among the hills. Who has
described an Italian summer better than Horace ? We almost
seem to rumble past the Coliseum and out on the tomb-lined
Appian Way ; in the chariot with his careworn senators, flying
the smoke, work, and war-cloud ever brooding over Imperial
Rome ; bound for the sparkling villas built into the resounding
sea at Baja, there to forget the daily alarms of street fires and
toppling houses, to bathe in quiet, listen to the seamews, and
eat serpent-like muræna and tasty cuttle-fish. And such will
ever be unsophisticated Italy. What of Epicurean sentiment is
there not in yon strolling banjo-player, reciting in the mediæval
strains of Tasso " Mia amica " and the " Campo santo " ? and
what of Catullus in the Pifferari ? nay, the very tone of triumphal
marches can be caught in yon petite bouquetière, as she lands
from a barquetta, singing as she goes " Margherikima e un
Tramvai." And then the drowsy street-calls, " Acqua ! dolch !
fresch ! "

In warm weather the great enemy of mankind is sleepless-
ness ; neither is the sultry glare which beats the livelong day
on the façade of an Italian house—causing the employment of
painted windows in the north and flat-roofed Oriental construction in the south, those *ingentes moles* of antiquity—at all
friendly to sweet forgetfulness and a state of physiological rest.
Many times have I risen at midnight and thrown open my
window looking out over the shallow torrent of the Po, and there
found delight in listening to the melody of the water, blending
with the hollow *shrill* of the Leaf Crickets (*Conocephalus*) from

the tufted banks, as though it were the river god breathing his wreathed shell until morning should go forth. How little do we know of this life intense, and what strange mythological fiction has it evoked in bygone time!

I had long heard echoing from inaccessible woods and suburban avenues a note which made me think of some large tropical cricket, although the producer of the sound remained unknown to me; one day, however, reclining about noon in a valley near a vineyard, I caught the same noise produced just over my head, and on asking a man, taking a siesta hard by, he assured me it was the Cicale. Finding a pathway winding up to the spot, I mounted and examined the vine poles, where I caught in a stroke of the net a large black Cicada, powdered on the body with a substance resembling mildew. On returning home and referring to my notes, I read " Scutello apice bidentato, elytris anasto-mosibus quatuor lineisque sexferrugineis;" which, with help of a more detailed description, clearly determined the insect to be the *Cicada Plebeja* of Olivier, and *Fraxini* of Fabricius ; but whether this or an African species is the kind named *Plebeja* by Linnæus appears disputed.

Though the choice recitative of this classic entity scarcely commanded the deafening hum of the mill-wheel, to whose clatter it loved to sit and sing, and much less the roar of the cataract, when once within harbour of the light acacia sprays, he proved no mean orator. The poets call him αὐτοφυὲς μίμημα λύρας, and by taking a few lines of Virgil, and reciting them with a good broad Neapolitan sing, no bad idea of the rattle of his notes is conveyed : *A-shee! a-shee! a-shee!* . . . *whee! whay!* The last honeyed harp-like refrain, *Whee! whay !* breaking from the leafy cover as the voice of a dryad ; as something indescribably strange and heavenly. During the rehearsal the classic Cicada may be observed crawling crab-like backwards and forwards over the bough, ever beginning anew this summery tune ; and then, when the notes have fairly died away, there arises from the flowery meadows around a whisper as of a sleeper breathing heavily ; an ominous sound I once traced to a heavy-gaited green frog con-cealed in a grassy tuft.* " By Jove," says Plato, in the opening lines of " Phædrus," "what a charming place for repose ! It might

* This Cicada is sketched in the act of singing, on Plate III.

well be consecrated to some nymphs and the river Achelous, to
judge by these figures and statues. Taste a little the good air
one breathes. How charming, how sweet! One hears as a
summer noise an harmonious murmur accompanying the chorus
of the Cicada." No one, however, has sung of the brisk music of
the ἠχέας τέττιξ better than Meleager of Gadara, whose Syrio-
Greek poems are always delicious.

> " O, shrill-voiced insect ! that with dewdrops sweet,
>   Inebriate, dost in the desert woodlands sing :
> Perched in the spray-top with indented feet,
>   Thy dusky body's echoings harp-like ring.
> Come, dear Cicada ! chirp to all the grove,
>   The Nymphs and Pan, a new responsive strain :
> That I, in noontide sleep, may steal from love,
>   Reclined beneath the dark o'erspreading plane."

The word "Cicada" has been derived from *ciccum*, a thin
skin, and also from *cito*, quickly, and *cadere*, to fall, a compound
suggestive of their being short-lived ; while ᾄδειν, to sing,
is stated to signify intrinsically a sound produced by motion of
a pellicle. A Cicada in ancient days was the emblem of music
among the Egyptians and Greeks ; and according to Polybius its
effigy was struck on the coins of races who claimed superiority
in that art, as the Messenians in Arcadia, and Locrians in Italy.
In the rich collection of the Vivenzio family in Nola it is stated
there exists a vase of baked earth on whose exterior is depicted a
humorous representation of a poet placing in the flickering flame
of an altar his lyre, from the strings of which some Cicadæ are
springing ; an allusion, we presume, to the tale of Ariston,
respecting which Strabo, with what sounds like mock gravity,
tells us that there existed this peculiarity in regard to the river
Halex, which, gliding between deep banks, divided the lands of
Regium from those of Locris, namely, that the Cicadæ on the
side appertaining to the Locrians were vocal, but those on the
other side mute, and that the reason of this some deemed to be
that the latter were in a shady place, and therefore, owing to the
dew, their membranes were not distended ; while the former,
dwelling in an open spot, had dry and horny membranes, which
easily emitted sound. Another version of the story is that
Hercules, wishing one day to sleep on this bank, was so tormented
by the " sweet eloquence " of the Cicadas, that, furious at their

concert, he asked the gods that they should never sing there for ever more, and his prayer was immediately granted. The truth of the tale, however, one would be inclined to find in the fact that the Locrians were Greeks and aliens, and formerly conceited of their musical tradition. In the days of Strabo there stood at Locris a statue to Eunomus, with a Cicada sitting on his cithara. And the legend went that two musicians, Ariston the Locrian and Eunomus of Regium, were contending in song. Ariston claimed for the Sun God his seat at Delphi; but this Eunomus maintained did not suit the dwellers at Regium; and as he warmed to his argument the very Cicada on the bough ceased to sing. Ariston, however, was by no means pleased at this, and still hoped for victory. At this moment a chord in the lyre of Eunomus snapped. The free-born Cicada was not to be beaten in his argument, and, flying to the spot, supplied the native sound of the string.

> " Phœbus, thou knowest me—Eunomus, who beat
> Spartio; the tale for others I repeat.
> Deftly upon my lyre I played and sang,
> When 'mid the song a broken harp-string rang;
> And seeking for its sound, I could not hear
> The notes responsive to my descant clear.
> Then on my lyre, unasked, unsought, there flew
> A grasshopper, who filled the cadence due;
> For while six chords beneath my fingers cried,
> He with his tuneful voice the seventh supplied.
> The mid-day songster of the mountain set
> His pastoral ditty to my canzonet;
> And when he sang, his modulated throat
> Accorded with the lifeless strings I smote.
> Therefore, I thank my fellow-minstrel—he
> Sits on my lyre in brass, as you may see."

In Northern Italy the duration of the classic Cicada in the perfect state is brief. Emerging at the commencement of July, the rattling of the males is heard only for about three weeks, and but a few days longer than the more abundant Blood Cicada. About the same time emerges the Cicada of the ash, *Cicada orni*. It is distinguished externally by a smaller size and spotted wings, and Linnæus describes it in his brief method as " Elytris intra marginem punctis sex concatenatis; anastomosibus interioribus fuscis." In the south of France this kind is excessively abundant in the forests of maritime pines at Bayonne

and Bordeaux, where, according to Léon Dufour, there remains
not a trunk of ash where the perforations of its grubs are not
manifest. In Spain also it is common in the environs of Malaga,
where it resorts in preference to the olive trees and stalks of the
agaves. Near Turin, where the southern fig and olive-yards
are replaced by slim oak forests and vineyards, with trees of a
temperate character, this Cicada on first emerging betakes itself
to the sunny vine poles and tree trunks; but, soon scorning such
lowly haunts, the males fly up on to the high poplars that line
the streams, or nestle among the sun-chequering umbrage of the
solitary plane-tree walks, and here ring out a busy and monotonous murmur wide over the landscape, *Chip! chip! chip!*
There is a pretty Greek lay to an oak-frequenting Cicada of such
lofty habit in the " Anthologia "—

> " Aerial branches of tall oak, retreat
> Of loftiest shade for those who shun the heat;
> With foliage full, more close than tiling, where
> Dove and cicada dwell aloft in air.
> Me too, that thus my head beneath you lay,
> Protect, a fugitive from noon's fierce ray."
>
> GOLDWIN SMITH.

Since the music of the Cicadæ, like that of other diurnal
choristers, is intrinsically harsh, harmonious alone in sunlight
amid rural scenes, this in itself may constitute a reason why it
should be distasteful to the melancholy cast of the Italian, and
more accordant with the Greek and Germanic tongues. There
is nothing refreshing in the melody when contrasted with the
fitful twitter of a belated Leaf-cricket from a shaded willow,
whose note comes with the trickling coolness of a mossy spring;
it is essentially a type of Nature barren and dry, a creature of
the dog days. Certain is it when we pass from Greece to Italy
the music of the Tettiges is universally condemned. The song
divine that supplied a broken harp string in the lyric games
becomes a personification of conceit in " Phædrus," where the very
night owl hooting around the Capitol was weary of this monotony
of the Pincian.

> " If ways do not accommodate they're tried,
> And oft receive the punishment of pride.
> A Cicada was jeering of a day,
> A Night Owl in her harsh and noisy way;

Who, weary with the business of the night,
Within a hollow bough was slumbering light.
The Owl, grown testy with the clamour, sighed :
' Keep quiet, can't you ?' but the more she cried,
And then again requested, 'Stop it, or——'
But she fired up, and dinned, the more, the more.
The Owl now twigg'd there was no other way,
Her words despised, to stay the babbler's play,
Devised at length a means to take her in,
And sleep in freedom from the noisy din :
' Sweet are the notes thou deem'st that thrill thy lyre,
Sweet as Apollo can alone inspire ;
But come, I have a mind to drink a draught
Minerva gave me lately to be quaffed ;
Your throat is parched ; take counsel, drink with me.'
The Cicada, who thought that she could see
Her voice was praised, all artless to her went.
The Owl down swooping from the hollow pent,
Her trembling chased and packed to Lethe's wave ;
So that refused in life in death she gave."

But not alone were the pipings of the Cicadæ painful in
civic ears; in this opinion, we find, acquiesces Virgil, with a
most exquisite perception for the mournful sound of the foaming
waves and chill starlight on the sea ; the enchanted rush of the
homeward-freighted galley round Circe's cape; the idle and
glassy calm off Tiber's mouth; the whistling breeze and
howling tempest, with all those vicissitudes the weather-beaten
voyager encountered in olden time who hoisted sail from Italy
to polished Attica. Farmer Virgil, who possessed a feeling no
less intuitive for the echoing song of the grape-pruner, and
dream-inviting murmur of the April bees at his sallow fence,
nevertheless, as we have seen, thinks the ten o'clock melody of
service to the herdsman.

If we turn to Hellenic libraries, we find, on the contrary,
the deified Tettix a favourite of every bard, from Homer and
Hesiod to Anacreon and Theocritus. Prophet of the dog-days
and their white lilies ; delight of wayfarer, and solace to the
lover reclined in the fragrant shade of plane-tree or dreamy
oak—no less was it favoured in the City of the Violet Crown ;
for if its clear wayside minstrelsy had led a forgotten priest
of Egypt to portray it as a hieroglyphic of music, we may be
sure no scoff of Æsop's, that it piped away the summer hour
and might dance in winter, could deter the children of Athena

and lineage of Cadmus from coquetting its golden charm in
their flowing hair, when the blood had crimsoned the sacrificial
axe.

> " O ! thou of all creation blest,
> Tettix ! that delight'st to rest
> Upon the wild wood's leafy tops,
> To drink the dew that morning drops,
> And chirp thy song with such a glee,
> That happiest kings may envy thee !
> Whatever decks the velvet field,
> Whatever buds, whatever blows,
> For thee it buds, for thee it grows.
> Nor yet art thou the peasant's fear,
> To him thy friendly notes are dear ;
> And still, when summer's flowery hue
> Begins to paint the bloomy plain,
> We hear thy sweet prophetic strain,
> Thy sweet prophetic strain we hear,
> And bless the notes, and thee revere !
> The muses love thy shrilly tone ;
> Apollo calls thee all his own ;
> 'Twas he who gave that voice to thee,
> 'Tis he who tunes thy minstrelsy.
> Unworn by age's dim decline,
> The fadeless blooms of youth are thine.
> Melodious insect ! child of earth !
> In wisdom mirthful, wise in mirth ;
> Exempt from every weak decay,
> That withers vulgar forms away ;
> With not a drop of blood to stain
> The current of thy purer vein ;
> So blest an age is passed by thee,
> Thou seem'st a little deity."

Or if desire arose to catch and imprison so vagrant a
spirit, and, placing it on rosy fingers, to drink in its floating trill
awaking summer's echo, it was then that the notes em-
bodied a sophistry, whispering to earth-born beauty of metamorphic youth and bloom in fields of asphodel, with eternal
banquet on celestial music, perchance on some ocean isle fanned
with the sea-breeze, and yet unknown to the mariner; or,
perchance, far away among yon stars, nascent in the dim
penumbra of the clear Grecian sky. There Leander might ever
court his Prote, or Faust his Marguerite, or the Turk be loved
hereafter by the black-eyed houris with green handkerchiefs.

But we also gather the noontide Cicadæ were not the

sole minstrels that fed the Grecian ear. Their painted atriums
were enlivened with the sunny creaking of diurnal Leaf-crickets,
and thrilled at the evening *shrill* of such as live darkling; and
many a neat little capricci has been engraved on gems by a
generation of forgotten poets scattered over the Archipelago
and in Sicily, to the transient guests of their basket-work cages,
whom they invested with a sanctity akin to that held by the
robin in England and stork in Germany.

> " Why, ruthless shepherds, from my dewy spray,
> In my lone haunt, why tear me thus away?
> Me, the nymph's wayside minstrel, whose sweet note
> O'er sultry hill is heard, and shady grove to float ?
> Lo ! where the blackbird, thrush, and greedy host
> Of starlings fatten at the farmer's cost!
> With just revenge those ravages pursue,
> But grudge not my poor leaf, and sip of grassy dew."
>
> WRANGHAM.

The poet Meleager, to attune his lyre, sought the golden
corn to capture the Locust sounding his sweet-speaking wings
with his feet; and as breakfast promises him garlic evergreen,
and pearly dewdrops that will melt in his mouth. Love and
resignation were alike learnt in these rural harmonies. The
maiden sitting in sunshine, listening to the chaunt of the Cicadæ
on the fragrant boughs, commingling with the rattle of the
Grasshoppers, forgets her lover and her tears; and one poet
deems death itself unrepulsive should the Cricket of the briar
raise over him a monument of imperishable strophes; but in
this he is rebuked by a mundane brother, who desires not to
dispense with the usual ritual. The autumnal death of the
performers themselves, and their flight to the dewy chalices on
the meads of golden Proserpine, were alike fantastically bewept.

Even now a lingering love for the music of nature per-
vades the South. The peasants of Andalusia hang the House-
cricket at the artificial summer of their Christmas fire, for the
sake of his pleasing but wearisome *Cree-cree*. The vernal
echoes of the Field-cricket have preference with the Turks, probably from their tendency to fill the mind with a train of
summery ideas of everything that is rural, verdurous, and joyous,
and therefore a suitable accompaniment to the western narcotic.

These are enclosed in paper cages, and fed on lettuce-leaves; but they also exhibit a penchant for the condiments of the pantry and store-closet, with an aptitude for domestication. Gilbert White, who made the experiment of hanging them in his sitting-room, tells us, " One of these Crickets, when confined in a paper cage set in the sun and supplied with plants moistened with water, will feed and thrive, and become so merry and loud as to be irksome in the same room where a person is sitting. If the plants are not wetted, it will die."

A rival candidate for the honours of the parlour is the Leaf Cricket, of which there are many species. One giant kind has ravished the untutored ear of the native of the Amazons with its midnight *shrill;* another is sold, enclosed in cages of fancy-work, in the streets of Shanghai, to fill the houses with its minstrelsy, and infuse a spirit of seclusion. There is an especially lively one in Italy that about twilight appears spectre-like on the dark acacia-tufts which shade the summer arbours, and from whence the male twitters stilly response to the strolling banjo, as once he may have hailed the lyre and barbiton ; so that were the ear adapted, the species had time to learn their modulations.

Nor is a music so full of poetry and so widely honoured wholly unknown to science. Many have been the attempts to render the songs of the Grasshoppers in music. Yersin, of the Vaudois valleys, who died young, was, I believe, one of the first to produce a score of the snatches heard among his Alps, and along the sunny Riviera. Brunelli, further back, was accustomed to keep a band of the Great Green Leaf Cricket in a cupboard, where they formed an orchestra, and whiled the day with recitative. The enterprising professor chirped a key-note, when at first a few of the boldest would answer, and gradually the whole choir struck in, and stridulated with all their might ; refreshing interludes were obtained by a rap at the door. Recently, a well-known author has testified to the pleasing nature of a solo from a select male of this species, confined under a glass on the table, which, as his music is only a little less deafening, might be preferable and more enjoyable than a Canary's.

Although Field Crickets and Leaf Crickets live well enough

in confinement—the only drawback to enclosure being an
irascible temperament, which causes them, like savages, to kill
and devour one another—it is not so with the House Crickets,
who, if they survive removal from the fire, require peculiar
management. Roesel recommends procuring a large glass,
filling it three-quarters full of earth, and covering the mould
with a piece of earthenware having a hole in it, so that the
individuals can burrow or not, as they prefer. And in some
parts of Africa, where such commodity has market value, the
natives find it possible to rear them in hot ovens. The Grasshoppers and, I conclude, the larger Locusts, will live a little in
a vivarium, and, if placed in the sun, chirp a snatch or two;
but they are very ephemeral, and only less so than the Cicadæ,
which are little adapted for pets, and how the ancients managed
to encage them is difficult to understand. May it be that
the play of light striking in through the aperture above the
impluvium, with the vicinity of fresh flowers and a sparkling
fountain, could place them in congenial conditions where those
dewy sprays might be had, said to be as necessary to a Cicada
as a bottle to a baby? They seem, anyway, quite to scorn our
ordinary breeding-cage, where they cling together like Bees,
and when placed in the window die of sunstroke within a few
hours : at least, such was the fate of those I collected among
the Superga Hills.

In conclusion, I may mention that by the first of August the
music of the Tettiges had faded from the groves in the neighbourhood of Turin; so, having learnt as much of their habits
as I wished, and found by actual dissection the part usually
termed the "mirror" in these insects is in reality an organ of
hearing, I determined on wending homeward. Several routes
were open to me for the return journey. I could either whirl
over the dizzy crags and snows of the Simplon or Great St.
Bernard, make my way by the Maritime Alps to Genoa or Nice,
and thence to Marseilles and Paris; or I could take the direct
route through the bowels of the Simplon, over whose snows
and mists the Honourable Lady Mary Wortley Montagu, in
the beginning of this century, was carried in a little seat of
twisted osiers fixed on poles upon men's shoulders, so that she
arrived that night at Lyons with a terrible fever. There is

nothing dreadful, however, in the tunnel, save the ordinary risk
incurred on the rail; and having settled on this route as the
most direct and economical, Macon and Paris formed but
pleasant breaks on the road.

The object, then, of the present work, the faithful companion
and solace of my peregrination, is not to teach science dog-
matically after the fashion of a complete text-book, but rather
to stimulate a taste for individual research by indicating the
drift of modern investigation. Indeed, it seems to the author
not a little surprising that so much should be known regarding
the physiology of the marine Mollusca, and that we should
still possess so little certain acquaintance with the corresponding
life-history and structure of Insects; and this must be the
more especially felt now that geology is taking rank as a
comprehensive pursuit, the contested theorems of which can
never be permanently established until specialists are found who
will devote their attentions to these much-neglected practical
subjects.

Inconceivable aid to a fuller knowledge of the biology and
rank of species would be afforded by experts and insect fanciers
if, when out in the fields capturing their specimens, previous to
rendering them mere bits of innate form and formate colour by
the traditional pinch, acid, prick, or immersion in spirits, they
would first note down the date and locality of capturé, and
placing them successively to the nose and ear, test their capabilities for music and odours. The results, if any, should be
recorded on a label attached to the pin, or, if travelling, on the
dogs'-ears and envelope corners to which butterflies and the more
delicate booty are usually consigned. A few hurried cyphers
made at the time speak volumes — "$^1/_v$/79, New Forest —
Stridulates, Ion mesothorax;" or, "Odour, musk." And afterwards a tally with fuller particulars and studies of habits, made
either on the spot, or in gauze-covered or glass-topped boxes, such
as are sold for geological specimens, or in an aquarium or
vivarium, together with such information as can be gained from
the lens, microscope, or other philosophical instrument, may be
inserted at leisure in a pocket-book or journal. From this data
a series of interesting cabinet labels may be easily constructed,
showing the stains of the secretions if not their analysis, or the

music and its modulations adapted to syllables fixed by intonations of the human voice. If the insect should, on the other hand, be reared from the larva or ova on a potted plant, or in the breeding-cage, or be forced under artificial conditions as a variety, invaluable reference will be afforded by attaching a label indicating its food plant, locality, and date of emergence, which will hereafter greatly tend to fix its identity. To fully record its metamorphosis, with its acquirement of lines and eyes, its changes of colour, and new members, a series of observations, accompanied with sketches, are required at each casting of the integument. Caterpillar skins may likewise be preserved by inflation, a method that has led to good results.

It is needless to say dissections of insects when freshly killed are easily made by means of a camel's-hair brush; to remove scales, a pair of nail scissors and a pocket lens are necessary. To examine the digestive organs it may be necessary to immerse the subject in water; and to examine the air-tubes, the insect may be cut open along the back, the skin pinned out, and the preparation set to dry. Even the delicate nervous system may be worked out to some extent by placing the insect previously in alcohol for a night. But different operators have doubtless different methods. The various structures are also capable of being prepared and mounted in the usual fashion for the microscope, but greater results will be obtained of their relations by using this instrument subserviently for the purpose of drawing up diagrams. Micro-photography would here be invaluable could it be only adapted to reproduce the delicate tissues and organs which are generally poorly rendered by drawings made under the camera lucida.

The present volume, exhibiting the scope and bearing of such occupations, may therefore find some recommendation, both as regards a novelty of arrangement and as regards such original matter incorporated; and the latter will, it is believed, be found to considerably extend the special illustration of such works as Dr. Darwin's " Descent of Man," Kirby and Spence's " Introduction," or Rennie's " Miscellanies." The concluding chapter on the variation of species will afford some information on a favourite topic with butterfly fanciers, and is in a measure an attempt to epitomise the labours of insect collectors and classifiers.

D

Space and inclination, however, forbid me to enter largely on the finespun theories of mimicry and protection now rife.

With the more curious and general reader, as treating of the music, dances, ornament, and distribution of a class whose ways contribute in no small measure to our enjoyment, and constitute an heirloom in our literature, this little work may also, I trust, find favour. How few, for example, not professedly entomologists, could tell you how a grasshopper sings or why a gnat dances ; and I despair not that even the physicist and mechanician may devise from a study of insect organs of music phonetic arrangements that shall hereafter be struck to charm anew the human auditory nerve, or that shall rank with the steam whistle as a herald of danger. The importance of this branch of natural acoustics to the electrician has already been manifested in the construction of the telephone.

Lastly, I have to tender my sincere acknowledgments to many leading entomologists, who from time to time during the last ten years have tendered me assistance, by naming my species, or by bringing my pursuits into public notice ; and authors and others to whose patient researches I am more largely indebted, will I trust, as far as literary compilation admits, find courteous mention in passing.

Guildford, 1880.

CHAPTER I.

Metaphysical Incentives to Reproduction and Distribution—The Germs of the Passions, Fear, Rivalry, Love, and Maternal Care, shown to exist in Insecta, and their Expression in Contractions and Secretions—Corresponding Senses of Touch and Smell.

No historian, poet, or philosopher, devoting himself to the study of mankind, has omitted to comment on the ruling bias of the passions, whether as regards the loss caused by their unbridled indulgence, or the incentive they afford to all commendable actions. Arising directly at the intimations of pleasure or pain communicated by the senses, they precede thought, reason, judgment, and other qualities of the mind, and in the savage divert the very members from rational machines to watchful weapons of defence. The emotions or stimuli of these metaphysical passions alone have their source low in the animal kingdom, and these we shall show pervade *Articulata* coexistently with the origin of complete organs of sensation in a multiple form.

Voluntary actions evincing these incentives are, it has been said, of two kinds, as they proceed from design or propensity; and in performing one of these kinds, the mind itself has an object in view, and is properly the source whence they originate. But in the other, the mind is merely a secondary agent, acting under the influence of stimulants, and frequently not aware of the consequences, or, although aware, often so infatuated as not to regard them, however fatal. It has been also long known to the naturalist that not a few of these propensities arise from the form and structure of the body, from the manner in which the optic nerve is affected by colours, the olfactory by smells, the gustatory by taste, and the auditory by sounds; from the different ways in which the fauces are affected by thirst, the stomach by hunger, and the genital parts by orgasm. Besides

these and other propensities which operate as stimulants in the
system itself, the naturalist has found that light, heat, and
moisture, in various degrees, from absolute darkness, coldness,
and dryness, act as stimulants upon living bodies; he has
experienced that electricity is a general agent, that several
plants emit flashes, and that some animals even give shocks
resembling the electric, while galvanism produces, and, as many
consider, originates, both muscular and nervous action.

Nothing exercises a more despotic power over the fortunes
of lower life than habit and instinct. Witness in proof of this
statement the perfection and unerring nature of the acts and
duties in which the bee, wasp, or ant engages from the first
moment of existence. The very excellence of the acts performed
by these unreasoning creatures is, as Dr. Carpenter has remarked,
a proof of a non-intelligent nature. Insects likewise from
organisation, one would infer, must be especially a prey of
propensity or instinct, since the double ventral chord receiving
the several ramifications of their *nervous system* at the rings of
the body, in a series of knots or ganglions, has its first swelling,
to which we commonly assign the name and notion of brain,
either little different or so much smaller than the rest, which
equally receive their groups of nerve branches, that we might,
in a majority of cases, suppose it rather formed to be guided by
their indications than suited to control them. The reader will
readily understand this by referring to Plate VII., where the
nervous systems of various insects are sketched out. Then as the
insect grows and develops to its last adult condition, the knots or
ganglions take corresponding change. In the moth (Fig. 3) we
see the first in order are larger than those of the caterpillar
(Fig. 5); they are, in cases where they are also marked in the
larvæ, commonly less numerous, and, in the present instance,
two of the larger appear to have approximated. Among insects
in their final perfect condition, this singularity is best seen
in the nervous system of the Cicada shown at Fig. 1. But with
this change in the nerve-knots, change in the habit and instinct
may be associated; and in the pronounced case of the crawling
caterpillar and flying moth an almost complete revolution is
perceptible. An eminent French zoologist, Dr. Virey, has
compared the animal in this case to a hand organ, in which, on

a cylinder that can be made to revolve, several tunes are noted; turn the cylinder, and the tune for which it is set is played; draw it out a notch, and it gives a second; and so you may go on until the whole number of tunes noted on it have had their turn.

Lamarck, as is known, defined *Insecta* as sentient animals that obtain from their sensations perceptions only of objects—simple ideas they are unable to combine to form complex ones, which intelligent animals not only do, but retain notions thus formed. Since the promulgation of this formula, fuller knowledge has resulted from the investigations of De Geer, John Hunter, Hubers, Kirby, Rennie, Burmeister, Goureau, Müller, Siebold, Westring, Landois, and others, as may be seen in the masterly *résumés* of Dr. Darwin and coadjutors. And consequent on completer recognition and partial localisation of the sensorial organs of touch, taste, smell, and sight, in caterpillars, and other larval forms of insects, and of hearing in *imagines*, their action on the nervous system has become patent, as determining emotions cognate with the passions of fear, love, rivalry, and maternal care, whose biological operation on the organic being manifests itself in reproduction and distribution, inducing variation and selection.

We likewise discover the intensity of these stimuli commensurate in their spontaneous indication as contractions, secretions, battles, display, and dances; vocal music, wing beatings, and instrumental music; or in migrations; characteristic of the third or perfect state of insects—and despite of the disproportionately larger posterior ganglions, we should be predisposed still to infer them determined in the capital knot. And this conjecture is strengthened by the impressions of sight, taste, and odours directly conveyed here through the nerves, as by the bites and antennal caresses lavished on the part; although, on the other hand, strange to say, disseverment asserts functional volition resident in those segmental ganglions immediately connected with the other centres of nerve ramification, which also receive afferents from organs of sense. For if the tales of wasps who have continued to imbibe sugar-and-water after decapitation be well authenticated, no less is it certain that there resides in the other rings a power to perform special actions. I remember observing the body of a yellow

Hover-fly I once by accident decapitated over night, standing
immovably on the window-sill the ensuing morning, and occupying itself, while I was engaged in writing, between the hours of
nine and twelve, in cleaning its wings with its hind legs and its
three pairs of legs by rubbing them in a determined manner
together, raising its forelegs vainly in air as if searching for its
head to brush up.   Whether such vitality in insects be an
instance of nervous reaction, continued respiration, or a phenomenon of life, is not so clear.   All but vegetative processes are
held to cease in the animal with the suspension of circulation ;
and as regards circulation in insects I believe much remains to
be learnt.   The *dorsal vessel* we notice pulsating along the back
of a caterpillar was not held by Marcel de Serres to constitute
a true insect heart.   He describes it as a closed vessel with no
visible opening, composed of two membranes, one internal and
muscular, the other external and cellular, and pervaded by a
close interlacement of tracheæ or air-vessels.   When opened,
its interior presents a transparent coagulable liquid, which dries
rapidly, and then exhibits the aspect of gum, of a colour seldom
deeply defined, but sometimes greenish, orange yellow, or sombre
brown.   Its contractions vary singularly in different species.
Thirty-six per minute in the caterpillar of the large Emperor
Moth, eighty-two at least in grasshoppers, and a hundred and
forty in one of the ground bees.

Very various have been the subsequent surmises regarding
this vessel and its function, until Dr. Bowerbank in the old
Entomological Magazine, No. 3, p. 239, at length fairly established a true circulation existing in the immature state of a
May-fly.   "The blood, abounding in flattened oat-shaped particles, may be seen circulating in every part of the body, not in a
continuous stream, but at regular points, in accordance with the
pulsations of the great dorsal vessel.   The latter, which is of
great comparative magnitude, extends nearly the whole length of
the body, and is furnished at regular intervals with double
valves. . . . .   The structure of the upper valve appears to
consist of a duplication inwards and upwards of the inner coat
of the artery ; that of the under of a contraction and projection
of the like parts of a portion of the artery beneath, so as to come
within the grasp of the lower part of the valve above it."   The

writer inclines to the opinion that a much greater portion of the
circulation than we can clearly define is carried on within special
vessels, as the blood may be frequently seen flowing within
curved and other lines, as if confined within very narrow limits,
and in the caudal extremity the ascending and descending vessels
are seen, like vein and artery, to accompany each other, and at
the same moment that the fluid passes up the one with the usual
pulsatory motion it descends the other. There is, however, no
perceptible pulsation of these minuter vessels themselves, and the
motion of their fluids therefore results from the action of the
great dorsal heart."

Possibly our imperfect knowledge of the circulatory system
of insects may be due to a very natural objection to vivi-
section, which does not affect us in any degree in the case
of an oyster or whelk; for, on lately dissecting an example of
the Privet Hawk Moth I had killed in the usual manner and
had assumed to be dead, a very different arrangement presented
itself from what the elder anatomists had previously led me to
conceive. I traced first the well-known dorsal tube (Plate VI.,
Fig. 11, B), yet clinging by its length to the upper portion of the
abdomen, and marked throughout by a *green* tint, doubtless
derived from the fluid enclosed; and I noted down its waning
valvular movements. I then gently removed the intestines,
when to my surprise there appeared another quite similar tube
(c), ventrally, and of an *amber* colour, whose upper fold moved
in similar fashion to the valves of the dorsal vessel. I traced
this forward to the junction of the thorax and abdomen, where
it could be plainly observed to unite with the upper vein in a
dilatation (A), that constituted a distinct vessel of a flat-roundish
form. I next noted down a two-fold alternating pulsation in
this vessel, that indicated a circular flow of the fluid as shown by
the double-headed arrows; and which certainly impressed me
with the notion of a rudimentary heart composed of an auricle
and ventricle, such as exists in mollusca.

The two main tubes had besides several efferents, *a, a,** and

* Dr. Vitis Graber ("Die Insecten," 1877, T. 1, s. 328—345) also mentions the
ventral vessel which he has noticed in Dragon-flies and Grasshoppers, and he
is of opinion it was first observed by Réaumur in the Saw-fly of the rose. It
should be regarded in the light of an artery to a dorsal vein.

those to the lower one seemed to open each time the flap or fold
spasmodically moved upward; while a central cylindrical duct
(B) passed from the hypothetical heart (A) ventrally into the
thorax, where its rhythmical action could be at intervals seen
extending as far as the second annulation, although the forms
of its vessels were obscured, from the fact that circulation was
already partially stayed in this portion of the body. Lastly,
the lower tube and flat-roundish vessel continued to palpitate
vigorously long after the valves of the green dorsal tube had
ceased to move. However we may consider these facts, this pulsation of the roundish vessel and its afferents points, I think
it will be admitted, to the equivalent of a heart in insects, which
circulates and aerates the fluids manufactured by the stomach,
and affords those secretions necessary to the manifestation of
the simple emotions, the evidence of whose existence we have
now to consider.

Simple muscular contractions indicative of fear are witnessed
in larvae, pupae, and perfect insects, and may result from touch,
sight, or hearing. Caterpillars and other larvae when touched
will contract or convolute into a ball; pupae when complete with
flexible rings, as the chrysalides of Lepidoptera, wriggle the
segments. Many beetles on touch turn lethargic, forcibly contracting legs and antennae, often into grooves of various depth,
possibly habitually acquired. This is characteristic of the Pill
Beetles (*Byrrhus*), and the families Rhynchophora, Coprinidae,
and Coccinellidae, as of the more elongate kinds of Elateridae. The
latter, familiarly termed "Skip-jacks," turning on their backs,
escape when the paroxysm is past by leaping high in air
(Plate I., Figs. 2, 3). Some contract their legs and depress their
heads either slightly, as species of Necrophaga and Malacoderma,
or receive them into grooves, as the Weevils. The male alone
of a red and blue *Clerus*, frequenting Umbelliferæ, appeared to
me thus affected, while his female sought to escape. Some lamellicorn beetles, in lieu of contracting their legs, rigidly extending
them as the common Dung Beetle, porrect the front pair and
draw up the hinder as its congener *Sylvaticus*, or indifferently
contract and extend them as the Rose Beetles. The active
Ground Beetles again, only partially recede them on alarm;
crouching down or squatting like rabbits and partridges. The

Stag Beetles, when arrested at twilight on a pathway, stand motionless as if paralysed, gaping with their mandibles, and when seized fall theatrically on their backs. The Bugs and many Neuroptera on touch similarly contract their limbs; the latter, with many moths, retaining their wings also stiff and motionless. Among the Bees, the shining red and blue Chrysididæ, often noticed on brick walls, involute their abdomens as wood-lice on seizure; but the Bumbles and others turn on their backs, whine, and extend the four hinder-legs like a cat on the defensive. These spasmodic symptoms not uncommonly result from sight or hearing. Bees also raise their legs and larvæ jerk their heads when light is intercepted. A dark trailing cloud-shadow passing over a colony of the Orange-ringed Caterpillars of the rag-weed (*Callimorpha Jacobæœ*) causes quite a sensation of wig-wagging while it lasts. I likewise remember once, on approaching to pill-box a very minute vernal beetle near London, seeing its limbs yield, when down it came, like a dust-shot, ere I could reach it. It must have seen my shadow! Species of other orders, the smaller moths especially, thus elude the ardent collector among herbs and grasses.

Of the butterflies it is noticeable that the majority, when confined in a pill-box sink into torpidity, so that some indeed, as the Peacock and Tortoiseshells, with others of the genus *Vanessa*, may be afterwards actually taken out and handled without moving; and some others, as the *Melitæ*, on the imprisoning lid being opened, escape, flutter, and then drop down tragically, as though dead. As a corollary we find many of these Day-fliers, and the *Syrphidæ*, or Hoverers among flies, exhibit similar lethargic symptoms on a cloud supervening; when species before so wary and active in gardens and shrubberies may be boldly gathered from the leaves and blossoms with the hand. The metallic purple and olive-coloured flies of the sunny genus *Sargus* are especially light-apathetic. Hybernation, or living through the winter, exemplified in this country by the Brimstone Butterflies and species of *Vanessidi*, may be viewed as a prolongation of this phenomenon, since on the influx of a brighter atmosphere with gayer meadows these regain their activity. Benumbing cold, however, is doubtless favourable to the continuation of this transitory sleep, participated in by many wasps, bees, and our own domestic flies, which in October resort to coigne and crevices, or the Shard-born Beetles (*Geotrupes*), who then bury themselves in the ground, thereby showing protection is conjointly connived at. It is also no less remarkable that the most sensibly lethargic Lepidoptera, as the Peacocks, Admirals, Tiger Moths, and Burnets are likewise most conspicuous from beautiful colour.

Muscular contractions on touch or sight, accompanied by emission of liquids or odours from secretory glands, likewise indicate fear. The larvæ or grubs of certain beetles when thus stimulated drop saliva, or exude volatile fluids from two or more dorsal tubes; in the case of the Red Poplar Beetle (*Lina tremula*) smelling of naphtha, while those of the Stag Beetles strongly savour of guano. In certain lepidopterous caterpillars, as those of the genera *Danais* and *Papilio*, these scent-tubes are protrusible. The secretion of the caterpillar of the common Swallow-tail Butterfly, redolent of pine-apple or fennel, is given off powerfully when by pressure between the finger and thumb it is induced to protrude its forked excretory tentacle from the second segment; and that of its congener *Podalirius* emits under similar circumstances a scent which has been compared to that of ripe pears. The caterpillar of the Puss Moth we find even attributed with a voluntary power of communicating electricity. A correspondent to the *Magazine of Natural History* relates, "On being taken from a young poplar the Cerura showed decided symptoms of irritation, which particularly drew my attention. It began to contract its body, drawing itself closely together, and by degrees elevated and extended its bifurcated tail. Then were slowly protruded from out of the points bright red filaments, and irregularly bent to one side. In a short time I felt a sudden tingle along my arms, which made me stop with surprise, and shortly another shock, which made me almost involuntarily throw the twig with the creature upon the ground." The larvæ of the genus *Porthesia* have two ordinary sac-like tubercles, one on either of the penultimate segments. The very beautiful scarlet-lined caterpillar of the common Gold-tail Moth (*Auriflua*), has these tubercles contractile, and above perforated by a keyhole-shaped opening,

in which a clear caustic fluid wells up, and this applied to the
tenderer portions of the human system, as the cheek or eyelid,
produces irritation and inflammation. When the caterpillar is
touched it draws in its hinder segments, and this fluid is transmitted by ejection or evaporisation to the hairy armature, which
it bathes in a clammy and baleful dew. Other hirsute palmer
worms are similarly irritating; many examples occur among the
Saturnidide and *Bombycidæ*, but in the case of the Gold-tail
Moth, as Mr. Meldola has pointed out, we have a singular
example of a caterpillar feeding on the non-poisonous plants of
our hedgerows—sloe, apple, oak, &c.—elaborating a noxious
secretion by a chemico-physiological process.

Some vegetable-feeding beetles on seizure exude rich amber
drops that conglobe at the leg joints. We notice this in the
portly Oil Beetles *Proscarabæus* and *Meloë*, that in spring
consume the spotted arum at the wayside, as also in the
aphidivorus pied Ladybirds, and ash-feeding Spanish Flies,
and in the latter these buttery drops have the noxious roast-
beef taint of the yellow-flowered Sand Mustard. Others of
these beetles drop saliva, as the Bloody-nosed Beetle, so common
on hedgerow bedstraw. These drops have a rich crimson
colour, due, it has been suggested, to alizarine derived from the
food plant; and in another Timarcha, somewhat smaller, I
found in the Isola di Capri, they were of a clear amber tint.
Other kinds exude a milky, glutinous fluid from the pores of the
dermis at all parts of the body, as the genus *Brachynotus*, found
on maple stumps in Canada, or our own willow-loving Musk
Beetles, with certain allied long-horned dryads, although some
say, curiously enough, the smell of tea roses or musk in these
waspish creatures proceeds from metathoracic glandular organs.
One of the lamellicorn beetles (*Osmoderma*) perfumes the trees
it has crawled over with a smell of Russian leather, and the
scarce, bumble-like *Trichius*, with the much-alike Burying
Beetles, are redolent of musk; but other carrion beetles, as the
Silphidæ, disgorge as we handle them nauseous brownish saliva,
in the fashion of vultures, with the taint of sulphuretted
hydrogen.

On alarm, communicated by touch or sight, Ground Beetles
and Staphylinidæ eject volatile corrosive secretions from long

excretory ducts opening into erectile anal tubes, sometimes
vaporising with explosion. The Bombardier Beetles, according
to Leon Dufour, conceal themselves beneath stones, and when
disturbed launch from the anus a white pungent smoke resembling nitric acid in its properties. *Brachinus displosor* (Duf.),
common on Spanish highlands, will furnish twelve such discharges,
but subsequently explosion with noise is replaced by the emission
of a yellowish or brownish fluid which gives out bubbles of air as
if it fermented. The mobility of the posterior abdominal rings
which are not covered by the elytra allows this coleopteron to
bend its abdomen to the point of irritation, whether below or
above; in the latter case the elytra after the explosion become
sprinkled with sulphurous dust. This phenomenon is seen in
either sex. Our native species (*B. crepitans*) is sometimes
gregarious, and then when one individual is disturbed the whole
discharge in unison; but after a few puffs, twenty according to
Roesel and Rolander, they only emit a white fluid. Its chief
enemy is the handsome *Calosoma inquisitor*, Rennie tells us,
which hunts it without mercy. " As it finds it impossible to
escape by speed of foot, it stops short and awaits its pursuer;
but just as he is about to seize it, he is saluted with a discharge." The fugitive beetle and pursuer are portrayed on
Plate I., Fig. 6.

The odour diffused by the little nimble beetles of the genus
Pterostichus, so common near damp walls, well-mouths, and
refuse heaps, flavours of vinegar, smelling-salts, or pyroligneous
acid; and in the small and brassy *Loricera pilicornis*, the
smelling-salt flavour is even more pronounced. The tree-beetles
Calosoma and foreign species of Callidium have a scent of
ratafia. The secretions may likewise be disagreeable to our senses,
as in Cychrus, or they are perfumes pleasant and musky, as we
find them in the meadow Tiger Beetles, of which a sort (*Cicindela
metallica*, Bois.) found in New Ireland adds a balmy fragrance
to the warm air as it flies. The larger ground beetles of the
genus *Carabus*, when handled, eject a saliva which is burning
and caustic if applied to the cheek and tenderer portions of
the skin. With the beetles with short wingcases, the ejection
of the scent from its reservoirs is accomplished by reversion
of the abdomen (Plate I., Fig. 9); and in a small species of

Lathobium (.'), common on pathways during early sunshine of the year, it flavours strongly of ratafia; while in Staphylinus we detect, again, the savour of musk, or in the large " Cocktails " (*Goerius*) we have a drenching black bottle of *vinaigre*.

Herr Erné, in an interesting memoir on the habits of *Velleius dilatatus*, a largish short-elytra'd beetle, parasitic in Switzerland in hornets' nests, where it sucks the honey and destroys intruding vermin, notices a sharp musky odour of such intensity that five or six of these forbidding earwig-like prowlers will sensibly perfume a room.

In aquatic beetles, and the tribe of the Heteromera, whose posterior feet or tarsi have only four apparent joints, the secretory ducts are similar in position, and the odour is nauseous, as in the surface Whirligigs (*Gyrinus*), and in the black and lucifugal larder beetles, Blaps; or the glands are situated at the cephalothoracic joint, as in the water Dytiscus; and their effluvia resembles sulphuretted hydrogen.

Passing to the light and airy bees, we find certain kinds on touch diffuse pungent scents. Though it must have doubtless been observed by the ancients that when an ant-bed was raked up with a stick a strong acid smell arose from its atmosphere, the existence of formic acid would appear to have been noticed first by Dr. Hulse in his correspondence with Mr. Ray, the doctor informing him that certain ants, if irritated, give out a clear liquid that tinges blue flowers red. This acid has been subsequently obtained in quantity sufficient for analysis by bruising ants and macerating them in alcohol, and then distilling over, when an acid liquor remains which, saturated with lime, mixed with sulphuric acid, and itself distilled, yields a liquid possessing all the properties of acetic acid. Another more simple method consists in bruising the insects and distilling, or merely infusing, them in water. In either case the resulting acid is said to possess the following properties : —It reddens blue flowers, flies off in the form of a vapour smelling like musk, is decomposed by a great heat, and forms salts with alkalies and earths called formiats, which are crystallizable and not deliquescent. It resembles acetic and malic acids, but is nevertheless distinct in character according to the investigations of Margraaf, Thouvenel, Lister, Arvidson, and Oehu.

Many ants are pregnant with this formic acid of chemists, but some are offensive with a smell of odour, as the *Formica fœtans* of Fabricius. The hairy Solitary Ants and the Wild Bees (*Andrenidæ*) have a taint of garlic; and some of this order that spin moth-like cocoons, as Bembex and Sphex, emit sudden puffs of ether, as do the little wasp-like species of the genus Crabro; but Cimbex emits a smell of musk. The large yellow European Scolia flavours of certain red aromatic cachous lozenges frequently seen on the counters of apothecaries' shops; and the gall insects of the oaks have, I believe, a disagreeable odour. The secretory glands of the Bugs are situated exterior to the insertion of the posterior legs, and emit fœtid effluvia on seizure, smacking of ratafia; or of that undefinable bug odour that annoys us so much when gathering raspberries; but many of the flat Plant Bugs take besides a very distinct smack of fruit essences—*Capsus tricolor* one of black currants, and Heterotoma is thought to flavour of sliced cucumber; one of the Pentatomidæ smells of the fresh hillside thyme, and the otherwise disgusting scale-like *Enoplops scapha* (Fab.) has a faint scent of the autumn ripe peach; another smells of hyacinth. In the Reduviidæ, or Musical Bugs, we find the ordinary hemipterous odour quite exchanged for that acid vinegary perfume found in so many insects.

The aquatic species of *Corixa* possess the bug odours, and the Water Boatmen, that oar on their backs, smack of the large Water Beetles. The odoriferous organ of the Land Bugs, Leon Dufour found to consist in a rather large pouch, rarely two, situated at the base of the abdomen, just below the digestive viscera, most often of a yellow or orange colour; its raised orifices lay on the thorax between the insertion of the second and third pairs of legs. Seize the common Grey Pentatoma, and plunge it in a glass filled with clear water, he says; "arm your eye with a lens, and you will see a number of little globules arise which, on breaking at the surface, diffuse the easily-recognisable and noxious odour of the bugs. This vapour, essentially acid, will both irritate the eyes, and leave a brown mark on the skin similar to that of a mineral acid."

Certain Cicadæ, according to Mr. Bates, unite the three stimuli of music, colour, and secretion, and the last is employed when frightened or as a defence. He tells us that in the neighbourhood of Para every tree was tenanted by Cicadæ, the reedy notes of which produce that loud, jarring, insect-music which is the general accompaniment of a woodland ramble in a hot climate. "One species was very handsome, having wings adorned with patches of bright green and scarlet. It was very common, sometimes three or four tenanting a single tree, clinging as usual to the branches. On approaching a tree thus peopled, a number of little jets of a clear liquid would be seen squirted from aloft. I have often received the well-directed discharge full in my face; but the liquid is harmless, having a sweetish taste, and is ejected by the insect from the anus." Grasshoppers and Leaf-crickets, on handling, drop brown saliva or emit scent from small tubular anal glands as the crickets and cockroaches; and among the net-veined insects, or Neuroptera, the duller of the Lace-winged Flies, that flutters about blighted garden apple-trees, is especially offensive, emitting the effluvia of odour with a flipping movement of the abdomen when retained by its ample wings. Some butterflies and the Tiger Moths exude clear white or yellow liquid on seizure at the prothorax; others, as the Death's Head, diffuse scents from expansile pencils of hair or "fans," and most of this order, with certain flies, drop solvent red, orange, or white saliva on emerging from the cocoon; a circumstance in the case of the Nettle Tortoiseshells that has given rise to disastrous blood prodigies in some parts of the Continent. Lastly, a small Bumble I retained at close quarters with a House Spider expired in a quarter of an hour—as I conclude, from fright, or possibly poisonous exhalation (Plate I., Fig. 7).

Fear dawning on the nervous system as the experience of pain, and yielding in reaction to desire, merges to anger, prompting rivalry, and producing struggles for existence or effecting a distribution of the species. Simple anger insects, both mandibulate and haustellate, manifest when provoked by touch. Beetles and Orthoptera seek to assail with their jaws, and the warlike bees, turning on their backs and extending the hinder pairs of legs, threaten with their sting or ovipositor. Bugs, often incautiously handled when gathering fruit or flowers, prick little less sharply with their needle-like rostrums. Rivalry again, essentially a masculine attribute, appears also in

savage females, and, prevalent in Insecta, finds its vent in
cannibal battles. On enclosing Orthoptera in either the immature
or perfect state, the males and females fight in an indiscriminate
mêlée, and then when all is over the female survivors subsequently gorge the slain, until, bloated and hideous, they can
scarcely stir from the spot.

It is thus the species of Mantis, or Walking Leaves, poor in
defence, find ready weapons in their raptorial forelegs, adapted,
in common with those of a Neuropterous and Hemipterous
aquatic genus, to forceps for catching flies as food; but with
which, when at close quarters, they guard and cut as Hussars
with sabres, so that, at a stroke, one will often decapitate or
cleave another through. The well-known entomologist and
miniature painter, Roesel, on enclosing together a male and
female of the European Prie Dieu, observes : " They threw up
their heads, brandished their forelegs, and each waited an attack;
nor did they remain long in this posture, for the boldest, throwing open her wings with the velocity of lightning, rushed at the
other, tore in pieces and devoured him." The leaping Orthoptera are equally vindictive with their mandibles. The *Gryllus
monstrosus*, with its curled wing-cases, is a dire cricket, once
seen not easily forgotten. Dobell, in his "Travels in Kamts-
chatka," relates, when the Chinese wish to enjoy a Cricket
fight, they place two males in an earthen bowl of six or eight
inches in diameter. The owner of each tickles his prize-fighter
with a feather, which makes them run round the bowl in different
directions, although they frequently meet and jostle each other
as they pass. After several such meetings, they at last lose
their tempers, and ere long, becoming greatly exasperated, they
fight with such fury that both are literally torn limb from limb,
The same occurs if males, male and female, or larvæ of other
species of Cricket, Leaf-Cricket, Cockroach, or Earwig are
confined together ; while night or gloom, the ancient nurse of
fear and mother of all things, predisposes to the conflict. The
larger Locusts fight in like fashion, and likewise kill themselves
by gluttony during their maraudings.

So wears out the spark of life in all such leaf-like insects,
who seem imbued with an animus near allied to that of many
a lurking spider, who is known to secretly devour her lover in

her snares. Less retiring, but little less moody and fierce, longicorn beetles, when enclosed together, amputate one another's legs and antennæ with a treacherous snip of their powerful and trenchant mandibles, as do several of the soft-skinned beetles, or Malacoderma, who then eat the slain and maimed in the fray. This circumstance does not escape the quick-eyed Chinese, who, in their old-world love for excruciating sport, keep bottles for matching beetle-fights, receptacles in which our school-boys sometimes engage the steel-blue Telephori that appear on hedgerow plants with the May blossoms. There is a tale told of the late Mr. Frederick Smith, who, having captured a number of examples of a rare Oil Beetle near Margate, as they were crawling near the nest of the bee on which they were parasitic, put them into a box, thinking no harm of them; but found that on the second day of their captivity, a free fight had taken place among them, the result of which was that some were killed and reduced to fragments, the greater number of survivors had lost either legs or antennæ, or both, and out of two dozen beetles only four escaped without injury.

Many pachydermous beetles with small jaws, and consequently bad fighters in a state of nature, notwithstanding, evince a coleopterous disposition to the combat. This is recorded of the tree-boring Scolytus, and of certain snouted weevils (Leptorhynchus), that appear in May and June in the Aru Islands. Regarding these, Mr. Wallace tells us, " I once saw two males fighting together; each had a foreleg laid across the neck of the other, and the rostrum bent quite in an attitude of defiance, and looking most ridiculous. Another time, two were fighting for a female, who stood by busy at her boring. They pushed at each other with their rostra, and clawed and thumped apparently in the greatest rage."

The Yellow Wasps, the symbol of war, do not seem mutually pugnacious in the ratio of other Aculeata and ants, who fight, combine in battle, or attack intruders on their communities. The queens of the hive fight, and their sterile maidens destroy the males in summer, those proverbially idle drones. Yellow ants of the kinds termed Emmets, or Mymica, attack one another when on the move (Plate I., Fig. 4, *b*) ; others combine in war, as the common Wood Ant (*Formica rufa*), and the larger

E

Herculean Ant of Europe. This seems a jealousy among females;
but love is the spring of action in the case of the solitary kinds,
whose winged males when contesting a female often finish by de-
vouring rather than relinquish her. Butterflies and some moths
attack their sex or kind, rising in air, and striking or breaking their
wings; the Painted Lady, in its jealousy that the sunbeam should
fall on its beauty, darting from its seat after the passing birds
and intercepting shadows. The males of a chivalrous Dolichopus
dash like fighting-cocks in one another's faces, and the common
Crane Flies are likewise said to engage. The emission of odours
that accompanies the paroxysm of fear may also express anger,
as in the case of two Rove Beetles meeting (Plate I., Fig. 9);
or in that of the ants seen fighting by Huber, who seized each
other, and, rearing upon their hind legs, mutually spurted their
acid.

These combats, it has been urged, tend towards a selection
of healthy males, which is doubtless true; but volition, even as
regards the social kinds, would rather seem to be presided over
by a natural law of masculine priority in appearance, submitting
the males to each and every sublunary influence, and rendering them
inured to manifold terrestrial strife previous to propagating their
kind. This is established in respect to various insects, Stag Beetles,
Saw Flies, Gall Flies, with others; and is rendered especially
conspicuous as regards butterflies, by the appearance of showy
gallants, such as the ringed Apollos on Swiss pasture, and in
this land the male Marbled Whites and Browns fluttering in
their summer haunts, ere yet a solitary female is to be seen.
It is likewise the common experience of those who watch their
moths emerge in the breeding-cages, and should be generally
true in respect to insects.

Love in insects seeks similar expression to fear, and secretions, it is to be believed, furnish suitable means of intercourse,
especially in fragrance-loving creatures like butterflies. Quite
lately Dr. Fritz Müller, who is prosecuting entomological
research in the virgin forests of Brazil, stated that he was
brought to consider as sexual organs for the diffusion of odours
the various pencils, tufts and manes of hair, and the chalky,
silky, or velvety spots of peculiar scales, as well as the recurved
margin or other pouches enclosing pale buff or white down,

which distinguish the wings of many of the brilliant exotic male butterflies. And in a general way this assumption is by no means improbable; for although instances occur where these insignia appear in the opposite sex in measure, and, as evolutionists would state it, acquired by the males, they have been transmitted in the lapse of time to their consorts, yet otherwise, in a majority of cases, they seem truly masculine. In the Papilionidæ, for instance, the front wings of the males have sometimes an oval chalky spot on their under side, opposite to which there is a dark brownish spot on the upper side of the hind wings, both spots emitting a strong disagreeable odour. At other times, the scent spot on the fore wings is replaced by a brush of hair, and that on the hind wing is chalky and fragrant of musk, as may be seen in the common canary-coloured butterflies of the tropics, Callidryas. Here likewise the females show on either side of their anal organs a shining spot odoriferous of a volatile acid. A third example we may find in the male of our English Clouded Yellows, where the chalky spot is seen without the brush, the pot of rose-water without the sprinkler. In other genera certain "plumules," or club-shaped scales, fringed with fine hair, have been accused of possessing scent properties. Our small White Butterfly, belonging to one of these, has a male, according to Kirby, redolent of thyme, and with some of his exotic fraternity the scent is described by Dr. Müller as delicious; but in such cases a caution, as Professor Westwood has noticed, is doubtless needful, for may not these bright beings that bathe in sunlight and vapoury distillations the livelong day, gather odour as they fly about. We have many moths, likewise, with raised scales on the wing-surface, but I am not aware this is the source of scent attraction. Lastly, organs of perfume are said to exist in the sub-family of the Swallow Tails along the anal margin of the hind wings of the males, which is then recurved. They are evinced by a tuft of hair with a disagreeable odour.

In the Nymphalidæ, or great group of butterflies with rudimentary forelegs, we find some of the common cosmopolitan Danainæ have dactylate hollow processes at the abdominal extremity, which are furnished with hairs, and on protrusion emit a disagreeable odour; and sexual pouches, possibly in

E 2

connection, exist on the first median nervure of the hind wing in a few species; both being seen in the tawny Danaïs, that extends its range to Southern Europe. Others of the Ithonia group have a tuft of long hair near the anterior margin of the hind wings, sometimes common to the sexes, and flavouring of vanilla, while the males have a brown spot covered by the tuft. The spot and brush of the Yellow Callidryas are reproduced in one of the Satyrinæ, and in some of the Nymphalinæ; while in the male of the blue *Ageronia Arethusa* a rather strong odour is emitted by two large brown spots, situated between the wings, which appear absent in others of this musical genus. The glorious blue satin Morphos, the pride of Brazilian forest trees, have protrusible hemispherical anal appendages, which sometimes are odoriferous of vanilla; and in the Brassolinæ, another bulky South American progeny, pencils of erectile hairs or spots of scales are present. So, too, the peculiar long-winged Heliconinæ of the glades have scent organs in the male, situated between the anal valves, and in the female these are placed on the dorsal side of the abdominal extremity; but some have also a scent-brush, if conclusion may be drawn from a figure in the volume on "Foreign Butterflies," belonging to the Naturalist's Library series.

But *Didonis Biblis*, according to Dr. Müller, is, so far as odours are concerned, the most interesting of all these fairy-formed and many-coloured things of the tropical bush. On seizing one of these butterflies of either sex, it pushes from between the fourth and fifth segment of the abdomen two hemispherical protuberances producing a disagreeable odour; and the male has a second pair between the fifth and sixth segments of the abdomen, while a different musk-like odour is produced by a black spot beneath the front wings. In the tailed Hair Streaks the sexual spot that may be seen on the disc of the front wings of the female of our green sort, is said to be odorous; and if so, this is a singular instance of a scent secretion supplanting a fixed wing colour.

The Skipper Butterflies, diurnal over our northern hills and woodland, amid the scorching tropical glare are known as evening visitors, stealing mysteriously to hang at the garden lupins in purple dusk, before a reflex light has awoke a second

and chaotic dawn. They neither in colour as a rule bear the
burnish and talismanic impress of the sunlight, or in form do
they attain the feathery etherealness of the day fliers; their
antennæ and various organs seem fairly heavy and moth-like,
and moth-like also are the expansile scent pencils of hair that
garnish the hind tibiæ of the legs of many of the males.
Passing to the Burnet kind, a like incarnation of the union of
night and day, the males of exotic Glaucopidæ have two long
retractable filaments, generally beset with hairs, on the ventral
side of the abdomen, sometimes emitting a strong odour.
Bristling scent pencils, however, are most characteristic of the
moths, and these usually distinguish the males, although, being
commonly retained in their pouches, their presence is little sus-
pected until they have been pricked out with a needle-point;
but in the twilight Sphinges, especially quick of nerve, they are
sometimes found present in both sexes, and sharp fear, as we
touch or grasp the flower-hoverer reposing in diurnal lethargy,
will cause their momentary expansion. Thus when the male
Death's-Head Moth is on the point of squeaking in our fingers,
as its abdomen inflates, a sessile pencil of yellow hair starts from
a fold between the dorsal and ventral arcs at the base (Plate V.,
Fig. 7B), expanding to a stellate form with swift whirling
motion, like a trundled mop (Fig. 7A); and immediately there
arises an oily volatile effluvia, resembling the scent of jessamines,
but soon becoming nauseous. This fluid aroma, secreted near
the insertion of the "fans," and traversing by capillary attraction
to their extremity, stains them bright orange at the glands,
shaded into yellow at their tip. Similar "fans" in the Convolvulus Hawk convey the sensation of amber or musk, and their
anatomy has been investigated by Prof. Targioni, who describes
the hyperdomal tissue opposite the insertion of the "fans" as
composed of hexagonally nucleated pyriform cells, from each
of which proceeds a tube to the base of the several hairs of the
fascicle, which are inserted by means of a process; so that the
unicellular glandules secrete the odoriferous matter, and the
hairs become their excretory ducts. These "fans" are likewise
said to exist in the male Privet Hawk Moth, and may characterise the genus Sphinx.

In the dull Noctuina, with their lined and branded grey

wings, the group that flocks to our sugar-pots and evening
lanterns from the long grass and flowery shrubs, we find the
" fans," with little exception, in the males, although, for
some reason unknown, they are sometimes altogether wanting.
They may be discovered by means of a needle, and pricked out
from a pouch beneath the first five of the dorsal arcs of the
abdomen, when they will be seen to be not sessile, but attached
to a muscular arm about two inches long, and have their hair
pencils where they lay in the fold opposite the fourth segment
stained by a deep saturation revealing the position of the
secretory glands (Plate IV., Fig. 9, f; Plate V., Fig. 8, x).   The
nature of this emanation, however, varies, as may be seen by
procuring in summer some of the pot-herb kinds that repose on
our garden palings.   In the abundant grey Cabbage Moth the
fans will be found to be orange, and flavouring of vinegar;
in the Dark Arches they smack of turpentine, and make us
think of the deal boards whence we have dislodged this heavy-
looking cornice-sleeper ; and in the evening Wainscot Moth of
flower plots they are said to be redolent of ratafia ; but in the
male of the Angle Shades they, on the contrary, are black,
although the secretion does not differ from that we observe in the
Cabbage Moth.   In the autumnal Red Underwing Moth of
willowed canals and poplar clumps the position of the fans is
quite different, for they are found on the legs at the upper part
of the second pair of tibiæ, which are grooved for their reception
(Plate V., Fig. 9, A. B.).

In the Erebidæ, a family remarkable for including the
Great Owl Moths of South America, and which, like the Atlas
Beetles, have but one representative assigned them in Europe, the
heavy fans are likewise on the hind tibiæ of the males, and
one brownish species takes its name after their odour, so we
often find them situated in the great family produced from
looping caterpillars, which includes genera that approach these
capacious wings in form.   But in other kinds of Geometrina we
may find the scent organs on the alar tracheæ as in their prototypes the butterflies, and here they occupy a position on the
submedian vein of the forewing (Plate V., Fig. 5).   Their
secretion is orange, smacking of turpentine and vinegar, or of a
white colour.   In Herminia, a genus of Pyralidina, including

brushwood-flying yellow-brown moths, nicknamed "Fan Feet," the male carries a pair of fans on each foreleg; one is inserted at the lower end of the tibia, and the second beneath the first tarsal joint.

As regards the derivation of the scent, it may be remarked of Heterocera generally, that in those kinds where the secretion smacks of turpentine, either the moth frequents deal boards, as the root-feeding Dark Arches, or its caterpillar feeds on fir needles, as the Tawny-barred Angle. Then if we regard the scent organs as a distinctive masculine feature, it is manifest they should act as an allurement to the female when the male is on the quest; and once when some Wave Moths were gathering in the silence of a Perthshire gulley around their wan consorts, sitting on the fringe of blaeberry bushes, I noticed against the last streaks of twilight that these beaux kept their fans extended. In like manner the ardent fashion in which the male butterfly or moth runs over his partner with his snuffing antennæ is an inducement to consider perfumes to be a charm in a new-born female, who would probably be heightened in esteem by having a stronger fragrance than flower-dust gathered among the grass tops.

Maternal care in insects superficially resembles what we observe in birds, but the various traits of nidification, brooding, and attending the young, are commonly united in the same species. Nor do insects, as a rule, indulge in monogamy, and coo and bill in pairs. The females oviposite, and directly or indirectly provide for an offspring in most cases posthumous. When the sustenance of the species is identical throughout life, as in the case of many crickets, bugs, and beetles, this need involve no anomaly: the female simply lays on or near her food; but where the perfect insect lives by suction, and its larva by mastication, as in the instance of bees, butterflies, and flies, this harmony rarely exists: the food has then to be sought by the pregnant female, who displays for a time the instinct of its larva and revisits its former haunts, a circumstance often involving nest-making and other phenomena we are accustomed to witness in the pupal metamorphosis. In the bee tribe, which forms a passage between the two groups, we witness a most marvellous example of this.

In placing their eggs some insects are manifestly guided by

smell, as the Blue-bottle Fly, who not unfrequently ov</br>iposites
on certain hothouse blossoms redolent of carrion, her proper
food and nidus. Many moths, according to Dr. Knaggs, will
not lay until presented with some strong nostrum; but other
sorts seem to be influenced by sight, as Trichoptera, whom I
have known to attach their ova to a fly's wing, evidently mis-
taking it for the reflection of a glassy brook. Nest and cocoon
making, with insects, is evidently a provision against the fluctuations of season, and especially the periodical visitations of
winter's cold and dearth. No perfect insect, however, spins
as caterpillars previous to metamorphosis, but both natural
endowment and economy are employed. Provident for the
larvæ, some Orthoptera, as cockroaches and *Mantidæ*, lay their
eggs in a case, like the skate and dog-fish. Some Lepidoptera,
as the moths of the Liparidæ, possess a dense woody tuft of
coloured hair around the anus, which they snip with pincers to
cover and protect them. European kinds, either like the Brown
and Gold Tail Moths oviposite on leaves in summer, or, like the
Gipsy Moth, place their eggs on tree-trunks at autumn. Other
moths glue their eggs together in the form of a necklace
around twigs, and Gnats thus construct a floating raft on the
surface of the water. The female of the Coccidæ, a tribe including the cochineal insects, excludes a cottony down with her
eggs, as the American blight, or simply dies with the ova in
her body, which then forms a shelly cover.

To insert their eggs, Cicadæ and Saw-flies cut grooves into
twigs with the serrations of their ovipositors; and other tribes
that have this external oviduct similarly indurated, as the Gall-
flies and Leaf-crickets, use it more or less as a drill. Fossorial and wild bees, with some lamellicorn beetles, on the contrary,
employ their mandibles or forelegs to excavate holes in friable
earth or wood, and form a simple or compound nest. As
regards providing for the young, bees are not equally proficient,
and it would almost seem as if we could trace various degrees
of intelligence. Some, as the Phytophaga, resemble the moth
kind in their transformations; others, as the Biophaga, which
includes the ichneumon flies, deposit their eggs in certain
insects, in whose bodies the young undergo their transformations; and others instead maim these insects, and drag them

to the hole where they oviposite, leaving their carcases as food
for the future larvæ. Others, again, live in a community of
three sexes, and are "child-nurses," as bees, ants, and wasps.
The sterile Formicidæ watch over the sun-hatching of their
nymphs, and matron bees tend the young singly (Melinus,
Epipone); or sterile females do so in company (Apis, Bombus);
but much of their economy is probably pleasant fiction.

Secretions likewise play their part in domesticity. The
flocculent coccus blight and summer cicadæ perpetually suckling
frothy plant-juices, often come to hand powdered over with a
clammy and resinous bloom exuding from the acicular pores of
their dermis; and we know the very similar tough yellow,
white, or black wax of bees, passively extruded as Huber
establishes, in thin light scales from the under-joints in the
hind body of many a sterile worker and solitary matron, has
its employment in the fabrication of those native and multifarious cellular marvels, seemingly constructed to that rule
proper to the dimensions of the bee body; as caterpillars and
grubs, on transition to their assigned sleep of torpor, eject
from their convoluted reservoirs and weave around them
houses of silk, resins, glues, or dusty matter. So that it would
seem but fair to assume that honey, wax, and propolis have
severally become secretions subservient to maternal sentiment,
while they exhibit a wider and structural origin. Nor can this
offcome of gluttony be ignored by mankind himself. Other
resinous secretions akin to these are recognised as producing
pleasant scarlet, rose, and crimson hues that command a value
in the market, such as constitute the famed Kermes, scarlet
grain of Poland, cochineal, and lac-lake, all of which dyes are
the produce of the mealy apterous coccus blight, which elaborates
them from a great diversity of plants, and which seems to hold
affinity in this respect with other staining matter, such as the
common writing-ink, extracted from the galls and tumours induced by insect agency on the plants themselves, or with the
tints and ciphers on the bodies and wings of species.

A gradation is further noticeable in the nesting of bees.
We observe some that line their holes with certain leaves,
wood-raspings, or a kind of glue; and then, after laying each
egg, close it in successively, as in a cocoon, having first

deposited some pollen and honey beside it. In all this the bee
can be but the humble instrument of Providence, exhibiting
the instinct of its tribe; for can we suppose she is knowingly
providing, or has a distinct vision of posthumous larvæ? So,
too, in regard to the more complicated phenomena of hexagonal
cell-building. This, Edward Newman, for instance, informs us,
is nothing more than takes place chemically in hexagonal crystals
and basaltic columns, and its popularity in the case of bees is
due to the interest imparted by imaginative and florid writers.
The views entertained by Dr. Darwin are similar. Those of
the bee kind, he notices in one of his works, which labour
solitary, as the species of Bombus, form cells of an ordinary
cocoon shape; while those which work in company in a confined space find their oval structures must become hexagons
by the law of equal pressure. Mr. G. R. Waterhouse says,
" Supposing a wax block were excavated in one of its sides
into the greatest number of equal-sized cylinders that it would
admit of, it would then follow that each cylinder would be
surrounded by six others, this being the only number of equal-
sized circles which may be placed round one of the same magnitude; and by removing the wax from the interstices, each
of these cylinders would become hexagons. Again, supposing
this block to be a flat mass, the ordinary thickness of a comb,
this block being cut into cylinders of equal diameters on both
sides, and the back of each cylinder being exactly over parts
of three opposing ones, when the wax is cut away at the
interstices, as at the sides, it follows that the bottoms of the
cells will be composed of three equal rhomb-shaped pieces.
This lesson Nature appears to have taught the Hive Bees.
The economy of White Ants differs from that of ants, in that
their larvæ are the workers, and there is but one male and
female in each nest, the latter of whom is inflated with
innumerable eggs."

Of females sitting on their ovæ and protecting the young,
Kirby and Spence mention three instances : the common Earwig,
a Van Diemen's Land Saw-fly, the June Birch Bug, and to
these some authors would add the Mole Crickets. With regard
to these examples appertaining to the Orthoptera, Hymenoptera,
and Hemiptera, we have accounts which, if unvarnished and

unadorned, might lead us to suppose the affection of the warm-blooded animals had here a reflex. The female of *Perga Lewisii* (West) was observed by Mr. Lewis, of Hobarton, to sit upon a leaf into which she had inserted her eggs, about eighty in number, till they were hatched. This takes place in a few days; and afterwards she carefully feeds them in the larva state, in which the brood keeps together, whether eating or sleeping, in an oval mass, sitting upon them with outstretched wings, shading them from the heat of the sun, and protecting them from the attacks of parasites for a period of from four to six weeks until her death. On June 4th, 1871, the Rev. Mr. Hellins noticed an *Acanthosoma griseum* on one of the lower branches of his birch-tree at Exeter, apparently engaged in extracting some nourishment from the catkins. On the 6th, at 3 p.m., she commenced to lay elongate eggs on the underside of a leaf, placing them in a rough diamond figure just about the size of her own body. The outer eggs were laid on their sides; the inner ones stood up on end. The mother then took her stand over them, but not apparently touching them with her body. On the 29th of June the young bugs were all hatched, clustered under their mother amongst the empty shells. These follow their mother about, and De Geer once having occasion to cut a branch of birch peopled with one of these families, observed the mother never stirred from her young, but kept beating her wings incessantly with a very rapid motion. The earwig, Kirby thinks, still more nearly approaches the habits of the hen in her care of her family. Frisch notices she sits on her eggs; and De Geer, having found an earwig thus occupied in the beginning of April, removed her into a box where was some earth, and scattered her eggs in all directions. She soon, however, collected them one by one with the jaws into a heap, and assiduously sat upon them as before. The young ones as soon as hatched creep under the mother, who suffers them to push between her feet, and will often, as De Geer found, sit over them in this posture for some hours.

Having briefly instanced the evidence which muscular contractions, battles, nidification, and provision for the larvæ afford of the salient impulses latent in Insecta, we will now proceed to consider the further manifestation of the emotions, fear or anger, love and rivalry, in the dances, music, and migration of insects. will then appear that, while music and display present parallel characters chiefly exhibiting love and rivalry and promoting the union of the sexes, migration, aerial and terrestrial, very generally takes origin in the same phenomena, and arises from the same stimuli.

The principal phenomena which mark the reproductive period of insects as we have observed are dances and music. We have shown that the battle of males for the female that burns so fiercely in higher organisations, which gave antlers to the stag, horns to the bull, spurs to the cock, and incisive weapons to the fish, smoulders yet more intensely in mandibulate insects of the Orthoptera, Coleoptera, Hymenoptera, and Neuroptera, many of whom bite and devour one another—a temper that extends to the females of many spiders; but, generally speaking, in a state of nature this influence is more gentle, especially in the case of suctorial kinds, as Lepidoptera, Diptera, Hemiptera, Homoptera, and many Neuroptera, which are, with the exception of the Bugs, where the cibarial organ is no longer for licking, but pricking, unarmed. And thus we find the fundamental stimuli of anger and rivalry not confined to battles of which the sex is the immediate object, but taking general expression in phenomena of dances and music, which are inspired by an emotion in cases common to the sexes. Thus we shall find that, while some males at a certain season form a dance and are then joined by the females, others pass their third stage in these performances, and others merely an early-winged period of the perfect state; while some, again, collect and pursue the female. or the female, on the contrary, collects the males.

Again, the feathered tribes, which in many cases exhibit similar sports at the pairing season, have a twofold method of communication, namely, a vocal music produced by the larynx, generally a mere jealous chirping, but which among the males at the breeding season acquires most exquisite modulation in call-notes like those of the nightingale and mocking-bird; also a faculty resident in the male of emitting drumming sounds to attract his female by means of the wing or tail, which are generally given out in air, but sometimes produced by solid percussion, and have been termed instrumental music. Both sorts of notes are often the call for the males to assemble, and thus originate fierce conflicts. The music of Insecta will, I hope to show, admit of parallel treatment; the buzz of the bee and drumming of the cicadæ, alike traceable to the agency of the tracheæ, representing the vocal kind, and the song of the cricket, produced by mechanical means, the instrumental, the latter music here taking precedence; and, assuming the power of modulation, we witness the former kind susceptible of in Aves, agreeably with the low bronchial stage of the trachea, which as other animal organs, has its origin in a multiple or complex form. This instrumental language has been termed Stridulation by Colonel Goureau.

We shall find, by careful consideration of these societies that gather at the accents of the emotions, that with some certainty it may be postulated that either, A, the males at a certain season form a more or less pronounced choral dance in air, and are then joined by, or seek, the females; or, B, the males become musical, generally in company, and attract the females; or male and female each sing, and own a mutual attraction. The first of these societies appears generally due to light or colour display, and in some measure may often be the result of vocal music; the latter may be due to the workings of instrumental or vocal music. The orders of insects in which these two laws find expression may thus be tabulated:—

A.

1. Neuroptera (May-flies, &c.).
2. Diptera (Flies).
3. Lepidoptera (Moths, &c.).
4. Hymenoptera (Bees, Ants, &c.).
5. Coleoptera (Beetles).
6. Hemiptera (Bugs).

B (a).

1. Homoptera (Cicadæ).

B (b).

1. Orthoptera (Crickets, &c.).
2. Coleoptera (Beetles).
3. Lepidoptera (Butterflies, &c.).
4. Hymenoptera (Ants).
5. Hemiptera.

Here the rank in which the Order stands is an attempt to express the ascertained tendency of the insects composing it to form dances in air, or an indication of the prevalance of the organs for music. The Homoptera, B (a), although the males of the Cicadidæ congregate by a music of the vocal kind, have as yet afforded no dancers, and the Diptera no instrumental musicians. A third and singular instance, probably a modification of Rule A, is seen in the swarming or migration of the Hive Bee, where the

neuters or barren females follow a fertile matron to a convenient branch of a tree or nook suitable for placing a new comb,
a proceeding probably due to the workings of rivalry, and which
many have thought guided by the hum of these insects. It may
be termed a celebration of a nuptial dance of barren females.

But did not other attraction prevent display and music in
coercing insects into flocks to further reproduction and distribution, it is probable the smaller kinds would be extirpated. Insects
generally are aggregated by a common food, and in this case the
organs of scent and taste are the agents implicated; others, as
bees and crickets, unite in nidification; and others are brought
together by secondary sexual characters. And, indeed, if we
examine the pageants in which love and rivalry promote the
reproduction and distribution of Insecta, and transmit these
forms of organic life in successive generations, we shall find the
sensorial organs implicated are those of hearing, sight, smell,
and touch, corresponding to the attractive qualities afforded by
sound, light, colour, shape, or odour, between which organs and
secondary sexual characters there is often correlative balance of
development. Thus the most musical groups, or those attracted
by odours, are usually dull in colour; and those bright in tone
have generally less efficient musical organs, or attractive secretions are absent. But groups may be specialised by one or more
of these attributes.

The employment of smell is markedly influential and widely
disseminated in insects ; and it may either be generally involved
in enticing the sexes to the emanation from flowery wastes,
fragrant leaves, fetid decay, or other pabulum ; or it may act
individually through a provision of specific or sexual secretions,
as those diffused by the "fans" of Lepidoptera. There appears
also reason to suppose odorous glands exert their spell to conduct
invalid females to the momentary attractive upas plant of
oviposition, often alien from their accustomed food as regards
smell.

The position of the organs of smell is yet in measure an
enigma. While some, with Cuvier, locate them at the mouth of
the spiracles, where, we shall show, there exists in flies certain
structures that may be subservient, others find them in the head,
especially in the antennæ. Huber, Kirby, and more lately Dr.

Vitis Graber, place the sense in the cavity of the mouth; and
Christian considers insects exercise it in relation to distant
objects through their antennæ, and in regard to near ones by
adaptation of their palpi. However, the primary function of the
antennæ and palpi is pre-eminently that of touch, especially in
terrestrial locomotion; and it has been inferred that these organs
direct the flight in species likewise aerial, on account of the
irregular path described by such insects when deprived of them,
as might occur by their being adapted to receive impressions of
odours, attractions, vibration, or radiation.

But as has been remarked they probably serve some further
intent than touch, for their shortness in flies and Cicadæ does not
accord with that usage. But apart from their employment,
naturalists came to attribute to them a sense corresponding
to smell or hearing, and up to the date of the discovery by
Müller, in 1829, of organs of greater structural adaptation for
the latter function, opinion on this topic ran even. Since that
period the idea has become more prevalent that they are organs
of smell, and it is certain insects employ them when searching
for nectar, and some which deposit on odoriferous matter have
them well developed. I conceive it also a fact that it is in kinds
which fly rapidly among obstructions in search of food that
we notice these parts lengthened, pectinated, or lamellated.
In the beetles, for example, the antennæ become prolonged
in the Longicornes, serrated in the Serricornes, clubbed like
those of a butterfly in the Clavicornes, Weevils, and Short
Elytras, and leaved or lamellated in the Lamellicornes.
In the scavenger ground beetles and lowly leaf-feeders they
are generally simple. In this case we find the club and
leaved adaptations confined to vagrant species that live on
excrement, flowers, or sap, and the serrated form in those
of various habit. The species also that present dilatation or
extension of the antennæ often employ these organs in a manner
to suggest adaptation for scent. The Rose Beetle of our gardens
when presented with flowers, according to Mr. Slater, erects its
antennæ and opens the leaflets of the club; and Sulzer remarks
when lamellicorn beetles settle on blossoms they alternately open
and shut the leaflets of their antennæ. The Longhorns employ
them as tactors, and thus track the female at the pairing season.

Butterflies are distinguished by antennæ with a dilated
club-like termination, which in the Skippers appears to pass into
the stout comb-like organs of the Sphinges; and the Painted
Lady Butterfly, when its haustellum discovers no nectar in a
clover tube, may be observed to depress the knob of an antenna
to the blossom. The plumy or pectinated antennæ we see often
confined to the males of the silk-spinning moths; and in the
Geometrina and night-fliers, where these brushes have de-
generated to mere pubescence, crenulation, or tufts, this character
yet exists, and must indicate some corresponding faculty. It is
evident, too, that as regards such kinds as the Gipsy Moth,
Vapourers, and Eggars, where the males carry the proudest
plumes, the feminine presence is not revealed by sight alone.
For when we expose a sedentary spinster of one of these with
not a speck visible in the sky, she fails not to draw her suitors
at the nuptial hour; and then, as the eager male alights within
a few inches of his cynosure, he remains yet obtuse as regards
her presence; for, moving round and round like a dog on the
scent, or a bee ejected from a pill-box, we often observe him
unconcernedly move off in the opposite direction, or even again
take wing. Nevertheless, to the human pituary membrane, with
some exception, as in the case of the Gipsy Moth, that on pupation emits a piquant effluvia of nitric acid, these insects and their
cocoons appear perfectly inodorous. The female lepidopteron
has also employment for her antennæ in connection with reproduction. One dull afternoon, on the 4th September, during
the wet season of 1879, when migrating Painted Ladies and
Gamma Moths were swarming everywhere, my eye was arrested
by the pretty dappled wings of a female of the large Magpie,
then little less abundant in our country lanes, who was flying
most purposely from leaf to leaf along a hedgerow. She successively visited a reddening bramble, a hawthorn, clematis, and
guelder rose, fruitlessly touching over their glandular surfaces
with a quick, alternate vibration of her black antennæ, in search,
as I at first supposed, of honeydew. The crisping foliage of a
thorny sloe finally arrested her, and seemed to confer satisfaction on her tactile perception; for raising simultaneously her
feelers and crawling on to the centre of a leaf, she hung on at its
upper surface, elevated her wings, and by curling her abdomen

round its apex, began to methodically attach her oval and shagreened eggs to the underside, close to the midrib. She could distinguish Souchong from Pekoe.

The males of the small sylvan dancing moths, like their watery trichopterous relatives, have long and thread-like antennæ. The crickets also have them elongate, and in the grasshoppers they are sometimes clubbed; so also of the Neuroptera. In the flies and the Cicadæ they are short, and the former have them dilated. In the bees they are often elongate, and appear to be used for smelling. The wasps when feeding vibrate tacitly their thick antennæ, and Ichneumon flies, as *Fœnus jaculator* (Fabr.), when searching a nidus for their ova, also vibrate their much longer organs, and apply them to the holes where they afterwards insert their ovipositor. Lehmann has further attributed to the antennæ a faculty of aëroscopy, or power of discovering the state of the weather; and Kirby tells us, on the authority of the poet Southey, that keen scent in crickets informs them of the well-known odours of mother-earth, even when wafted across the salt and dreary ocean. Thus on the approach of the ship *Cabeza de Vaca* to the coast of Brazil, the proximity of land was inferred, and, as the result proved, truly, from a Ground-cricket which a soldier had brought from Cadiz then beginning again to sing.

But in order that the antennæ should be olfactory organs a special organisation is necessary. Straus-Dürckheim in 1828 remarked their constant presence, and alludes to the nerve-branches that enter them, fitting them for the localisation of a sense; and Newport in 1838 notices these nerves, and also the tracheal offshoots which penetrate them and ramify at every joint, adapting them, as these authorities consider, for organs of hearing. Erichson next, in 1847, observes them to be covered with variously-disposed pores, closed at the bottom by a thin membrane; and Vogt, 1851, notices the pores and closing membrane, which appeared to him clothed with numerous hairs. These authorities differ from the former in attributing to them touch and smell. Then Lespès in 1858 observes the pores, and mentions otoliths in connection, which Claparède, in 1858, says are tubes. Dr. Hicks, lastly, refutes Lespès as to the presence of the otoliths, but still considers the function of

F

the antennæ auditory (Plate VI., Fig. 3). He describes the
pores, *p*, as including a fluid and severally closed by a membrane,
sometimes thin and sometimes thickened and raised to a hair-like
eminence, while a nerve, *n*, entering the antennæ sends a branch
to each, which ramifies on its inner side.

Sir John Lubbock, in his volume of "Scientific Lectures,"
published in 1879, notices these structures in the antennæ of
ants as "certain curious organs which may perhaps be of an
auditory character." He says : "There are from ten to a dozen
in the terminal segment of the small yellow Meadow Ant, and
indeed in most of the species which I have examined, and
one or two in each of the short intermediate segments. These
organs consist of three parts—a small spherical cup opening to
the outside, a long narrow tube, and a hollow body shaped like
an elongate clock-weight." It should be remarked, however,
that the author has previously stated his conviction that an
organ of smell is located in the antennæ of ants, and he details
how they can be arrested by a camel's-hair pencil scented with
lavender-water, or how an antenna was withdrawn when this
nostrum was presented to its tip.

Then, finally, we have in immediate relation to muscular
contraction a specialised sense of touch resident in the mandibles,
palpi, filiform antennæ, or fore tarsi, which with the anal appendages take masculine modifications for the purpose of securing
the female in the chase. Thus the males of beetles are recognisable and often distinguished by the number of dilated or
flattened joints of the tarsi ; and in the aquatic beetles such
modifications sometimes assume the form of conspicuous suckers,
and their females have fluted elytra to increase the hold.
Singular dilatations of the feet are likewise found in the males
of the little wasps Crabro, an Alpine grasshopper, and certain
small black flies, *Hilaria* (Plate I., Fig. 8). But this sexual
dilatation must not be confounded with a notch and scale on the
fore tibiæ, which, comparable with the comb on the claw of the
Goatsuckers, is used by butterflies, moths, and some beetles to
clean their antennæ, which are repeatedly drawn out between
the two as a seamstress passes the thread she is about to wax
between the thumb and fore-finger. Other kinds of insects
that want this modification substitute their mandibles, as the

crickets, grasshoppers, beetles, and bees, or employ a palpus, as the Phryganidæ, or the spine terminating the fore tibiæ, as some others. Most insects in cleaning the body, head, and wings employ either the tibiæ, or tarsi of their legs, which we conclude must have a power of feeling the gathered dust and defilement invisible to the immovable visual organs.

This extreme sensitiveness of parts to touch has a yet higher expression as a telegraphy or language between the sexes. Insects with mandibles make known their presence to the newly-emerged beauties by a rough bite at the head; the males of butterflies, bees, and Neuroptera caress fondly with strokes of the antennæ, and the predaceous meadow-flies of the genus *Empis* toy together with the forelegs much as children play at "Baker's man," each seeming to desire to have theirs uppermost (Plate I., Fig. 5). In certain insects that live gregariously, as the social bees, ants, and beetles of the genus *Bembidium*, antennal touch takes its greatest expression as a means of communication; and here it is not alone an exponent of sexual desire, but expresses recurrent emotions in the economy of life, such as the presence of food and alarm on the disturbance of a foraging file (Plate I., Fig. 4, *a*). Ants likewise throw out their lines on the plan of scouts that advance, return, and touch antennæ, and then the stream moves onward in the track. The scouting, however, as Sir John Lubbock has endeavoured to establish, is carried on by antennal scent, and is furthered by that singular property of surface conduction of odour known to the lower creation, and doubtless due to the deposition of the odorous dust by aerial currents. I well remember how a little black bee, common on hawkweed heads on the Grisnez chalk-cliff, on being brought into a room behaved in an ant-like fashion, walking backwards and forwards to the freshly-gathered blossom, touching as she went with her antennæ, and now and anon curling herself up as she would tuck in to sleep again in the feathery down. Yet the fashion of a bee on the scent differs from that of the ants in one respect—they work outwards in concentric circles, and not on the linear system; and of this the late Mr. Frederick Smith brought forward two examples in the Hunting Wasps. One was introduced to his notice by a gentleman at Barrackpore, who having taken a wounded Field-cricket

F 2

from the grasp of a *Chlorion*, threw it some two or three feet clear of its tunnel. The matron, robbed of the provision for her eggs, first examined the nest five or six times to make sure the cricket had gone; the ground immediately round the nest was then carefully scrutinised, but not meeting with any success, she commenced working in a circle, which gradually increased in size till the circumference took in the body of the cricket, which was then brought back. The other instance was of a Hunting Wasp in Greece, which similarly captured a spider in a room.

BIBLIOGRAPHY

ON THE SENSES RESIDENT IN THE ANTENNÆ.

* 1761. Sulzer, " Abgekurzte Geschichte."

* 1770–82. Drury, " Illustrations," p. xxiii.

† 1790. Bonsdorff, " De Usu Antennarum."

* 1779. Bonnet.

† 1791. Brunelli, " Comm. Acad. Bononiens," VII., pp. 199–200.

* 1797. Dumeril. Millin, " Magas. Encyclo.," t. 2, pp. 435–446 ; " Dissertation sur l'Organe de l'Odorat et son Existance dans les Insectes."

† 1798. Luczot, F. M. S. (avec Charles Nodier), " Dissertation sur l'Usage des Antennes dans les Insectes, et sur l'Organe de l'Ouie " (Besançon, Briot), iv., p. 12.

†   Göze, " Natur Menscheril und Voresch," v., 389.

1800. Lehmann, " De Antennis Insectorum. Disser. post. Londini et Hamburgi," pp. 15–47.

1815–26. Kirby and Spence, " Introduction to Ent."

* 1819. Samouelle, " Entomol. Comp."

1822. Blanville, " De l'Organisation des Animaux," p. 565.

† 1828. Straus-Durckheim. " Considérations générales sur l'Anatomie Comparée."

† 1831. J. Rennie, " Insect Miscellanies," pp. 105–117.

† 1832. Burmeister, " Handbuch d. Entomologie."

1835. J. Duncan, " Natur. Library Ent.," Vol. II. (Beetles), Edinburgh, p. 101.

† 1838. J. W. Clarke, " Mag. of Nat. Hist.," Ser. 2, Vol. II., pp. 393–395 ; Mr. Newman's " Remarks on the Antennæ of Insects," ditto, pp. 472–476.

† 1839. W. Ericheon, " Archiv für Naturgeschichte," p. 284 : " Das in den Antennæ das Gehörorgan der Insecten zu suchen."

1840. Newport, " Trans. Ent. Soc. of London," Vol. II., pp. 229–248 ; " On the Use of the Antennæ of Insects."

* 1845. Robineau Desvoidy, " Ray Society."

* 1846. J. W. Slater, "Zoologist."

* 1847. Künster, " Ray Society."

* 1847. E. F. Erichson, "Dissertatio de fabrica et usu Antennarum in Insecta " (Berlin).

* 1851. Vogt, " Zoologische Briefe " (Frankfurt-a.-M.).

† 1858. Lespès, " Annales des Sciences Naturelles" (Paris). t. 9, pp. 225-254.

1858. Claparède, Jean Louis, " Ann. sc. Nat.," Ser. IV., t. 10, pp. 236-250; " Comp. Rend.," t. xlviii., pp. 921-922.

1859. Carl Clauss, " Müller Archiv.," pp. 552-563 ; " Ueber die von Lespès Gehörorgane bezeichneten Bildungen der Insecten."

† 1859. Hicks, " Trans. Lin. Soc.," Vol. XXII., p. 147.

* 1873. A. Paasch, " Archiv. für Naturgeschichte " (Wiegmann, Berlin), b. 1, Taf. 39, pp. 248-275.

† 1874. A. M. Mayer, " American Naturalist," Vol. VIII., No. 10 ; " Experiments on the supposed Auditory Apparatus of the Mosquito."

* Authors who assert the antennæ to have the sense of smell there located, in addition to that of tact.

† Authors who assert the antennæ to possess the sense of hearing.

Scarpa, Schneider, Bockhausen, Oken, Carus consider the antennæ adapted to organs of audition.

Linnæus, Lyonnet, Latreille, Christian, consider the antennæ adapted to organs of smell.

SECRETIONS OF LARVÆ, OR IMMATURE INSECTS.

	EXCRETORY DUCTS.	APPROXIMATE SCALE OF SCENT.	INCENTIVE CAUSING EJECTION.
COLEOPTERA.			
Elater	Two rows of nine contractile dorsal tubercles	Oily, Box leaves	Irritation.
Lucanus cervus	... ...	Guano	
LEPIDOPTERA.			
Papilio	Forked protrusible tentacle on second segment ...	Pine-apple or fennel, *P. Machaon;* pears, *P. Podalirius*	Seizure.
Danais ...	... ...		
Porthesia	Two contractile tubercles, one on either of the penultimate segments...	Caustic and inodorous ...	Irritation.
Saturnidae			
Bombycidae			
HYMENOPTERA.			
Cimbex	... ...	A greenish liquid	Irritation.

PERFECT INSECTS.—COLEOPTERA.

	EXCRETORY DUCTS.	APPROXIMATE SCALE OF SCENT.	INCENTIVE CAUSING EJECTION.
GEODEPHAGA.			
Cicindela ...	...	...	
Bra-chinus ...	Two erectile anal tubes	Secretion vaporises with explosion on ejection	
Cychrus ... (Carabus)..	...	Eject caustic saliva	
Lorlcera ...	...	Secretion vaporises with explosion on eject half ...	
Anchomenus (?)	...	Resembles smelling salts or pyroligneous acid	
Pterostichus ...	...		
HYDRADEPHAGA.			
Gyrinus ...	Situated at the cephalothoracic joint	Nauseous	
Dytiscus ...		Resembles sulphureted hydrogen	
NECROPHORUS.			
Necrophaga ...			
Silpha ...			
LAMELLICORNIA.			
Trichius ...		Musk	
Osmoderma ...		Russian leather	
LONGICORNIA.			
Aromia ...	The pores of the dermis of the body	Musk	On seizure.
Dorcacerus			
Ceragenia			
Lophonocerus			
Acanthoptera			
PHYTOPHAGA.			
Timarcha ...	The pores of the dermis?	Red or yellow saliva	On seizure.
Brachynotus ...		A glutinous fluid	

PERFECT INSECTS.—COLEOPTERA (continued).

	EXCRETORY DUCTS.	APPROXIMATE SCALE OF SCENT.	INCENTIVE CAUSING EJECTION.
HETEROMERA.			
Proscarabæus	At leg joints	Amber coloured, rancid smell	On seizure.
Meloe	At leg joints	Amber coloured	
Cantharis	At leg joints	Amber coloured, flavouring of *Sinapis tenuifolia*	On seizure.
BRACHYELYTRA.			
Staphylinidæ			
Goerius			

LEPIDOPTERA.

Scent Organs in Butterflies, extracted from Paper by Dr. Fritz Müller, in "Trans. Ent. Soc.," 1878, p. 211-221.

	POSITION OF EXCRETORY DUCTS AND THEIR ADJUNCTS.	APPROXIMATE SCALE OF SCENT.	INCENTIVE CAUSING EJECTION.
BUTTERFLIES.			
Papilionidi (Swallow Tails)	In males, along the recurved anal margin of the hind wings with a tuft of hair	Agreeable	
Pieridi	In males, front wings sometimes with a chalky oval spot at base, beneath		
Leptalis	Also a brownish spot on upper side of hind wings of males	Disagreeable	
Theruesia	Also a brownish spot on upper side of hind wings of males		
Callidryas	Spot on fore wings replaced by a brush of hair, that on hind wing chalky	Musk (acid from anus)	
Colias	No brush		
NYMPHALIDÆ.			
Itana, Lycorea, Danais	A protrusible dactyle hollow process at the abdominal extremity, furnished with hairs, and sexual pouches on first median nervure of hind wings in *Danais Erippus* and *Gillipus*	Disagreeable	
Dircenna, Ceratinia, Ithomia, Mechanitis	A tuft of long hair near the anterior margin of the hind wings	Vanilla in *D. Xantho*	
Thyridia, &c.,	Found likewise in females (more rudimentary)		
Antirrhea	Spot and brush as in Callidryas		

LEPIDOPTERA (continued).

	POSITION OF EXCRETORY DUCTS AND THEIR ADJUNCTS.	APPROXIMATE SCALE OF SCENT.	INCENTIVE CAUSING EJECTION.
BUTTERFLIES (cont.) **NYMPHALIDÆ (cont.)**			
Morpho	Protrusible hemispherical anal appendages	Vanilla...	
Brassolinæ	Pencils of erectile hairs or spots of peculiar scales		
Heliconiæ	In male, scent organs situated between anal valves, in female placed on the dorsal side of the abdomina [extremity]		
Acræinæ			
Mysecdia, Epicalia, Didonis	As in Callidryas brush black		
Agronia	Brown spots between wings, not in A. Amphinoae and Feronia	Disgusting	
Didonis	Two protrusible hemispherical protuberances between the fourth and fifth, fifth and sixth, segments of abdomen, producing a disagreeable odour, and one of heliotrope; also a musky odour given out by black spot beneath front wings		
Thecla	Sexual spot on disc of front wings in many males		
Hesperidæ	Expansile fans of hairs on tibiae of hind legs in many, as in moths		
MOTHS. **(Compiled from various sources.)**			
Glaucophie Gryptole-bia	Two long retractile filaments beset with hairs on ventral side of the abdomen in males, odorous in Ilelemaia insaurota, Cryptolechia...		
Acherontia	Expansile fan of hair, contained in a pouch between the dorsal and ventral arcs of the abdomen, common to the sexes	An orange secretion smelling of jasmines, but nauseous	Irritation.
Sphinx	Fan in male alone	Musk or amber	
Macrosila	(?)		Irritation.
Mania (Mania)		Vinegar	
Xylophasia		Turpentine	
Leucania		Ratafia...	
Acronycta	Fans in male, attached to a muscular arm, and concealed in a pouch beneath first five dorsal arcs, secretory ducts at fourth, secretion orange	Vinegar	
Mamestra Philogophora	Fans black, secretion white?	Vinegar	

LEPIDOPTERA (continued).

	Position of Excretory Ducts and their Adjuncts.	Approximate Scale of Scent.	Incentive causing Ejection.
MOTHS (cont.)			
Atamea ...	Fans at anus?		
Catocala ...	Fans at upper part of second pair of tibiæ, which are grooved		
Erebidæ ...	Fans on hind tibiæ of males, Ex. Erebus odorus, Lin.		
Boarmia, Tephrosia ...	Fans on hind tibiæ of males	Vinegar	
Macaria ...	Fans on submedian vein of fore wing	Turpentine, secretion orange	
Charia ...	Fans on submedian vein of fore wing in prunata, testata, and populata (larger sex)	Secretion white	
Acidalia ...	Fans on hind tibiæ	Secretion white	
Herminia ...	Male with two fans, one inserted at lower end of the tibiæ, and the other beneath first tarsal joint	Secretion yellow	
Ennychia ...	Fans on fore tibiæ	Secretion white	

HEMIPTERA AND HOMOPTERA.

	Position of Excretory Ducts and their Adjuncts.	Approximate Scale of Scent.	Incentive causing Ejection.
Cimicidæ ...	Situated exterior to the insertion of the posterior legs	Bug odour	Seizure.
Pentatomidæ ...	Situated exterior to the insertion of the posterior legs	Bug colour or cucumber	Seizure.
Capsus capillaris ...		Fruit essence	Seizure.
Heterotoma ...		Fruit essence	Seizure.
Lygæus Hyoscyami ...		Thyme	Seizure.
Enoplops scapha, Fab. ...		Flavour of peach	Seizure.
Syhirus bicolor ...		of head Nettles	
Reduviidæ ...		Vinegar	Seizure.
Scutelleridæ ...			
Cicadidæ ...	Fluid from anus		On alarm.
Cicadidæ ...	(In many) a waxy secretion exudes from pores of dermis		
Coccidæ ...	(In many) a waxy secretion exudes from pores of dermis	(Cochineal, Pela, &c.)	

HYMENOPTERA.

Cimbex	Musk	
Cynips (lignicola)	Disagreeable odour	
Scolia	Cachous lozenges	
Sphex	Ether	Irritation.
Formica	F. rufa, formic acid; F. foetans, excrement, Kirby	
Mutilla	Garlic ...	
Bembex	Palter	
Crabro	Palter	
Andrenidae	Garlic	
Apis, Bombus	Wax, a resin with basis of a fat oil	Nesting.
Pimpla	Unpleasant odour	Irritation.

ORTHOPTERA.

Blattidae	Small tubular anal glands	Smell of cockroaches	
Gryllotalpa	Small tubular anal glands		
Saltatoria	(Brown saliva)		On handling.

NEUROPTERA.

Chrysopa (spec.)	Anal?	Excrementitous, C. perla, phylloch-roma, septempunctata (McLachlan)	Seizure.

CHAPTER II.

DISPLAY AND DANCES MATERIAL AGENTS IN REPRODUCTION AND DISTRIBUTION—CORRESPONDING SENSE OF SIGHT.

THOSE refreshing visitations of sleep, coincident with our passage through the discordant shadow cone projected spaceward by our revolving globe, are not participated in by the rest of the organic creation, which in part have normal characters and sensorial organs sharpened to render them nocturnal; and as they approximate, the invertebrata become also less constitutionally capable of extended exertion, so that the Articulate and Vegetable Kingdoms, in their periods of recurring energy, mete the hours of the earth's revolution, and in the term of their existence record the brief seasons of its circuit. These hours and seasons of transient life are thus alike rendered conspicuous and grateful in the many-coloured births of flowers, and by the gay masquerades, music, and dances with which Insecta effect their reproduction: pageants that surge and wane as the golden chariot-wheels move in their fictitious track around the bright zodiacal lights—now kindling into tumult at the sun-god's balmy approach, now subsiding to melancholy as he sinks chilly from the zenith.

So, too, at the momentary interception of the melodious sunbeam by a flitting cloud, brilliant flies on the hedges, and the frailer Dragon flies of the marsh land, faint as they sit on the leaves, so as to be readily captured in the hand; and the flitting butterflies, with other suctoria, remain motionless, as enchanted; but no sooner does the source of light, heat, and chemical action pass from its dull shroud, than both suctoria and mandibulata renew their activity, and traverse with fresh murmur the scented air. Indeed, it would seem as if the very sense of fear originated in contractions induced by touch and gloom; and that sensitive plants, whose leaves fold and droop in darkness, or on irritation, preluded those sombre, inexplicable forebodings paralysing animal nervous systems. It is thus that emulous life, in its desire to perpetuation, ruled by the undulatory emanations from the central orb, instinctively shuns this reign of chaos; and those organisms that awake their gambols and melodies on the borders of the night show by increased delicacy and the employment of their sensorial organs, they are but adapted to people a diffuser day. With these even the white and frosty lunar reflections exert little potency, and are shunned, and the fainter planetary light appears alien.

The sleep or lethargic periods of flowers, as Sir John Lubbock remarks, exert an influence on the recurrent activity of insects. Those visited by bees close at nightfall; and, on the other hand, flowers fertilised by nocturnal moths, often pale in colour, open and become fragrant at dusk; those that close at noon, and throughout the day, should also harmonise with the periods of suctoria. It is thus manifest a majority of insects are dependent on the solar beam, either in respect to its light-giving or heating rays; and as a continuance of dull weather alone might retard exclosure, or prevent the union of the sexes, we find traces of parthenogenesis, or the hatching of unfertilised eggs, in Lepidoptera and Homoptera, and in many kinds retardment of development, or a capacity for hybernation.

But light in the abstract exercises its spell. Gnats, moths, beetles, bugs, may-flies, and even Orthoptera, flock to the midnight lamp; and in tropical countries these vagrants arrive of a calm evening in such numbers as to constitute a domestic nuisance. I may cite my personal experience with a small moth when I was stationed in the Mauritius, near Mahebourg, that used to choke up the candles successively in the course of the evening, while scathed individuals, dropping down, covered the supper-plates like small dust. It multiplied, I think, in the Fil-la-haut or Tamarisk trees that lined the shore. Moths, then, are pre-eminently dazed and drawn by light, especially those of the heavy, dull-coloured group of our night-fliers, which stand out from the Lepidoptera on account of their exquisite organs of scent and hearing. Many of these, as entomologists are aware, may be procured for the cabinet by merely going the round of street lamps at the fall of evening, or by placing a light at

an open window; and, next to the Nocturna, the butterfly-like
Geometrina are most frequently so attracted. But light exerts
an opposite or repulsive effect over other insect species, agreeably
with regard to its paralysing power. A notable instance of this
is found in an individual of the same group—the Copper Underwing. The manœuvres of this moth, when surprised by the
autumnal lantern on the sugared tree, are adverted to by Mr.
Stainton in his Manual; and it is really curious to note how the
insect avoids the bull's-eye of the collector by sidling round the
trunk into the shade, and yet continues slyly probing the decoc-
tion with lengthened haustellum.

Passing from light, which has its source in chemical union to
that emanating from the sun's photosphere, diffuse daylight is
accompanied with kindred manifestation. Diurnal insects, re-
leased in a room, generally make for a window, to flutter
against the panes they cannot pass; but others, on the contrary, and among them most nocturnal kinds, flee the rays,
seeking any obscure shade or crevice which offers partial concealment. In the crepuscular Dung Beetles, we may witness
the two qualities blended, for when let loose from a box the
more active males invariably wing for the light, while
the more passive female seeks to avoid it by burrowing.
Or should we desire an illustration from the light, sensitive
group of Lepidoptera, we may find it in a congener of the
Copper Underwings, the Flat-winged Mouse (*Amphipyra trago-
pogonis*), which, on being dislodged of a bright September
morning from its concealment at the window-sash, leaps fran-
tically, like an *Ægeriid*, or Cotton Grass Moth, falls theatrically
on its back, and after remaining motionless a few moments,
recovers itself, and, "mouse-like," runs hopping backwards and
forwards along the shadow of the wainscoting, intuitively seeking a dusty crevice to wedge into, a proceeding that invariably
abrades the scales from the wing-base, so that specimens are
seldom found perfect. But this lucifugal character in insects,
with its attendant muscular contractions indicating fear, I have
already sufficiently touched on; love, on the contrary, I shall
notice as being fostered by the calm of darkness and gloom—
a phenomenon doubtless ruled by the circulative system.

The compound kaleidoscope eyes of insects, obsolete in

gloom-frequenting Cave Beetles, and in nocturnal moths phosphorescent, are then uniformly sensitive to light and its spectral
elements. Most species move towards an excess of luminosity;
some become torpid on its withdrawal, or are lucifugal; a
majority are drawn by colour. Whether or not coloured objects
might be employed with success as a means of ensnaring specimens does not appear; but it is at least certain chromatic
attraction enters into the economy of the species, of which
examples in nature from time to time occur. Some kinds
associate colour with the taste afforded by their habitual nourishment. We sometimes notice the Humming-bird Moth and
Drone Fly abscond from garden-flowers when the sun shines,
to probe red or purple dabs on wall-paper in our sitting-rooms,
which are likewise occasionally visited by the Blue Meat Fly,
in mistake for carrion. An ingenious analysis of this attraction
was afforded by Sir John Lubbock, who, on baiting paper slips
of various colours with honey, and then shifting them, discovered certain bees he had marked invariably returned to the
tint they previously selected. But besides an indication of
food, insects regard colour as a call to reproduction, or an incentive to battle. Female Dragon-flies whose males are blue
have been known to collect around a blue fishing-float; or the
males of the little Meadow Butterflies (*Polyommatus Alexis*) to
fight a piece of paper of that hue. So, also, a piece of glowing
rose-paper has attracted the Brimstone Butterfly, a piece of
green paper the Small White, and an umbrella-top a Wood Fritillary, where it may be remarked that one butterfly has orange
spots and produces a red variety, and another possesses silvery spots.

Fear is also intimately connected with colourisation, and butterflies and moths with rich white or red wings are those especially
where we find alarm produces paralysis, instead of stimulating
escape. In the moths this becomes marked in Tiger Moths,
Ermines, and others, comprising the families Zygænidæ, Chelonidæ, and Platypterigidæ; and even the small white-speckled
hedge *Hyponomeuta*, when we near a pill-box to the grass
blade where they so indolently sit, suddenly conceive fear of
discovery, and, contracting their members, drop among the roots
and tangle, a proceeding where gravity is not a little abetted
by a wedge or arrowy form. And thus, in the case of insect-

painting we discover the sentiments of love, rivalry, and fear
have been writ in various hues in ages gone, and have become
associated with special feeling, red and white betraying the
tenderest natures, and yellows and metallic blues acute light
sensitiveness.   Indeed, the colouring and patterns of insects
may be broadly regarded as a design for mutual attraction
of the kinds and sexes subservient to reproduction, or as a
means of affording protection to the species; and as regards
the individuals, it becomes an expression of pleasure or pain,
a language, as I hope to indicate, to shield the various forms
for a period against that extinction which has removed successive eons of geologic life from our terraqueous surface.
Nature likewise imparts virtual permanency, by imbuing each
more or less with the hues and shapes of the evanescent objects
where they resort, and thus we commonly find an insect tinted
and marked according to the various parts of a tree, the lichened
twigs, mossy trunk, spring verdure, or brown horror of autumn,
where we may chance to discover it; among such harmony, we
doubt not, it may alone exist, crawl, or fly during its season of
activity, or hide in its lethargic hours.   And, in fact, we discover the insensible cutting down of forests, drying of marsh
land, and cultivation of heaths yearly affords us evidence of a
gradual extinction of species, in measure due to this removal
of mimetic shelter, although other causes may combine, such
as a simultaneous disappearance of food-plants, or other intrinsic
circumstances.

On penetrating further into this presiding phenomenon we
begin to recognise it is those surfaces in insects which are adapted
for coverings, or exposed in repose, that are especially protected,
while the parts they conceal often assume bright and striking
hues.   Deceptive mossy and licheny tints characterise the upper
surface of the fore-wings or their representatives in moths,
caddis-flies, grasshoppers, bugs, and beetles; and in the instance
of butterflies we notice these appearing beneath and often confined to the under-surface of the hinder wings, behind which
the primaries may be withdrawn when the wings are shut in
repose.   Here, however, a South American genus (*Ageronia*)
forms a notable exception.   These butterflies rest with their
wings flat, and like many moths present superiorly mottled

colours, assimilating in hues the tree-trunks on which they are
fond of alighting. In all orders species occur where the protected portions of the wings and body assume leaf-like shapes,
and of this the raptorial Orthoptera, and Indian butterflies of
the genus *Kallima*, whose under-surface resembles brown paper,
are familiar examples. These pendants of phanerogamic vegetation are also certainly sufficiently omnipresent to come in the
way of any vagrant insect traversing the earth or air; but
should it be desirable to realise the deceptive excellency of the
wide-spread lichen, moss, and carpet patterns, one has but to
stroll the woods among the primroses and violets of spring, when
this realm of terrestrial Thallogens is laid bare in hoary grandeur,
and its frost-work will be then noticed everywhere, bearding the bare
soil and torpid tree-trunks, like seaweed whence a tide has ebbed.

From constantly observing insects reposing on the surfaces
that resemble most their colourisation, we intuitively perceive
such must exert a constant attraction for their ocular organs,
and that kinds resembling one another in tint should flock
together, a law which also appears to explain why mimetic
species of diverse orders are often noticed flying on the same
spot, as Hover Flies with bumbles or other bees they imitate.
Cases, too, may occur in which the strictly protective hues
serve to bring the sexes together. A male of a Small White
Butterfly who had been eluded by a female he was fluttering
about in a hedge, I observed visiting the dry leaves of an
Erysimum, which, as will be found, nearly match the specific
protective colour beneath the hind wings; and the female of the
Green Hair Streak closes her wings, exhibiting the bright inferior green, when found by a male. It may also be remarked
the spots and lines beneath the wings of butterflies, and on the
primaries of moths, sometimes take a metallic glitter, like the
scales of fish, or nacre, conspicuous in obscurity. The following
genera will afford examples :—

	BUTTERFLIES.	MOTHS.
Silvery streaks or spots	Anthocharis. Argynnis. Megalura. Charaxes Cinadon.	Xylina. Plusia. Chryso-tista.
Gold streaks or spots	Erycina?	Plusia. Argyresthia.

G

Generally speaking, sober browns, blacks (Plate IV., Fig. 12), chlorophyle tints, and aerial purples are the pigments we notice on protected surfaces, and this is well exemplified in butterflies and moths. On the other hand, those more florid hues specially designed to evoke the passions consist mainly in white and its prismatic elements. These, which may be regarded as primitive when compared with the more composite or protective colouring, are not always strikingly present; and we might surmise that individuals possessing such beauty in the highest degree require, on account of their apathy, this greater stimulus from the solar beam to fulfil their drama in the economy of life, or that they stand higher in the scale in respect to organisation. The first hypothesis, I may remark, appears true, in regard to the brighter-coloured Lepidoptera, as these seem certainly to want auditory organs, or to have them less perfectly formed. We may also regard such primitive colour as in measure an index to the ocular organs and nerves that receive its impression, for why, abstractly, should one insect be excited by red, another by blue, and a third by yellow, or white, unless predisposition for such be present? Fritz Müller, indeed, assures us certain butterflies will only visit blossoms of a certain primitive tint; and if this be correct, we should be inclined to consider it a manifestation of this capacity and preference for a ruling colour ray.

Primitive and attractive colouring, then, in Lepidoptera is present in a greater or less degree in the species and individuals replacing the more composite grounds; while, as regards differentiation of the sexes, it characterises males in butterflies, though not invariably, and in moths it is often most marked in the female sex, an anomaly evidently resulting from a design to render them visible at dusk in the majority of cases. In these groups sometimes the wings are wholly different in the sexes, as for instance, female white, male brown; or female greenish, male fulvous or yellow; or female yellow, male inclining to orange; and *vice versa*. We will tabulate a few striking cases instancing the genera or species where they occur.

COMPLETE COLOUR DIFFERENTIATION.

		BUTTERFLIES.	MOTHS.
A	{ Male, brown	{ Triphya	{ Hypogyma.
	{ Female, white		{ Spilosoma.

COMPLETE COLOUR DIFFERENTIATION (*continued*).

		BUTTERFLIES.	MOTHS.
B.	{ Male, fulvous { Female, greenish	} Cethosia. } Argynnis.	
C.	{ Male, yellow { Female, white	} Gonopteryx.	
D.	{ Male, orange . . . . . . { Female, yellow . . . . . . { Male, yellow . . . . . . { Female, orange	} Genera of the Geometrina and } Bombycina. } Euthemonia, Eunomos, Angu- } laria.	

In the last instance, D, the inversion is not alone one of colour, but also of size; the females of a British representative, the Clouded Buff Moth, being notably smaller than the males. A second case occurs in diurnal Lepidoptera, where the males or females are more or less distinguished by a suffusion of primitive or rainbow colour, as white or yellow grounds marked in the male with red, and olivaceous tints replaced more or less by blue, orange, purple, or green in either sex, but characteristically so in the male. We will tabulate a few instances.

INCOMPLETE COLOUR DIFFERENTIATION.

		BUTTERFLIES.	MOTHS.
A.	{ White or yellow { grounds marked { with red in the { male.	} Anthocharis.	
B.	{ Olivaceous tints re- { placed by blue, { orange, purple, { or green.	{ Thecla. { Chrysophanus. { Polyommatus. { Dynamine.	

Here likewise, in the instance of the Coppers, the males, as a rule, are more orange than their females, but in *C. zanthe* we find the reverse to be the case. This secondary tint again may be complementary and shot, arising from longitudinal bead-like striæ on the wing scales; and of this the Emperors (*Apatura*) and Brazilian butterflies of the genera *Morpho*, whose yellowish-brown wings, placed perpendicularly to incidental light, become interferential with rich purple, afford illustration.

In Orthoptera, Hemiptera, Homoptera, and Coleoptera the disposition of the colours is as in cryptopterous Lepidoptera.

Dermal wing tints, in the case of the rapacious grasshoppers, sometimes possess a smooth beauty like stained glass, enhanced by a transparency to light which serves to throw the reflections from the crimson and yellow scales on the handsome hind wings of moths quite into shadow. In either case likewise this bright colouring is associated with protected areas. The elytra or forewings of the afore-mentioned locusts, which they employ to conceal their charming colour-wings, mock the dark patches of fibrous moss they frequent, while their equivalents in the nocturnal moths are branded on their superior surface with the protective lichen-marks and tints of the group. But although bright colouring or metallic marking is not rare in this group, the Orthoptera are not often so sexually distinguished, a circumstance probably owing to their endowment with excellent organs of music; a few cases nevertheless occur, as one of the Saddled Leaf-crickets (*Ephippigera rugosicollis*, Ser.), where the male is described as fawn-colour and the female is green; and that of a North American cricket (*Œcanthus niveus*), where the male is ivory-white and the female yellowish or greenish.

Beetles are distinguished among insect orders by the high degree of induration of the integumental parts exposed in repose, and we perceive the external cuticle frequently chitinised or horny as if it had been exposed to the influence of a petrifying spring or hoar-frost. This process of hardening may be viewed in various stages of completeness, in the semi-terrestrial grasshopper and bug tribes, which only occasionally use their organs of flight, and everywhere harmonises with the habit of the species, conferring, as it does, a heavy movement where momentum is acquired at the expense of agility. In this latter respect, its influence on the anterior members of aerial locomotion is noteworthy; these present every nuance, from the limber wings of the light butterfly to the simply expansile wing-cases of the beetle, which, in their lowest form, are united together and abbreviated so as to appear but a modification of those excrescences so frequently noticed on parts of the insect thorax, into which the trachea has pushed a vesicular ramification. It is consonant, then, that the alar colouring and patterns should reappear on the elytra, and these are sometimes resident on the dermis and sometimes are blazoned on a covering of pile, or again are present on feather scales, the homologues of those that clothe the wing-membrane in Lepidoptera. The interferential colours of the green Diamond Beetle (*Entimus*) are ascribed to these scales, which possess a laminated surface similar to what we find in mother-of-pearl; elsewhere, a play of colour in Coleoptera is imputed to a similar integumental modification.* Rarely are colours and patterns of beetles sexual; some lamellicorns and longicorns may be adduced as example, nevertheless, and among the former we find the glistening *Hoplia cærula* with a male bright blue and a black female.

The posterior alar organs of grasshoppers and bugs are often colourless; those of beetles, I believe, are universally so; and as we descend the chromatic scale, we witness in the membraneous wing-groups of diurnal Neuroptera, Hymenoptera, and Diptera pure colour and patterns still less marked on the alar organs, while these acquire often prominence on their bodies which assume a maximum brilliancy. Yet even here exceptional cases occur where alar colouring, protective or attractive, extends to the wings, and is there even sexually marked, as, for instance, in flies of the genus *Dolichopus* (Plate V., Fig. 1), where the male has black wing-tips; or the Dragon-flies of the genus *Calopteryx*, where the male has bluish organs of flight. In the Dragon-flies, again, as a group, males are not unfrequently distinguished by bright primitive tints, as for example—

A. { Abdomen of male red, } Libellula ferruginea.
     { in female olivaceous }   ,,   sanguinea.

B. { Abdomen of male blue, } Libellula depressa.
     { in female olivaceous }   ,,   cærulescens.

C. { Abdomen of sexes discriminated } Æschnina.
     {   by lines and spots } Agrion.

With bees, as flies, bright colouring of the abdomen is universal, and it is either resident in the dermis or diversifies a covering of hairs; having in the first instance often a metallic lustre. The humble bees present likewise sexes differentiated by markings, and in the case of *Anthophora retusa* we have a male bee black, with an orange female; this of course refers to abdominal colouration.

* The play of colours on a Dove's neck is due to a bead-like thickening of the fibrils, and not to impressions, as has been stated.

When we stroll or drive among those sequestered nooks in
Europe where Nature yet asserts her fairy reign, a little reflection shows us that with anthropods generally it is among
the sun-frequenting life that bright patterns occur; with others
haunting cooler shades, or active only during the low temperatures of night, licheny or mossy designs are universal; and some
indeed are wholly black, bleached, or colourless, a feature yet
more marked in the subterranean hordes.   Species that inhabit
the dark pools and streams, also, are dingy or dull as compared
with the wanderers of the air.   Then if we consult a cabinet as
regards terrestrial climate, we shall further extend this law, and
find that the rays of the sun impart a richness of tint varying
with the ardour of his beams, and that tropical species which
are diurnal have a gaudiness compatible with the languor of
a clime that becomes their guardian, expressed in the opaque
paint-like pigment that imparts a varnish to their dermal tints,
with a heaviness to their external coverings.   In the Brazils,
for example, all colours, whether of birds, insects, or flowers,
are brilliant in the extreme.   Blue, violet, orange, scarlet, and
yellow are found in the richest profusion, and no pale faint
tints are to be seen.   Even white seems purer, clearer, and
deeper than the white of other countries.   Here, as elsewhere,
the butterfly kind takes precedence in regard to colour, and it is
also possible these colours are distributed as to district; for
example, metallic spots are very frequent on insects from the
Cape, a locality which also produces the well-known silver tree.

Protective colouring extends to the pupal and larval stages,
and markings exceptionally brilliant in caterpillars may effect gregariousness, although sexual phenomena are here of course latent,
and the springs of action less comprehensible.   The larvæ and
nymphæ of *Mantidæ*, are, it is stated, sometimes coloured like
flowers for another object, that of attracting the insects on
which they prey, an end equally gained in the protective phyto-
mimicry of the *imagines*, which allows unsuspecting approach of
the victims.   It is also affirmed that as a general rule caterpillars which are dull-coloured and have a smooth skin, or those
which are nocturnal in habits, are greedily eaten by birds; and
on the other hand spiny and hairy Palmer-worms are spared and
often brightly coloured.

Many insects have the planetary power of rendering themselves visible at night, by means of a diffused light emanating from various corporeal parts, and physiologists who have investigated the matter agree that this is generated by a soft yellow secretion, intersected by tracheal ramifications, and that it is essentially connected with the act of respiration; but further the subject is perplexed. In Kirby's day it was a question whether the light should be considered a substance, a quality of matter, or the effect of vitality, and whether the chemical process involved was a slow combustion of phosphorus or hydrogen, or an emission of light which had been stored up. Although I am not aware the subject has materially progressed, or that any chemist has undertaken a satisfactory analysis of the aforesaid soft yellow substance, as was then suggested by Mr. Spence, it is now the fashion with the scientific to compare the radiance of the Glowworm either to the oxyhydrogen light, or to an oxidation of carbon attendant on respiration. Phosphorescence in insects does not appear confined to any order, being found in beetles, moths, caterpillars, coleopterous larvæ, and centipedes; and there is no reason why it should not occur in bugs, May-flies, and Mole Crickets; but the luminosity of the Lantern-flies is nowadays regarded as a fiction and travellers' tale, originating with Madame Meriana. As regards *imagines*, the luminous property-like colour secretions, though generally common, is sometimes sexually marked; the grass-frequenting female Glow-worm is an often-quoted instance. "It has been supposed," says Kirby, "that the males of the different species of *Lampyris* do not possess the property of giving out any light; but it is now ascertained that this supposition is inaccurate, though their light is much less vivid than that of the female."

Appended to this chapter will be found a list of insects in which a luminous property has been noticed, and it has been accredited to many others. In a paper just published in the "Transactions of the Entomological Society," the Rev. H. S. Gorham has recorded his observations on the structure of the Lampyridæ with reference to their phosphorescence, directing attention especially to the development of the eyes, which he finds increase in size, while the pectination of the antennæ is decreased, in the exact ratio of the luminosity of the various species.

The attractive quality of insects' colours from the foregoing appears nearly that presented to the human eye, and, utilised in sedentary or aerial display, originates phenomena of love and rivalry, battles, dances, and gregariousness in evident parallelism with those evoked by music. But this attractive virtue, which must be considered as stimulative, does not reside especially in either sex, as some at first sight might be inclined to assert; for we find conspicuous colourisation, though for the most part distinguishing the males, sometimes by a species of inversion appearing in the females; the sexes also are often very similar in hue. And the reason of this is that the females very generally attract the eager males by sedentary display, of which the moth kind affords notable instance. Here we may remark the paler hues of many heavy Bombycina females who exhibit on herbage, and the grey, white, or satiny shades of moths that repose on tree trunks, sexually marked in the Gipsy Moth, who is rendered in measure terrestrial by her limp wings. Others more or less apterous, like the Vapourers and Psychidæ, owe what little chromatic attraction they possess in all appearance to their conspicuous cocoons.

To ensnare many of these kinds, it is only necessary to expose a female in a box covered with gauze; males of the Vapourers, Gipsy Moth, and Oak Eggars are all then readily enticed, even when none are visible in the sky. In the case of the Gipsy a pungent odour exuded during pupation doubtless adds to the allure, and to many such an effluvia has seemed necessary in all cases, seeing that the eager suitors are endowed with a keenness of perception approaching the marvellous, and come from a range it is difficult to assign to insect vision. But this matter is wanting in proof. In a state of liberty the smaller moths are quite as attractive for their males as these diurnal members of the silk-spinning group. White Geometrina, such as the species of *Acidalia*, may often be observed climbing the sprays of plants to display at nightfall; and these kinds assemble the males in numbers, although, singular to say, species of the Geometridæ that exhibit delicate tints of evanescent green prove no less attractive. At the call of passion individuals of tinier Tortricina and Tineina are often noticed surrounded with quite a circle of admirers.

Female Dragon-flies and flies with bright and metallic coporeal colours may often be noticed reposing on plants in the sunshine, attracting ever and anon the attention of some passing male, who, staying his course, remains for a while, as seized with an ecstasy, suspended over their charms like a hawk marking his quarry, and seeming as if dazed by the glow of pigment beneath him. This is very characteristic of the Libellula and Syriphidæ.

Let us pass now from colour to luminosity. That the apterous Glowworm's "love-illumined form" on her mossy couch is a beacon to the vagrant male, Kirby argues from the fact that the light, most brilliant in the female, in some species, if not all, is present only in the season when the sexes are destined to meet, and is strikingly more vivid at the very moment when the meeting takes place. It has also been questioned from negative evidence whether the male is actually drawn by the light; and in reply to this I can only mention the case of a Glowworm I found among the solitudes of the New Forest one summer's evening, which, on being brought into a lighted room caused several winged males to enter at the open window.

To further the union of the sexes, we also notice the male insect is usually smaller, more fleet and agile on the wing, than the female; and the wings of the latter, often unsuitable for flying, are in some bulky egg-bags so abbreviated as to be scarcely discernible. But in the heath-flying Clouded Buff Moth we have an example of the converse, and instances may be quoted in other orders, consonantly with which, perhaps, we notice the task of aerial locomotion after coupling is indifferently apportioned to the sexes; or at least, this can be affirmed of those light ethereal forms, the butterflies. For while the Whites, Clouded Yellows, and Meadow Blues spread their limber vans to bear away their brown or dappled loves, the Black Veins, crimson-spotted Thais, orange wood Fritillaries, and indigenous Browns, carry their admirer* through the liquid sky.

Aerial display, in which colour, reflection, or luminosity is rendered intermittent by rhythmical motions, originates in love and rivalry, and like sedentary display effects gregariousness,

* Trans. Soc. Ent. de la France, T. V., p. 79. Butterflies will not pair in confinement. Newport, Trans. Ent. Soc., V. IV., p. 60.

promotes reproduction, and favours migration. It finds its function
of variation in the dynamics of flight. Thus, as Dr. Pettigrew
has shown, the vertical projection of insect atmospheric paths,
arising from alternate rising and forward wing-pulses, are invariably curvilinear by the second law of motion, although in
flies it will be found (Fig. 1) the wing blur indicates a reciprocating twist, converse to that observed in butterflies (Fig. 2).
So likewise it is immaterial whether the wings hook and move
in unison, like those of bees, moths, and other insects, or act
similarly and simultaneously as those of Dragon-flies; whether

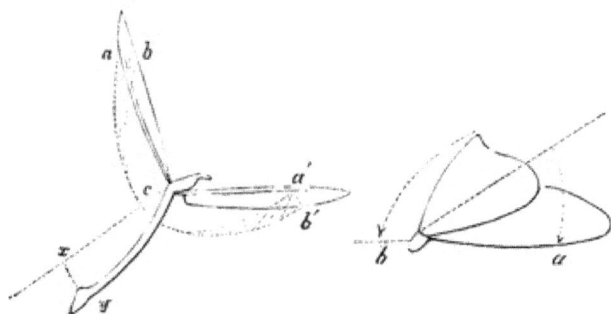

Fig. 1. *a, b, c,* forward impulse; *c, b', a',* rising impulse.
Fig. 2. *a,* forward impulse; *b,* rising impulse.

they feather as those of flies, or shut like those of butterflies, as
they move upwards.

The wings of an insect may be considered as levers, and
during flight they take two strains, a longitudinal and transverse. The first is provided for in the large and often prominent
longitudinal air-veins, which in thin-membraned wings increase
conformably in diameter, strength, and prominence, as we
approach the wing-base, where this strain attains a maximum.
The latter is received and sustained by the wing membrane and
cross veins, that unite with the longitudinal and transmit them
the strain. The position of the centre of percussion is approxi-
mately indicated by the intersection of the two chief longitudinal
veins. The wing is further an elastic propeller, and the impetus
of the stroke arises in great measure from its spring. In the
exotic genus of butterflies *Euplea* the pliancy, and in *Idea* the
flexibility, of the pinions are marvellous. The costal or front

margin is also strengthened in swift-flying insects. The un-
folding of the wings is accomplished by inflation of its veins,
virtually prolongations of the tracheæ; and their expansion is
performed by one set of muscles placed centrally and horizontally,
their elevation and depression by others obliquely vertical, and
situated laterally in the thorax. There are also rotator muscles.
The wing has an intricate articulation, and when this is membranous it is covered or protected by a little scale; check
ligaments are likewise present to regulate its action. The fore
and hind wings may be considered as representing in movement
the primary and secondary quill feathers of Aves. But a variation in outline we associate with a change of character in flight.

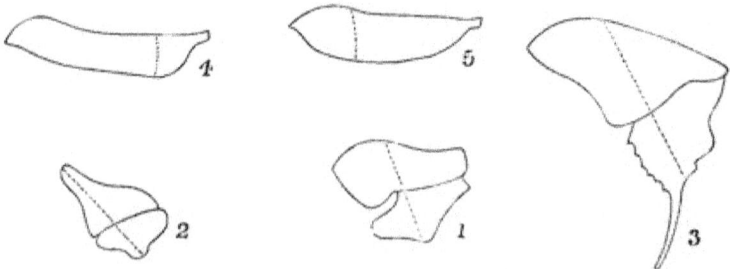

Fig. 1. Outline of the wing in Skimming Insects. Examples: Polyommatus,
Vanessidæ.

Fig. 2. Outline of the wing in Skipping Insects. Examples: Pieridæ, Tortricina,
Hesperidæ, Liparidæ, Bombycidæ, Saturnidæ, Ourapteryidæ.

Fig. 3. Outline of the wing in Soaring Insects. Examples: Papilio, Apatura,
Morpho.

Fig. 4. Outline of the wing in Insects that Hover. Examples: Syrphidæ,
Anthomyiidæ, Bombylius, Culex, Ceratopogon, Chironomus, Psychoda, Formicidæ,
Termites, Sphingidæ, Sesiidæ.

Fig. 5. Outline of the wing in Insects that Dance. Examples: Trichocera,
Heliconia, Hepialus, Adela, Nemophora, Ephemeridæ, Trichoptera.

Skimming insects (Fig. 1) have broad and rounded wings,
with ample sustaining area. Skippers (Fig. 2) are characterised
by a blunt or falcate wing, and their flight may be compared to
that of short-winged birds, but the Soarers (Fig. 3) have wings
of greater breadth and are more triangular, inviting comparison
with acute-winged birds; hollow lines, such as we see in the
Brimstone Butterfly, appear to favour concealment by allowing
the species to settle suddenly.

From this variation in wing outline, too, arise eccentricities of flight in males at the pairing time. Thus certain beaux of the Liparidæ and Bombycidæ, with comb-like antennæ and falcate wings approximating (Fig. 2), zigzag vertically in air, and hunt backwards and forwards, searching for the female, who, apterous, or with wings paler in livery, displays on her cocoon, the herbage, or a tree trunk. These when diurnal in habit pair at noon, on the advent of a cloud-shadow or summer mist.* Many butterflies similarly search for a partner, and then pair at noontide when the sky is overcast, love here acknowledging a kindred stimulus to fear and anger.

Vertical figure dancers, when examined, will be found to have a stereotyped wing outline, lanceolate, or lancet-shaped, with the maximum breadth nearest the tip, which sustains their balance (Fig. 5). They are widely distributed in the orders, with exception of bugs and beetles. As regards butterflies, Mr. P. Gosse tells us the South American *Heliconia Charitonia* before sundown aggregates in a dance after the fashion of gnats, and the individuals then sleep the night in groups of half-a-dozen on the leaves of the creepers. † The heteroceral dance is attractive, and promotes sexual intercourse in the case of the common Ghost Moth, whose silvery-white males swing pendulum-like over the grass of low damp meadows, until some yellow female careering wildly along collides with one, who immediately follows her, and they pair.‡ The orange-coloured males of the Golden Swift likewise dance under lonely hedgerows, and in more accurate curves, although the female of their darker congener the Northern Swift ascends at dusk a bracken frond on Scottish moors to allure its fleet-winged mates that dart around. In similar fashion the sylvan twilight is enlivened during the freshness of early summer by the brisk swinging movement of glossy-winged societies of the Long-horned Adelæ, which when silence falls and the shades of dusk creep on, yield in turn to the similar antics of the sickly yellow Nemphoræ.

The males of some Neuroptera dance and collect, and when joined by their attracted females they pair. Over low alluvial

---

* Trans. Ent. Soc. de la France.   † *Glosseia Charitonia.*

‡ Ent. Mon. Mag., V. VIII., p. 63.

meadows and along their willowed banks, where the water runs
brownly, the dashing males of the common May-fly glitter
seraphically in the afternoon sunshine towards the close of May,
rising and sinking about a foot, with rhythmical motion that
makes their pearly wings the envy of the speckled trout and an
habitual solace to the angler. The females, though mostly unobserved, are sometimes in air, when the flashing males seize
upon them, and they pair* enshrined in light. The smaller
Caddis Flies leave their watery grottoes to similarly weave the
dance of amour in the last rays of sunlight, and I remember, on
the 7th of September, 1874, when wending homeward over the
bridge that spans the ornamental water in Regent's Park, at
about 6 p.m., having noticed a yellow *Hydropsyche* (?) organising
a dance in the fashion of its lepidopterous prototypes. The
larger and paler sex floated fitfully and almost playfully in and
out among the groups of smaller males, who rose and fell in
misty clouds, and these swooping down from time to time
selected the fair who presumed, the male clinging on with the
tarsi of his fore-legs to the abdomen of his suitress, and by
reason of his unusual weight compelling a precipitate flutter to
the grassy margent or a ducking (Plate I., Fig. 8).

Anxious to verify the fact whether the dances of the small
Crane Flies were a nuptial ceremony, I watched the performers, who turned out almost invariably to be coteries of males
attracted by each other's performance; these would dance until
wearied, and then, perching on the buds and leaves, sip honey-
dew. However, at Guildford, the 15th of the ensuing March, I
scrutinised at sunset the gambols of the so-called Winter Gnat,
that arising from garden heaps of mouldering leaves in November, enlivens the dead season by its chori among evergreens.
And I then found that when the glittering males, who were
springing up and down in the frosty sunshine, crossed a female
in figure of 8, the rush of wings grew brisker, and a couple,
separating from the throng, fell to the earth. A slender species,
specially designed for its more elegant semicircles and quadrants,
is the swinging Pendulum Crane Fly, that commences at noon,
just when the April foliage and balmy airs arrive, a dance of

* Kirby and Spence, Lettr. XVII.

rivalry alone or in company, in our apartments, porches, of
gardens; and the autumnal brood of this species I have observed,
like the Winter Gnat, to pair of an October evening. All these
vertical curve dances seem similarly to be commenced by the
males, who as they flash out their colours or interferential glitter
in the air, attract their kind to the sport; but the coupling I
have observed only to ensue towards the hush of evening.

Insects that hover on the wing have wing-outlines resembling that of a blade of the horticultural shears (Fig. 4), definable
as elongate and triangular, with a maximum breadth near the
base, and a costal edge rounded at the tip; and Humming Birds
and other Sphinges have an expansile tail of hair they expand
when they poise. In the wing configuration, and resultant
specific motion, we seem to trace a transition from the preceding
group, for while the Orange-belted Hover Flies poise, humming
singly, on open spots near swamps, woods, and bushes, now and
again attracting their kind to wheel in air, others, as the groups
of Yellow-ringed Gnats and their mute water-born allies, band in
gossamer clouds that lightly lift and slowly fall. As regards
these gregarious dancers, simmering evening flocks of blood-
thirsty gnats, midge, or mosquito, constituting the one Linnean
genus *Culex*, are only too familiar to the cottar on the borders of
extensive fens, rivers, and forests, and more than one alarm of a
rural cathedral being on fire has arisen from their smoky volumes
rising and rolling above the district steeple. The clouds of
plumy *Chironomi*, that spring from those scarlet worms so frequently noticed in our kitchen water-barrels and tanks, are in
some places scarcely less dense; and the pairing of one species,
among the reedy plash at Calais, I witnessed towards sundown one mild afternoon in the beginning of stormy November. The scene was extremely beautiful, the red
autumnal sun shining brightly on the silvery elfish masses
that lay along the old town moat like heavy marish damp,
and as each couple paired they struggled to gain the sere and
seedy plants bordering the bank, an attempt, however, in which
a majority failed, and dropping down were whirled away on
the sluice. The males and females of the ants emerge but to
commingle an hour of their hasty lives in these nuptial clouds,
both sexes twisting off spontaneously their pantomimic wings

when the giddy delight is past, a circumstance rendered the more singular, as the sole mundane idea of their ever-apterous neuters, born from the same parents, is nesting, provision of food, and attendance on the eggs, larvæ, and pupæ—two instincts which if amalgamated in one would doubtless confer high rank on the possessor, had not Nature, in producing the hymenopterous workers, cast them sterile from her laboratory. In their filmy dances, often noticed in this country on those warm and sultry afternoons that occur from July to September, the males, rising with general impulse, zigzag obliquely; and as the whole column lifts and falls they select their heavier females, who follow and float among them suspended head to wind. Sometimes, according to Kirby, the swarms of a whole district unite their infinite numbers, and, seen at a distance, produce an effect resembling the flashing of an aurora-borealis. The noise emitted by myriads of these creatures does not exceed the hum of a single wasp. The representative White Ants, belonging to the distinct family of Neuroptera, similarly dance, but the males chase and contend for the females after wing-shedding.

Many dull-grey flies of the group termed Authomyiidæ, or flower-frequenters, near relatives of our domestic species, move obliquely in bi-sexual dances, during which the males often encounter or seize the female. Thus is enacted the merry sidelong ballet of *Homalomyia canicularis* over the testy head of the country visitor, as he lies of a bright morning shaping his dreams and schemes in unison with the sweet air and stilly echoes; and thus, too, hum its congeners *Scalaris* and *Anthomyia meteorica* around the leafy umbrage, and anon over the head of the sweating horse during an afternoon's round of aristocratic visiting. What charms would not the country mansion be robbed of if deprived of these habitual gambollers in its cheerful rooms and peaceful shadows, and how much wonderment has been unconsciously bestowed on their swinging dance of rivalry and love!

But other species, naturally hoverers, or with wings more rectangular form horizontal circular dances. Thus, while the Alpine *Hilaria* of Loew. has been described as zigzagging in August athwart sunbeam and shadow in the fashion of the

Anthomyidæ, with a brilliant white or silvery reflection, due to
certain oval, opaque, film-like adjuncts to the hind legs ;* another
of this genus, the little black *Cilipes*, in May and June swarms
and wheels in horizontal circles, with a fitful rush over stagnant
water.   During July, on grassy ditches in the neighbourhood
of London, the male of the long-legged *Dolichopus nobilitatis*
constantly approaches his female, expanding and vibrating his
wings; he then takes repeated aerial circles around her with
a buzz, or should a rival alight, the males approach and dash
in one another's faces (Plate V., Fig. 1).   The males of
Cattle Flies (*Tabanus*, *Gastrus*, and *Chrysops*), with certain bees
(*Bombus*, *Anthophora*, and *Melitta*), circle in company, and the
latter grapple and contend for the female.   According to Huber,
the female Hive Bee, at the end of June, when about five days
old, issues forth at noon after the males, and, soaring upward,
circles horizontally in the air.   Male beetles of the Chafer kind
(*Melolontha*) collect in circular dances over trees; and the
Midsummer Chafer (*Rhizotrogus solstitialis*) thus pairs at evening.   Luminous Serricorns of the Elateridæ and Malacoderma
form social dances, or collect in spots, and the light is intermittent or responsive, indicating love, rivalry, or fear; for species
become luminous on touch, collect at a flambeau, or wing to the
torch of a female.   Certain Dragon Flies (*Calopteryx*) live in
company; the blue dapple-winged males of the Demoiselle (*Virgo*)
chase backwards and forwards along the same spot at the water's
side in pairs, and the females follow and settle beside them with
their wings raised, spectatresses of the sport.

Lastly, rivalry has momentary expression in the dances of
the white or highly-coloured butterflies, who may be seen rising
in air, circling each other vertically, battling or toying.   Chief
of warriors, the purple-shot males of *Apatura* fix their throne in
the heats of July on the zephyr-fanned summits of the oak wood,
from which, on sunny days, these Emperors may be seen mounting
upward as far as the eye can reach, battling in air, and eventually
descending to the identical sprays from which they started, often
with their wing-tips broken by the fray.†   Other terrestrial species,
as *Polyommatus*, *Chrysophanus*, *Pieris*, and *Lasiommata*, appear

* Kirby, Lettr. XVI.; *Ent. Mon. Mag.* Vol. XIV., p. 126.
† *Ent. Week. Intel.*, 1859, p. 139; Curtis, *Brit. Ent.*, Vol. VII., p. 338.

jealous and pugnacious in relation to the bravery of their hues. This temperament has its maximum in summer on warm, moist spots, or as we approach the tropic zones, where torrid skies drive and locate the various species on the borders of pools and streams, and where, collecting in thirsty groups, they rise ever and anon in air, and whirl in variegated garlands round and round, with rapid rush of wings, like the rustling autumn leaves. Such butterfly waltzes form alike a prelude to pairing and migration.

Of aërial dances now noticed, vertical figures, or hovering, where species are virtually mute, seem specially to illustrate conditions of attraction by reflection, colour, or luminosity; and in this way the pale colour of the longhorned nocturnal *Nemophoræ*, and sexual differentiation indicated in the genus *Hepialus*, the metallic gloss on the wing-scales of *Adelæ*, or opal refractions on the membranous wings of May Flies (*Trichoptera* and *Tipulidæ*), whether arising from notched scales, hair, or wing surface, or emanating from dermal glitter of gem-like abdomens, as in the case of sun-poised Hover Flies, and sedentary Dragon Flies, all severally fulfil the conditions of display, as does in like measure the leaping spangle of the Fire Flies. But in the dances of bees, flies, and beetles we witness commingling another element, and this especially as their curves of motion approximate horizontality, so that cases occur where the tensive qualities of sound appear to replace those of light. The circular dances, which especially exhibit this feature of vocal music, if primarily dependent on wing outline, in many Lamellicornes appear exaggerated by the horizontally-extended hollow elytra, which seem to render intrinsical that wheeling or curvilinear horizontal path we see in Dung Beetles,* a feature by no means general, for others of this tribe hardly expand the wings, as the Rose Beetles, and some raise them vertically over the back, as the Burying Beetles. Aërial display exhibits one common feature—it is enacted in the wind's eye; aërial migration, it can be shown, on the contrary, takes direction from the wind.

The sense directly corresponding with display is sight. The

* They can, however, fly in a straight line.

visual organs, like those of hearing, take rise in Arthropods, in low complex form. They are either compound (*oculi*) or simple (*stemmata*), and one or both kinds may exist in the same insect. The stemmata or eye specks, placed centrally in the head, in Myriapoda sometimes number twenty, in Arachnidæ six or eight. Among insects, they are generally present in bees; they exist also in diurnal grasshoppers and crickets, in butterflies and moths, and in terrestrial bugs, being here absent in nocturnal and aquatic species; they are also seen in many flies, and some beetles (*Omalium*, *Anthophagi*), as in many lepidopterous and coleopterous larvæ. They are usually three in number, hemispherical, triangulated, adjusted for upward vision in Bees and Lepidoptera, or front vision in Orthoptera, and Libellulidæ; but in Geocorisæ there are only two, as also in the Fulgoridæ. In the larva of *Dytiscus*, according to Dr. Grenacher, the skin is slightly swollen to form a lens; and the vitreous humour and retina, with its pigment, are differentiations of the hypodermis or cellular layer. The larva of *Acilius* has a more convex lens, with the vitreous humour more decidedly differentiated. In Arachnidæ, again, the retina is more strongly differentiated, as in insects; but in the latter the vitreous humour is not much developed.

The faceted eyes are generally two in number, variously shaped and directed, and situated laterally in the head; but in some Longhorned Beetles they are more (*Tetrops*) or less divided by a lap of the integument, as in some aquatic beetles (*Gyrinus*). These may be virtually said to possess four oculi. The compound eyes structurally may be considered a contiguous aggregation of the stemmata in the form of faceted six-sided prisms—and, according to Professor Müller, they act simultaneously in the formation of a single resultant image. In section each element presents a cornea (*a*, p. 99), in the common Stag Beetle of extraordinary thickness, and in *Libellula* thickened in the posterior facets. Behind the cornea in diurnal insects is a ring of black pigment (*c*) (choroid), acting as a diaphragm in diminishing the quantity of light which passes to the crystal-like cone (*c*), and consequently wanting in nocturnal cockroaches and darkling beetles. The crystal cones are of great length in the Libellulæ, and at their extremity, where they become suddenly slender, they are continuous with the nervous filaments

(g) which pass to the optic ganglion. The faceted eyes are classed by Dr. Grenacher according to the development of the crystal-like cone, as follows: Aconic eyes, in which the cellular structure of the crystal-like cone is distinct—Nematocerous Flies, Cimicidæ, Earwigs, and Coleoptera, with exception of the Pentamera; Pseudaconic eyes, where the crystal-like cone is represented by a conical space behind the facet of the cornea filled with a clear fluid, and by four nucleated cells posterior, as heteroceral flies, *Tabanus*, and *Musca*; Euconic eyes in which the cellular structure of the crystal-like body is alone traceable posterior to the cornea — Pentamerous Beetles, Crickets, and Grasshoppers, Bees, Cicadæ, Phryganidæ, Dragon Flies, and Butterflies. These are thought to have the most perfect organs of vision.* In Crustaceans the structure of the faceted eyes is in the main that of the euconic in Insecta.

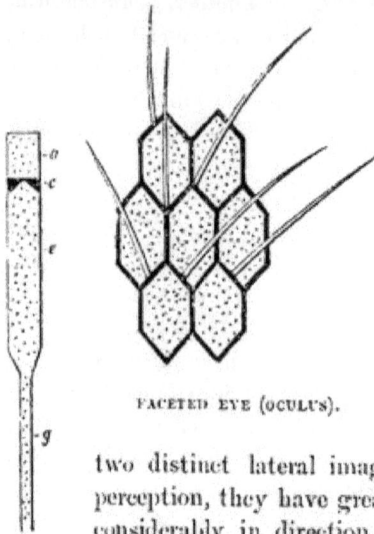

FACETED EYE (OCULUS).

The faceted eyes are, of course, immobile, and form two distinct lateral images, and thus, if deficient in perception, they have great range of vision. They vary considerably in direction. In aquatic beetles and the *Leucanidæ* they are somewhat triangular, divided by an angular lap of the integument, and directed downwards and upwards. In Flies and Dragon-flies they are often large and conical, directed upwards and forwards, as in the Cattle and House Flies. In Orthoptera they are ellipsoid, vertical, and directed sideways. In Longicorn Beetles they are directed backwards and lunate, encompassing the tactile antennæ, which anteriorly supplant their office. In some Flies and Neuroptera they occupy projections of the head. The maximum perfection of the compound eye, according to Dr. Lowne, is attained in the

* The ramifications of the tracheæ surround the oculus and commingle with its nerves.

H 2

multiplication of the compound ocellulæ and fusion of the percipient cellular structure of each into a single organ connected
with a single nerve filament. The eyes of the diurnal butterflies
will thus take rank among the most perfect, both as regards con-
vexity and number of facets, which in species amount to as many
as 34,650 ; as also in relation to the homogeneity of internal
structure. The dragon-flies, with eyes less elevated and circumambient, are not far behind, 14,000 facets having been determined
in the head of one species. These are vagrant groups, with the
sexes often differentiated by bright colour. On the other hand,
sluggish species, gregarious from a common food, have generally
less development of the visual organs, and rarely complete
chromatic differentiation, although brightly coloured, and distinguishable by pattern. The Plant Bugs are an example. Selection by colour and luminosity is thus in direct relation to the
quality of the visual organs ; which, nevertheless, from their
enlargement in the raptorial dragon-flies and suctorial flies, conjointly with an imperfect development in Longicorn Beetles,
indicate variation dependent on food as well as display.

A LIST OF LUMINOUS INSECTS.

(Extracted in great measure from Vol. XV. of the "Entomologists' Monthly Magazine.")

	EMITS LIGHT.	REFERENCE.
COLEOPTERA.		
Elater noctilucus, ignitus (American Fire Flies.)	Chiefly from two yellow tubercles placed at the lateral margins of the thorax	Kirby and Spence, Introduction, Lettr. xxv., Macartney. "Phil. Trans." 1810, p. 291.
Lampyris noctiluca, L. splendidula, L. (Glow-worms) [hemiptera]	Female, from two luminous spots beneath the anal segment; Male, from two luminous points on the penultimate abdominal segment. Larvæ and pupæ luminous	Kirby, and Spence, Introduction, Lettr. xxv.; Ray, "Hist. Ins." 81; "Illiger Mag." iv., 195; De Geer, "Mem." iv. 49.
Drilus flavescens, Oliv.		
Luciola Italica (Italian Fire Fly.)	From last two ventral segments, which are opaque white in colour	"Bull. della Soc. Ent. Italiana," vol. ii., p. 177-180.
Melanactes (Elateridæ)	Larvæ	"Proc. Ent. Soc. Philad." 1862, p. 123, tab. I. f. 5; and l. c. vol. iv., 1865, p. 8.
Nyctophanes	Larvæ	
DIPTERA.		
Chironomus tendens	From the thorax and abdomen	Menizin, "Deutsche Ent. Leitschr," 1875, p. 432; "Ent. Monatsblat," Dr. Kraatz, 1876, p. 41; Pallas Kleine "Notizen," 1781, iv. p. 396; "Zool. Rec." xii. p. 170.
Tipula ?	Larvæ	"Mag. of Nat. Hist." N.S. i. 549.
Ceroplatus sesioides	Larvæ	Wahlberg, "Ofïv. Vet. Akad. Forh." 1848, pp. 128-131.
Thyreophora cynophila	From head	Robineau Desvoidy, "Essai sur les Myodaires," 1830; Macquart, "Suites à Buffon," ii. p. 197.
LEPIDOPTERA.		
— ?	Larvæ, from the rings of the body	Bigg, Withers, "Pioneering in South Brazil," 1878.
Mamestra oleracea	Larvæ	Boisduval, "Silbermann's Rev. Ent." i. p. 266.
Noctua (Polia) occulta	Larvæ	Ginmertinl, "Bull. Soc. Nat. Moscow," 1829, vol. i., p. 140; "Ann. Soc. Ent. France," 1832, p. 124.
APTERA.		
Geophilus electricus, L. phosphoreus, L. (Centipedes.)	From the rings of the body	Kirby and Spence, Introduction, Lettr. xxv.

CHAPTER III.

INSTRUMENTAL MUSIC CONSIDERED AS A MATERIAL AGENT IN
REPRODUCTION AND DISTRIBUTION.

" Insects have, therefore, neither voice nor speech. They produce a sound,
however, by the internal but not by the external air. For no insect respires,
but some of them utter a humming sound, as bees ; flies and others are said to
sing, as the cicadæ. All these, however, produce a sound by the attrition
(vibration) of the internal air, which subsists within a membrane under the
division called the transverse septum, as, for instance, the kinds of cicada ; and
this is likewise the case with flies, bees, and all the rest, in consequence of
raising and depressing that membrane when they fly. For sound is produced
by the attrition of the internal air. But grasshoppers produce a sound through
the attrition of the legs by which they leap."—ARISTOTLE, chap. ix., Book IV.

As we have already observed, among the first aspects that strike
those who study the economy of Nature are the various and
alternate seasons of rest and activity of which it is susceptible.
These are marked dial-like in the limber flowers or leaves that
during the day periodically expand and close, in the lethargy
and flight of winged insects, no less than in the times of
repose and wakeful music of others. We have likewise here
not alone to treat of that harmony which confers a voice on the
summer earth and air, for neither does the insect population
beneath the crystal element remain wholly impassive at the
incense of the seasons ; and, considering the faculty of hearing
in water has been estimated to be four times as great as in
air, it must not altogether surprise us that the water insect
also swims and creaks purposely in the murmuring depth of
our parlour aquarium.

Broadly speaking, Stridulation, or Instrumental Music, in the
Animal Kingdom, proper to Articulata, extends to Insecta,
Arachnida, and Crustacea. As regards Insecta, it characterises
the *imagines,* although exceptionally occurring in the pupal or
larval (?) stage. The musical organs sexually common in most

beetles, butterflies, and moths, as in a grasshopper genus, assume
generally masculine differentiation in Orthoptera, indicating
dermal alteration and induration; they are either duplicate,
paired, and similarly situate as regards the bodies' median line,
or their development is single, as the alar organ of Leaf-crickets,
or quasi unique, as in a family of bugs, and the longicorn
beetles. Reciprocating stimulatory friction of articulate parts to
express emotion postulates adaptive acquisition, consequent on
assumed integumental tendency under attrition to determine
a smooth undulatory surface, and propagation by hereditary
transmission, supposing the theory applicable; any way, a
rudimentary structure of this description exists in the Stag
Beetle at the inferior and posterior extremity of the head;
and whenever a member or group of insects is capable of
music, we may establish a degradation of the organs almost
invariably in mute individuals of the opposite sex, or in
other members of the genus or family. Practically, the microscope establishes the essential constituent, the file or *lima*, to be
a dermal or skin excrescence, with a systematic exaggeration or
coalescence of its external callosities, wrinkles, tubercles, or a
protrusion of the spiral thread of the wing-veins or other
tracheal organ (Plate III., Fig. 7; Plate II., Figs. 4 and 7).
Theoretically, this active or passive source of sonorous vibration is a variously-placed more or less f shaped tumour,
provided with denticulations more or less regular, which are
vibrated and sounded diagonally over a narrow raised callosity
or ridge, on the chitinous integument or modified alar vein.
These latter, constituting the passive or active clasping organ,
assume the function of a violin bow or plectrum.

The musical organ's existence in four orders of Insecta has
been ascertained, and it attains a maximum, with structural and
functional perfection, in the mandibulate tribes. Thus while the
musical instruments emit but faint sounds, or monotonous creaking
indicative of fear, or at most of anger, in genera of beetles, butterflies and moths, bugs and bees, in some Lepidoptera and
beetles, and in most Orthoptera where they are accompanied
with membranous adjuncts and generally differentiated to the
males, they become capable of mutually intelligible modulation
by variation in the direction, force, or length of their stroke, so

as to be employed to express the more complex emotions of love
and rivalry, causing, at certain seasons, the music to assume the
character of a stimulus to reproduction and migration.

The culminating points of musical perfection in a group are
often indicated by other characters expressive of emotion: thus
with beetles the Death Watches exhibit love and rivalry in
music, and are most sensitive to touch; or the longhorns pre-
dominate in pugnacity. In Lepidoptera music is in direct relation to colour, sound to beauty. Musical Orthoptera and Longicornia exhibit pugnacity. This evidently indicates parallel
development of the sensorial organs.

The action of stridulation with a majority of beetles and one
of the bee group is a more or less rapid protrusion and contraction of the abdominal segments, a respiratory movement which
we shall show results from tracheal disposition in Insecta. In
some moths and grasshoppers music is implicated with a bladdery
inflation of the skin; but in other insects it is not directly
dependent on respiration. With some the action is a sharp nid-
nodding, performed by the elevator and depressor muscles of the
prothorax or head. Many butterflies and the crickets produce
their music by wing friction, resulting from a rapid movement
of the extensor and deflexor muscles; and the grasshoppers to
the same end employ the subtile elevator and depressor muscles
of their agile leaping legs.

The music of some crickets, beetles, butterflies, bugs and ants
may be synthetically studied by means of the vivarium or
aquarium; but that of others, as grasshoppers, must be observed
sub divo. A box covered with gauze answers well as a cage for
Orthoptera; but beetles that employ their mandibles should be
placed under glass or wire gauze. The method here adopted in
notating insect music is that of rhythmic syllables composed of
consonants and vowels, which may be carried to sufficient perfection to cause grasshopper species individually to respond. Notes
produced by musical instruments are also in vogue, but these
necessitate a technical knowledge, and even then little more
than the comparative mellowness, sprightliness, or plaintive
nature of the harmonies can be caught. The most acute ear, and
that best attuned to agrarian music, notably fails in the task of
reproducing its fine quirks and light echoes by means of our

conventional scales. Certain passages in the recitative of crickets evidently correspond with our intervals: but the greater portion of all such music is much too rapid, and it is also so uncertain when the performers may stop, that we cannot extemporise any musical bar whatsoever. Then the pitch is rarely attainable by our shrillest and harshest instruments, and the intervals employed are so minute that they differ greatly from the gross periods into which we divide the octave. The eccentric development of the musical phenomena on European areas, no less than the irregular distribution of species, speaks of forms extinct or of outlying groups.

STRIDULATION OF THE NEUROPTERA.

The stridulation of the Neuroptera has been inferred from the probability of an extinct insect found by Dr. Scudder in the Devonian strata of New Brunswick, being musical and appertaining to this order, and on the story of the beating, presumably stridulation, at Upminster, of a louse-like insect identified as the Museum pest, *Atropos pulsatorius*. The latter is stated to tick for hours together on paper like an antique pocket-watch, with sudden shakes of the body; to answer its pulsations artificially produced; and after a fortnight thus to pair with the female. I find no subsequent confirmation.

STRIDULATION OF THE HYMENOPTERA.

If, with Goureau, we consider the sound given out by the Sand-wasps to be vocal, we shall have only the Formicidæ to treat as stridulators among the bee tribe. And apart from the hissing of questionable origin given out at times by ant-streams, the solitary species of the parasitical genus *Mutilla* have been long known to possess the power of producing a shrill frictional cry, sibilant in tone, and common to the winged male (Plate III., Fig. 9♂) and wingless female (Fig. 9♀). This sound Goureau considers perceptibly due to their rubbing a shining surface on the third abdominal segment beneath the inner border of the much-elongated second segment (Plate VI., Fig. 8, *a*); and in this he is corroborated by Herr N. Westring, who remarks this shining surface is a little flat dark shield, which, as seen beneath the glass, is transversely and finely rugose (Fig 8, *l*).

The Solitary Ants are partial to barren, sandy soils, where the males frequent flowers and the females lurk in holes until the hymeneal hour, when they may be seen borne by the former through the air. Their subsequent oviposition in Bumble Bee nests decimates the constructors, one establishment evolving seventy-six of the European Solitary Ant and only two bees. The species attain a maximum in warm climates, thirteen inhabiting Kephalonia. While passing a fortnight on the island of Capri in the Bay of Naples, I was able to procure a few females of *Matilla hungarica*, F., some wandering on clayey spots, and others taken covered with honey, emerging from the cells of a small violet bee (*Chalicodoma muraria*, F.). When first seized these would produce a sharp sound—"Tip! Tip"—by drawing their shield half under the second segment and then pausing; this note was then lengthened by a complete movement of the segment forward, and then the action lapsed into a quickly reciprocating double stroke, causing the note to rise in pitch. On seizure they also threatened with an elongate sting. The musical organs of the Solitary Ants closely resemble those of the Longicorn Beetles, except in being placed on the abdominal rings in lieu of those of the thoracic division, and that the single lima is here the active agent.

At Guildford, towards the middle of July, 1877, I was in the habit of entertaining myself, by examining with a pocket lens some individuals of a small yellow emmet (*Myrmica ruginodis*), which had established a sparsely-visited domicile beneath a wall, and were invariably collected on the blighted spikes of certain neighbouring clumps of wayside thistles.* When the sun shone these thistle-heads became really objects of interest if thus scrutinised. You saw the aphides exemplified in their various stages, each with its rostrum plunged deep in the sappy cuticle; and gathered round the desiring ants, who, ever on the alert, moved from one to another watching for each slowly welling drop of honey-dew, which as it appeared at the anal tubes, was forthwith an object of eager contention, two often sucking the precious secretion together.

Wishing to reproduce this curious scene, which may now be seen

---

* *Cnicus arvensis* and *lanceolatus*.

indicated on Plate I., I cut off a thistle-top or two, and secured
them with the blight, drowsy in potation, to the mouths of small
phials of water, at the same time carefully separating the ants
and covering them beneath an inverted wine-glass. Now it so
happened the weather was hot and sultry, and these emmets
probably irascible; for they had not been left long to themselves when a puny-looking individual was observed placed head
downwards at the side, and near the inverted edge of the glass,
rapidly vibrating its abdomen vertically from the pedicle, and
simultaneously giving out a continuous singing sound; in colour
and intensity resembling the sharp whining of the little dipteron
Syrilia pipens (Plate I., Fig. 4, *c*).

Concluding the rhythmical motion accompanying the music
indicated this ant as a stridulator, I carefully studied its external
anatomy beneath the microscope, and found the abdomen contracting anteriorly with callosities, as though the skin was
drawn in; here produced, and movably inserted into the second
knot or articulation of the pedicle, moulded in the form of a
dark ring, traversed with more than twelve minute yet regular
annular striæ (Plate VI., Fig. 7, *l*). This was reproduced where
the second knot (*p*) articulated to its triangular antecedent, but
with the striation less marked; elsewhere, the exterior surface of
the epidermis was merely punctured or wrinkled. As the spiracles
of this emmet are minute, I would, therefore, ascribe the singing
of the puny (male?) individual to the friction of the first-men-
tioned striated ring. To Mr. F. Smith, of the British Museum,
I am indebted for the correct specific name of this little yellow
ant. Several observers have noticed a hissing to arise from
certain ant streams when molested, but I am unaware of its
nature. The Solitary Ants, besides music, have the attraction
of coloured pile; one, *M. cardinalis*, is of a bright scarlet,
others are distinguished by spots and lines.

The stimulus to music in the bee kind, we have noticed,
is evidently fear or anger; but some emit a sound perhaps
maternal, for an African fossorial Hymenopterous insect, termed
by Drury *Cœruleana*,* " makes a clicking noise when it flies, like
a racket, which may be heard at twenty yards distance." The

* *Pepsis cœruleus* (?).

species preys on grasshoppers, which it carries from bush to
bush. This alar stridor, which we shall again notice in the
butterflies and crickets, appears due to serration of a wing-vein.

STRIDULATION OF THE HEMIPTERA.

As we have musical fishes, whose drumming in the deep
river or ocean bed maybe gave colour to the ancient tales
of the mermaiden, so there are musical water-bugs (*Hydro-
corisidæ*), which utter a startling voice beneath the calm lake
or inky pool, over-starred with white water-lily and ranunculus.
When pondering on a remark by L. Frischs, that the male
Broad Water Bug, a species which from a rude accompanying
figure of nymph and imago appeared to be the common *Naucoris
cimicoides*, L., produces with its neck a fiddling noise like
the Longhorn Beetles, I dubiously experimented on some
individuals, late the tenants of a willow-hung, rain-water pool,
and finding a faint sound like the scratching of a needle-point
would at times ensue when the overlap of the delicate tense
and pellucid tergum of the prothorax was nodded to and fro
and slightly pressed on the upper surface of the mesothorax, was
led to examine the lateral front angles of the latter, where I
detected a minute lima of the *f* shape, thickly set with striæ.
I next investigated another member of the rapacious *Nepidæ*,
a handsome ruddy-winged Water Scorpion, very kindly sent
me for the purpose by the author of the "British Hemiptera,"
and finding the edge of the prothorax, if depressed and worked
over the mesothorax, elicited an ominous sharp click or crackle,
proceeded to inspect the superior lateral angles, and here discovered two faint, triangular, rounded, striated surfaces, resembling
in all respects but position the lima of many Longicorns.

In *Corixa Panzeri*, Fieb., a little, light-cared bug that basks
on its subaqueous sand-flats like a flounder, they are again seen
here *f* shaped and elongate. *Corixa striata*, Cur., its congener,
has been recorded performing in a parlour aquarium. At the
meeting of the British Association in 1845, Mr. Ball noticed the
fact of one of the Notonectidæ, *Corixa striata*, Cur., emitting loud
and powerful sounds, somewhat like those of a cricket. "These
sounds were given out while the animal was about two inches
and a half under water, and so loudly as to be distinctly audible

in an adjoining room through the closed door. The first observation of this fact was made about two years since by Miss M. Ball, who has since frequently verified the original observation. Mr. Ball stated that he had himself heard, on the 15th of June instant, this remarkable sound. It is probable the sound is emitted only by the male: it has as yet only been heard in the months of May and June." Lastly I examined a specimen of the Water Boatman, *Notonecta glauca*, L., and its melanic variety, *Furcata*, Fab., from a slimy farm pond by the roadside, and here also finding the projecting prothorax if moved over the mesothorax elicited a sharp, high creak, submitted the more rounded surface of the scutum to a high magnifying power, and thought to find the cause either as before in lateral limæ, or in an obliquely striated band lying along either side of the central lenticular depression.

On the other hand, the stridulation of certain terrestrial Hemiptera has been already established, but caution is needed in following up the investigation from a tendency in the skin of these insects to run to wrinkles where a frictional surface could not be supposed to act. Or again, species have been enrolled as stridulators, and stated to possess limæ on the ventral arcs of the abdomen played on by the hind legs, without appended notice of their being seen or heard performing.* Among our northern Land Bugs (*Geocorisidæ*), the stridulation of the strongly rostrated species of Reduviidae, as *Reduvius personatus*, and congeners, or of *Coranus subapterus* (Plate IV., Fig. 13), on seizure, by rubbing the extreme point of the rostrum (*s*), in the obliquely striated channel on the prosternum (*l*), has been most ingeniously analysed by Herr Westring and Herr Reuter. *Reduvius testaceus*, rare at Malta, possesses a very marked power of stridulation, and species of the genus *Pirates* are known to perform in the same fashion.† This music, in conjunction with the repulsive scents emitted by certain groups of the Land Bugs, from pyriform glands between the hind legs, intimates a perception akin to fear.

Our single native species of *Reduvius* is best known to hen-

* The stridulation of the genera *Pachycoris*, *Scutellera*, *Strctrus*, *Optomus*, *Coeloglossa*, *Arctocoris*, and *Psacata*, wants confirmation.

† Kirby and Spence, Lettr. XVII., p. 317.

keepers, from the sharp prick it administers with its rostrum. Douglas and Scott say of it : " Found occasionally in houses and fowl-houses, it also flies by night to lights in windows. The larva feeds on the Bed Bug, and other insects, and covers itself with dust." The problem as to how this insect produces a monotonous stridulation has been thus solved by Herr Westring. " I took a *Reduvius personatus*, Wolff, and remarked after this had been stuck on a pin it nodded with its head, and at such times the stridulation was heard. As I observed the neck was shining or smooth, I could not bring myself to consider that part an organ of sound. On pushing a fine brush between the sucker or rostrum and the canal, the sound ceased, although the movement of the head was continued, but on its removal the sound was again heard. This experiment was repeated with the same result. So I considered it settled that the stridulation was caused by the friction of the rostrum in the canal on the prosternum. This I established, on observing with a microscope that the canal in a certain light was covered with the finest transversal edges, starting from either side and running obliquely backwards, meeting in the middle at an obtuse angle. The active organ was the needle-fine point of the rostrum, which moves over the canal at an inclination of about 45°." Ray compares the cry of this bug to the chirping of a grasshopper.

Various species of Reduviidæ I lately procured performed in similar fashion. Placing the indurated or tempered termination of their short and thick rostrum in the lenticular striated groove extending from the front edge of the prosternum to the insertion of the first pair of legs, they then commenced to rub this angular point backwards and forwards by a nodding motion of head from its prothoracic articulation, the length and celerity of the movement perceptibly regulating the fulness and pitch of the notes, while the organic structures and frictional surfaces determined the gamut.

The first stridulator I met with, the pupal form of *Reduvius personatus*, was taken begrimed with particles of dust, within the folds of a sun-bleached muslin window-blind ; the second, *Oncocephalus notatus*, Ramb., was captured in a railway carriage near Foggia, both in the month of May.

During captivity these would perform somewhat reluctantly on seizure; the notes of the first had the musical timbre of the minute longicorn beetles (*Leptura*, Fabr.); those of the latter had a more rustling sound, which caused me repeatedly to think they arose from my having inadvertently crumpled the elytra. A third stridulator, the perfect winged form of the former bug, I captured on the 18th July behind a window-shutter; it performed readily, and its notes were sharp and distinct. This species has been said to emit a disagreeable mouse-like odour. A fourth example, *Harpactor iracundus*, Scop., taken on the banks of the Po at 6 a.m. one June morning, when engaged in sucking the juices of an earwig, was a more sturdy performer, with a sharp, creaking stridulation. And, although if retained for more than a second in the hand its music would often subside to a tone scarcely perceptible by the human ear, yet, if the insect was then allowed to slip just a little through the fingers, this action apparently conferring some sensation akin to pleasure at release, the rostrum was seen at once to elongate its strokes in the channel, and the notes came out again loud and clear. When suddenly seized it had also the power of emitting a strong vinegary scent. The power of stridulation does not, however, appear to exist in the larval state. All species of this family of which I am cognizant employ their rostrums to prick their captors sharply. Those Palæarctic kinds here observed to stridulate will amount to six; and various of the larger exotics have no less an appearance of being very decidedly musical and vindicative. Travellers might experiment on their capabilities.

The sounds emitted by the Water Bugs seem low when compared with those that can be reproduced in many Longicorn Beetles. I am not aware that any of the species stridulate when seized; and although many emit clicking sounds when confined in a glass vessel, I never could detect an instance in which their production was not accompanied by the head striking on the bottom or sides of the receptacle. It may be also remarked the bugs have hitherto evinced only one incentive to stridulate, namely, fear on seizure, although one species, like *Cryptorhynchus lapathi*, among the rhychophorus beetles, was most disposed to do so when the paroxysm was passing off; and, further, that it has yet to be determined whether the several species, viewed in regard to sex and stage of development, present gradations in their capability for music, or not.

STRIDULATION OF THE LEPIDOPTERA.

Let us now repair to the fragrant orange-groves of the New World, and listen to the rattling of love-wounded butterflies that there meet, to chase through noon-tide twilight. "The large and brilliantly-coloured Lepidoptera of Brazil," says Dr. Darwin, "bespeak the zone they inhabit far more plainly than any other race of animals; I allude only to the butterflies, for the moths, contrary to what might have been expected from the rankness of the vegetation, certainly appeared in much fewer numbers than in our own temperate regions. I was much surprised at the habits of *Ageronia feronia*. This butterfly is not uncommon, and generally frequents the orange-groves. Although a high flier, yet it very frequently alights on the trunks of trees, on these occasions its head invariably placed downwards, and its wings are expanded in a horizontal plane, instead of being folded vertically, as is commonly the case. This is the only butterfly which I have ever seen that uses its legs for running. Not being aware of this fact, the insect more than once, as I cautiously approached with my forceps, shuffled on one side just as the instrument was on the point of closing, and thus escaped. But a far more singular fact is the power which this species possesses of making a noise. Several times when a pair, probably male and female, were chasing each other in an irregular course, they passed within a few yards of me, and I distinctly heard a clicking noise, similar to that produced by a toothed wheel passing under a spring catch. The noise was continued at short intervals, and could be distinguished at about twenty yards distance."

Darwin is corroborated by Mr. A. R. Wallace. "This, the common Ageronia (at Pará), produces it remarkably loud, when two insects are chasing each other, and constantly striking together. One alone does not produce the sound in flying, and I have never heard it made by the small species *A. Chloe*, which is equally common with the other. I am inclined, therefore, to believe that it is produced in some way by the contact of two insects, and that only the larger and stronger-winged insects can produce it; but M. I. B. Capronier says, 'There have hitherto been some doubts if the noise referred to was peculiar to the male; but M. Van Volxem, who has had frequent opportunities of observing these insects enjoying their frolics, affirms that the noise is common to both sexes.' We also learn concerning these musical species, christened the 'Whip' butterflies, that they are common where their food-plants, *Dalechampia*, abound; that they have a short, rapid flight, and constantly alight on the trunks of trees; and that four species* produce in flying a sound which has been compared by a good observer to the rustling of a piece of parchment.'"

Mr. E. Doubleday says that "he had examined the butterfly described by Mr. C. Darwin as making a noise, and that he had detected a small membranous sac at the base of the fore-wings, with a structure along the sub-costal nervure like an Archimedean screw or diaphragm in the tracheæ, especially at the dilated base of the wing." Instigated by these observations, I have myself examined the same butterfly, and find the superficies of the wings that overlap are of considerable dimensions. On the surface of the hind wing the costal vein (marked *s*, Fig. 1, Plate IV.) is elevated, indurated, black, curved, and bare of scales for about 3''', and at first sight remarkably smooth and glossy. But if this apparent smoothness be observed obliquely with a strong magnifying power, in bright daylight, parallel indentations or slight striæ are seen all along its hinder surface, and under the microscope these develop into a fine file or lima. When the wings are expanded, this costal vein is received into a little concavity in the inflated, rounded, smooth anal vein of the fore-wing (marked *l*, Plate IV.) beneath and near its base, which can be traced, after the fore-wing has been detached, by the depression its great prominence leaves in the membrane adjacent. This concavity seems suited in every way to act as a clasp, sonorous when the wings are moved, while the whole adjustment presents, in elementary form (if we overlook a slight divergence as regards correspondence of the veins), the bristle and catch that lock the wings of moths.

* *Feronia, Fornax, Amphinome,* and *Arethusa.*

I

The music, then, appears common to the sexes, and to be produced from the stimulus of love or rivalry, but is perhaps also capable of expressing fear, for sometimes a short, clicking noise is made when one of these butterflies is caught in the net. Some other South American Rhopalocera stridulate. Dr. Fritz Müller relates :—" On October 30th, 1876, at the mouth of the Rio Trombudo, a tributary of the Itajahy, I saw two butterflies chasing each other, which produced a loud clicking noise, and settled from time to time in the manner of *Ageronia*, with the wings expanded horizontally, on dry stems of bamboo.* I, of course, imagined them to be some species of this genus; but, after having succeeded in catching one of them, found that it was *Eunica Margarita*. I may observe the neuration of the wings of that butterfly bears a rather close resemblance to the above, so indeed it may be more nearly allied to that musical genus than is generally assumed. On February 21st, 1877, at the foot of the Serra de Itajahy, I again heard a noise resembling that of *Ageronia*, but rather louder, produced by two small brown butterflies, which I did not succeed in catching."

Among the Nymphalidæ, or butterflies with only four perfect legs, the African species of Charaxes, perhaps, utter a sound in flight, and several kinds stridulate in repose. The Purple Emperor (*Apatura Ilia*), when settled on the damp road or on cow-dung, may be observed to partially expand its wings, and move the fore back over the hind, as if producing a sound; and the species of *Vanessæ* will often, when disturbed during their characteristic northern hybernation, or during the similar torpidity induced by dull weather or enclosure in the dark, produce, as they sit, a high sand-papery sound, by repeatedly spreading out their wing horizontally, and rubbing the fore-wing's basal portion over the hind, which action and accompanying stridor will be repeated on touching the insect's wings, or by merely presenting an object at the performer, evidently indicating the impulse of fear or anger. In this genus colour and music take parallel development, as evinced by one of the most beautiful of stridulators—a welcome annual visitant to the hydrangea and dahlia of our flower-gardens.

* *Taguara.*

The Rev. Joseph Greene informs us that on the 8th of December, when out on one of his memorable autumnal diggings for pupæ in Buckinghamshire, he came to a beech-tree on a high bank, the roots of which formed an arch about a foot in height, and faced the north, the opening being quite exposed to rain, snow, &c.; and that, as he was on the point of inserting a trowel into the cavity, he heard a faint hissing noise, and to his surprise he found, in searching for heterocerous pupæ, he had startled a colony of Peacock Butterflies, wintering there with shut wings. He tells us :—" Two were attached to the concave part of the arch, the third was on the ground, and the noise I heard proceeded from it ; " and adds, " The noise resembled that made by blowing slowly, with moderate force, through the closed teeth ; and, while making it, the wings were slowly *depressed and elevated.*" Apparently doubting the evidence of his senses, Mr. Greene pushed off another of these insects, " which immediately commenced the same movement of the wings, accompanied by a similar noise." That it was the testy temperament of the performers that thus sought vent, as spoilt children cry when awaking from sleep, he next afforded an ingenious proof. Pointing the trowel at one of the performers that had expended its spite, and closed its wings again to slumber, he saw it immediately turn towards it, and recommence the noise and motion with renewed vigour; and he noted that, whenever this experiment was repeated, the same querulous manifestation ensued.

The late Mr. Hewitson, whose collection of butterflies was widely known, writing on the 28th of January upon the conduct of a Peacock Butterfly that had been hybernating since the first hoar frosts (we presume on the ceiling or wall of his sitting-room at Weybridge), says :—" They had been cleaning my room, and had driven it from its winter quarters. I had handled it rather roughly, which it resented by *spreading out its wings horizontally to their fullest extent, and rubbing them rapidly together:* it produced a distinct sound, like *the friction of sand-paper.* This it continued to repeat for some time, and seemed greatly exasperated." The Camberwell Beauty has been seen behaving in a similar fashion. " In 1872," says Mr. A. H. Jones, " a female *Vanessa Antiopa* came into my possession, in a hybernating condition, and in that state she would, when disturbed, partially expand her wings, and at the same time was produced a grating sound, which seemed to come from the base of the wings."

This braving our winter's climate is no chance occurrence, but is almost yearly observed with regard to the Peacock or the Tortoiseshells, and shows a remarkable accommodation of habit and persistence of species, when we consider two of this genus find their limit of northward distribution in the Cheviots. Doubleday describes the fallen boughs of the forest covered with protective horse-shoe fungi as their original winter quarter of the Palæarctic areas. " Last winter," he says, " some large stacks of beech fagots, which had been loosely stacked up in our forest (Epping) the preceding spring with the dead leaves adhering to them, were taken down and carted away, and among these were many scores of *V. Io*, and of the Tortoiseshells, *Urticæ* and *Polychloros*."

It may be conjectured that beneath pure Italian skies the portion of the year passed by these butterflies in hybernation would be brief; but in more northern countries the few species that gladden the landscape shrink from encountering the first visits of the benumbing airs of winter. I took advantage of a singular opportunity thus afforded, during a sojourn in the Highlands, to investigate the capability of the Small Tortoise-shell Butterfly for stridulation. On the 22nd of August, a dull day, when there had been a sudden fall in the temperature, a fresh brood of the Nettle Butterflies, newly sunning themselves at West Loch Tarbert, hastened in from the fields to shelter and remain torpid, perchance to dream. I detached one of these, a female, hanging on cobwebs in an outhouse, and seated her, still drowsy, on the palm of my hand. Then, with the other hand, touching lightly the tails of the hind wings, I induced her to depress and shut the wings successively. Each time she testily performed this action I heard distinctly, as the fore-wings were brought forward, when only the extreme basal portion of the wings was in contact, a sound soft and refreshing, like evening footsteps on the pavement, or grating sand-paper.

In the Small Tortoiseshell, then, certainly, and in the Peacock more than probably, it will be noticed, the sound produced by the vexed insect must have arisen from the friction of some hard parts at the basal portion of the wings. Prepossessed with this idea, I submitted specimens of the wings cut from the male and female of either of these to an excellent microscope. And I then found a minute file or serrature (lima) came at once to view, situated on the anal vein (Plate IV., Fig. 1, *l*) at its base, and running along it for one-third of its length, for which distance it is tumid, spindle-shaped, and bare of scales. In the case of the Peacock, and I believe also of the Small Tortoiseshell, it was much more strongly developed in the female than the male, and the vein had a blacker, firmer consistence. The vein that clasps this notched or filed one when the wings are rubbed together is not difficult to find in the costal vein of the hind wing (Plate IV., Fig. 1, *s*), recognisable by a raised surface, curved outwards, with a smooth bevel above where it comes in contact with this filed vein; but it likewise only presents this character at the base, for if we trace it outwards, we soon notice its upper surface to sink in a series of sharp notches beneath the feathery scales. But this is not all; in each of these butterflies there is an organisation which I would compare to the mirror of the males of the Leaf-crickets in structure and object, for we find at the fore-side of the costal vein the wing-membrane is bare in a little circular patch (*su*) which is embossed, a provision, I conclude, to impress the musical tremors arising from the friction of the filed vein on the air.

This genus, characteristic of the group of the Nymphalidæ, it will be seen has the organ of stridulation placed inversely from what we find it in Papilionidæ; in the Lycænidæ the adjustment seems the same. Mr. A. R. Wallace, in a paper on the Butterflies of the Amazon Valley, says, regarding the indigenous Theclæ *—"They have a very peculiar habit of moving the two lower wings over each other in opposite directions, giving an appearance of revolving discs." The phenomenon is also observed of certain Hair Streaks in the Isthmus of Panama by Mr. T. Belt, who relates that on his arrival at the small town of Tierrabona he saw, " on wet muddy places near the stream, groups of butterflies collected to suck the moisture. Among

---

* *Eudymion, Marsyas, Etolus, Pholeus, &c.*

them were some fine Swallow-tails, quivering their wings as
they drank, and *lovely blue Hair Streaks*. The latter, when they
alight, *rub their wings together*, moving their curious tail-like
appendages up and down."

Here we have corroborative evidence of the alternate motion
that certain of these little-tailed butterflies, when reposing,
impart to their hind wings, which forcibly recalls the stridulatory action in the genus *Vanessa*, differing in this respect, that,
whereas the hind wing is rubbed by the fore one in the one case,
it is the fore-wing that is rubbed by the hind one in the other.
Having become impressed with this analogy, I, one May, gave
my attention to the habits of the indigenous Green Hair Streak
Butterfly, to see if anything similar was observable in its
economy. Nor was I disappointed in this, for on two occasions when it was reposing sedately on a bramble leaf with shut
wings I was rewarded by seeing it alternately lift the hinder
wings, and in a leisurely fashion rub them backwards and
forwards over the fore ones, although I could catch no audible
sound. The movement may be also observed in numerous
species of the genus of " Blues " (*Polyommatus*) that flit over
our Palæarctic meadows. Regarding the presence or absence of a
stridulating organ in connection with this rhythmical action, it
will be found that beneath, on the overlap of the fore-wing in
various species of *Thecla*, or Hair Streak, the scales are supplanted by a patch of hair, and just above, the anal vein is bare
and raised. This bare raised portion in the Green Hair Streak,
submitted to a microscopical scrutiny, appears crossed at uniform
distances by pronounced striæ, which indicate internal dia-
phragms, and constrict the tube into a series of bead-like formations. *T. Ætolus*, one of the butterflies noticed by Mr.
Wallace performing, the overlapping fore-wing has its anal vein
decidedly denuded beneath, indicating friction, and, at the base,
it is also blackened, indurated, and striate. We thus see in
Rhopalocera or club-antennæ Butterflies the alar limæ intimate
vein induration, with abnormal protrusion or constriction of
their tracheal helix, while their music may express love,
rivalry, fear, or anger—certainly the latter.

Moths distinguished from butterflies by the absence of the
button terminating the antennæ have many of their groups

nocturnal; and thus inhabiting a colder atmosphere, we find their bodies are often more thickly clothed with hair and scales than the diurnæ. They also present us with a music, whose emotional springs remain now to be interpreted by metaphysics, as its mechanism by anatomy. Certain of these that present a passage from the sunny butterflies are said to possess musical organs similar in construction and position to those of diurnal Cicadæ, which they not remotely resemble. Of this, the genus *Glaucopis*, in the family of the bright little Burnets and the Australian *Hecastesia Thyridon*, afford examples, the latter being known to produce a sound. Generally the heteroceral music presents features that confer on it a doubtful appearance as to whether it appertains to the instrumental or vocal class.

The Death's Head Moth (Plate IV., Fig. 8), on certain spots planted with potato, occasionally emerges from the chrysalis state in no inconsiderable numbers, and although seldom seen flying, and scarce attainable but by rearing from caterpillar or chrysalis, is at times found in hiding in the daytime, and at night has been known to congregate round a lantern. But in the evening glow, just when the October sun is oversoon dipping beneath the horizon, the Death's Head is not unfrequently observed suspiciously fluttering around beehives, that, heated by the short noon, are effusing a thymy fragrance of nectar, which a short proboscis and heavy flight lead him to seek in specific quantity, in preference to any doubtful measure at the bottom of cups and tubular flowers where his kind hover. And when the shades of night have settled on the landscape, this sly marauder may again be observed stealing from the drowsy citadel of the bees, intoxicated and bathed with sweet honey, yet unharmed and unmolested.*

But another trait, and one with which we are more immediately interested, is a power this reveller has of emitting a shrill and plaintive squeak, scarcely distinguishable from that of a mouse. This flute-like note the moth has not been observed to give out on the wing as is usual with Lepidoptera, but only on the ground when disturbed and seeking concealment, or when touched or handled. A correspondent to the *Entomologist's Weekly*

* Kirby and Spence, Lettr. XVII., p. 317.

Intelligencer, after remarking these insects usually creep from their pupæ shells at midnight, or later, adds : " I have frequently allowed one or the other to have a fly around the following morning; they were soon satisfied, and would, while uttering their squeaking notes, hastily retire and settle in some dark part of the room."

The Death's Head (Plate IV., Fig. 8), with its mimicry of the macaws, like those screaming birds, lives well in confinement, but is apt to die from gluttony. A number of individuals kept together showed a disposition to pat one another with their fore-feet; and it has been shown that a touch of the delicate tarsus provokes a squeak. So that, with our previous knowledge of this as a weapon in courtship, we may infer that the sound of dread will eventually appear the same timid and nervous note as other instrumental music, and, no longer an echo of the menagerie or harbinger of ill to the peasant of Brittany, will have to be classed with the mysterious harpings that collect at the chimney-stack, the shriek of the barn-owl at the hayrick, the chatter of the bat at the eaves, or the baby cry of the hyæna.

When we turn from the object to consider the organism productive of the sound, we are able to deduce more satisfactory conclusions; for in addition to facilities for an examination of the anatomy of the terrible Sphinx itself, we glean much from the celebrated controversy in which Réaumur, 1742, Schröten, 1785, T. Vander Hoeven, 1859, support a proboscis-palpi theory ; Roesel, 1755, Ghiliani, 1844, Wagner, 1836, a proboscis expiration theory ; Lory and Nordman, 1838, the abdominal expiration theory ; and Passerini, 1828, Westmaas, 1860, the Passerini theory. The view here adopted will be intermediate between the frictional and expiration theories. I will begin by quoting the experiments of Landois in 1867, in confirmation of Réaumur, thus given in his little pamphlet on the "Ton und Stimmapparate der Insecten":—

"On first impression the singular piping made led me to think the note was produced by friction, because it has a resemblance to the note of many other insects which produce their music in this way. I could not approach the note instrumentally. One similarly finds it scarcely possible to reach the pitch of the note of the Longicornes.

" An attentive observation of the living insect invariably reveals a perceptible motion of the palpi (Plate IV., Fig. 8, *l*), the proboscis (*s*)—rolled up between them—at such times remaining almost without movement.

" The inner surface of either palpus towards its base is naked, but when submitted to the microscope this seemingly uniform glazed and shining surface is seen to be crossed by numerous fine grooves composing an apparatus resembling in construction the file of the burying, dung, and longhorn beetles" (Plate IV., Fig. 4, *l*).

Réaumur was the first who published an account of the note of the Death's Head Moth, and he was fortunate in possessing several living examples at one time, so that he was enabled as much as possible to make a clear investigation. His conclusion was the note of the moth might be produced " by the friction of the two bearded limbs against the proboscis." In this case, then, the two palpi carry the files, and the horny and short proboscis lying between them acts the part of a clasp.

" Since the files originate at the base of the palpi, the moth is only dumb when these are totally removed. When the proboscis is cut off the note naturally can be no longer effected, since the surface of this organ on which the palpi rub is also removed. If the files be rubbed against the proboscis each time, the note of the insect is produced. The files of the males have finer furrows than those of the females, which is in harmony with the fact that the stridulation of the females is deeper than that of the males."

Dr. Landois then gives a table of the denticulation of the palpi in seven *Sphingina,* and concludes the Death's Head stridulates audibly, " because in this insect the teeth are very strong and large." I may likewise add if one cuts off a palpus of a moth that has been emitting its cry for some time, a further proof of the implication of this part can be obtained, as the hair at its fore-edge will present the appearance of being rubbed short, or worn away.

The Death's Head accompanies its squeak with a singular phenomenon. As the male Hoopoe is said to call its vernal mate by inflating its beak and then rapping it against the bark, so this moth, as in mimicry, when it squeaks puffs out the first three

rings of its abdomen by collapsing those behind. The air, then,
it is supposed, leaving the large air-sacs, is forced as by a pair of
bellows into the end of the digestive canal "immediately before
the peculiar stomach," through a communication discovered by
Wagner; and traversing this, passes to the haustellum, issuing
by sixteen little perforations at the extremity of either of its
parallel tubes, as may be inferred from the air-bubbles noticed at
each squeak when the tip is submerged in a fluid. So may we
conclude the sound is really formed in the haustellum comparable
to two flutes joined side by side, in which nodes and loops
originate on expiration, augmented by strokes of the filed palpi,
while the sound waves issue at the sixteen round perforations.
By this self-same arrangement butterflies and moths suck in
their nectarious nourishment.

But the organ of crepitation is not restricted to our Death's
Head, *Atropos*. The stridulation of *Acherontia Lethe* is mentioned
by Sir Emerson Tennant in his "Natural History of Ceylon,"
and another species from Hindustan, *A. Satanas*, also squeaks,
according to Col. Gott. St. Pierre also mentions the stridulation of a kind found in the Mauritius. Both male and female
Death's Heads crepitate.

Passing from the Acherontiidæ to the Smerinthidæ, the late
Mr. Walker, once employed in cataloguing at the British Museum,
quotes a record from Natal that a Hawk Moth (*Basiana Postica*)
"gives out sounds resembling those of a longhorned *Lamia*
beetle for minutes together." *Langia zenzeroides*, another
Sphinx, is described as uttering a "faint stridulous cry, like
that of the Death's Head, but not so loud." Then among the
clear-winged *Sesiidæ* we have the musical *Sesia Pelasgus*, alias
Thysbe, with reddish-brown wings and hyaline discs. Regarding this, the Canadian Humble Bee Hawk Moth, Dr. George
Gibb says : "The sound is something like the squeaking of a
mouse or bat. I retained one I captured myself for some time
alive to hear its murmurs. This squeaking noise continued as
long as the creature remained alive, and was much louder than
in any other of the numerous Sphinges it was my good fortune
to capture."

In the silk-spinning moths of the Bombycina, next in review,
the rule appears general that the capability for stridulation is

greater in the male sex, the organs by which it is produced having in it the greatest development. Many species possess a high degree of susceptibility, and pertinaciously sham death in the net ; while the pervading brightness of their colour seems at first sight to extend the law pertaining to the butterflies and negative that pertaining to the leaping grasshoppers, where, as a rule, a dulness of hue indicates an increase in the capacity for music in a species. The organism by means of which these moths produce, or probably produce, a stridor, consists in a little triangular bladder (Fig. 11, Plate IV.), external, and formed, it would appear, by a vesicular dilatation of the integument of the episternum of the metathorax, over the surface of which runs a lenticular crumpling representing a lima (l) placed vertically, and lying invariably in the depression on its tense membranous superficies, that receives the inwardly bowed hind femur (s), the inner superficies of which is so directed as to suggest its effecting a stridor by friction. The complete apparatus invites comparison with that of the male *Pneumora*.

This vesicular bladder was first discovered by Solier in a Tiger Moth (*Chelonia pudica*) common on the Riviera, and advanced by him in explanation of a sound this insect was heard to produce on the wing. It was noticed to bear striæ on its anterior margin, both by Solier and Laboulbène, about sixteen to twenty in the male, and eight to ten in the female, some six of which are more elevated than the remainder (see Fig. 11, Plate IV.). The latter author, who ascribes the sound to the friction of the hind femora over the striæ, also mentions the vesicle as present and well developed in two other Continental kinds, *Chelonia matronula* and *Œrtzeni*, the former of which was observed by Czerny to produce stridulous notes, and in *C. flavia* he found the episternum, though denuded, not dilated. The bladder and striæ are perceptible in the indigenous Ruby Tiger; they are also present in the common red Rag-weed Moth, as also in *Callimorpha Hera*, a richly tulip-coloured kind, found as far north as the Channel Islands, though but rarely occurring in England.*

But this crumpled bladder only needs searching for to

---

* If a fresh specimen of *C. Hera* be taken, and, seizing the extremity of a hind tarsus, the femur be moved over the striæ, a distinct sand-papery sound like the stridor of the *Vanessæ* is heard.

establish its presence very generally among the lichen-feeding
Lithosiidæ, as well as in the Tiger Moths, whose palmer-worms
are herbivorous; its development attaining a maximum perhaps
at the tangential point of these two beautiful groups. Guenée
describes this identical vesicular inflation in several Alpine
species or sub-species of little orange moths (*Setina*),* some
of which he describes as producing ticking sounds like a
watch when placed to the ear. We may see it too in the
indigenous species of this family that propagates on the
lichened bark in forest glades, and trace it in other genera.
Dissections of the striated vesicle have shown it to be empty,
and divided into a right and left cavity by a membranous
partition. Fear and anger appear the stimuli to music in the
Heterocera now noticed; but the stridulation of others has
greater range of expression.

We will now proceed to consider the cry emitted by the
common green Silver-lines (Plate IV., Fig. 3, A). This insect,
whose arboreal life and propensity for uttering a queer squealing
on the wing, that, heard in the early spring evening, seems the
oracle of that mystery which history has ever shrouded in the hush
of its oaken shades, once had a very wide geographical distribution,
specimens coming to us even from Australia. It has likewise an
unstable position in classification. A green obliquely-striped
caterpillar, recalls the Hawk Moths; a boat-shaped cocoon, a
mobile family of the *Pyralidina;* and the moth itself has a
general resemblance to the *Noctuina*, *Pyralidina*, and Blunt
Wings, in all of which groups it has done duty. The partiality
of the male for flying around bushes, solitary or toying with its
female whilst uttering its cry, that has been compared to that
of the Woodman Beetle, and may be likened to the fitful and
rasping sound of the bat, taken in conjunction with the emerald
scales which adorn its primary wings, presents a striking micro-
cosm of the parrots among the palms of the tropics. The first
notice of its stridulation is found in Morris's " British Moths."
" I wrote," says this gentleman, " of my own knowledge; I
remember the time, place, and circumstance well. I was then at
Bromsgrove School, and was out ' hunting ' one evening; and I

---

* *Aurita, Ramosa, Roscida, Irrorca, Flavicans*, and *Andereggi*.

remember that it was very early, and before actual dusk, on a hill, or rising ground rather, some two or three miles from the town, near Stoke Court, where I saw many of these moths, the only time I ever saw them alive, flying up and down and very fast, and hard to catch, near or above the top of an old-fashioned high hedge, on the side of a wide, grassy lane. I could not help being struck by the curious stridulous sound they made as they flew." Here the gregariousness of the insects is noticed.

Next the noise proceeds from the male. Dr. Buchanan White, writing from Perthshire, says :—" On the evening of the 28th May, when mothing in the oak wood surrounding my house, I noticed what I thought was a beetle, flying round a small oak, and giving vent all the time to a sharp, quick sound, very similar to that produced by the Scottish longicorn beetle *Astinomus*, when held between the fingers. Though I failed to catch this individual, I was more successful with another which was behaving in the same manner. When in the net the sound ceased, and I saw to my astonishment that the insect was a moth." This proved to be a male Silver-lines. On the same evening Dr. White again went out to the oak wood, and captured another specimen in the act of "squeaking." "The sound was quite distinct at a distance of ten feet or more. Next morning I treated him (it was a male) in the same manner as I had the first specimen, and with a similar result. I found that a good imitation of the sound may be made by rubbing the point of a knitting-needle on the closed blade of a clasp-knife. I have since taken another specimen, also a male, flying round an oak, but not producing any noise."

Then the note is employed during courtship, and marks the pairing-time of the moth, as I had myself opportunity of witnessing some years since, when spending a summer sketching at St. Catherine's Ferry, Argyleshire. On one of those fine days after the protracted spring rains that occur in the beginning of June, wending home about sunset, on the skirts of a newly-leaved oak shaw, I was suddenly arrested by a novel and loud succession of twitters in the dusk air, and on looking up saw a male and a female Silver-lines Moth, which came fluttering down from the foliage, and were toying just in front of me.

Now, as to the production of this sound, analogy would refer

us to a lima and clasp. And it is further certain that a
note quite similar to that of the moth results either from
simply expanding the fore-wings, when the elbow at the base
(Plate IV., Fig. 3, *l*) leaves the indented edge of the lateral
and dorsal piece of the metathorax (*a*), round which it
clasps in the posture of repose, with a frictional sound, or from
chafing this elbow backwards and forwards over the hinder wing
surface. On submitting these parts to a high magnifying
power, I find the callosity (3, B), which in many moths locks
the forewings to the body, unusually developed, obliquely
striated, and placed at the inner edge of the elbow, while the
sub-costal nervure of the hind wing is raised to form a catch,
and minutely indented. When stridulating it has been observed
to inflate the abdomen and protrude a filament (fan ?).

One of the *Tortricina* (Plate IV., Fig. 2) shows a propensity
to stridulate. It was, I remember, one sunny afternoon I went
to pay a call at East Cowes, in the Isle of Wight. The lawn
was being mowed. My attention was attracted by a little ring
of male *Dicrorampha sequana*, paying respects to a female at the
edge of the long flowering grass. They flew around her, chased
each other, and went through various antics, always retiring
settling on grass-stalks in the immediate vicinity. But what
surprised me most was to see one exasperated little atom suddenly raise its two fore-wings perpendicularly, and pass them
rapidly over the front edge of the two hinder.

When the little society was on the point of dispersing, I
managed to pill-box a few, and subsequently submitted them
to a magnifying power to see if any trace of a file and clasp was
to be found to account for the strange proceeding witnessed. All
I have to say is, that, finding the front edge of the hind wing of
one a little rubbed, I denuded it, and then discovered it furrowed
perpendicularly to the edge, but whether this be sufficient to
enrol this diminutive creature as a stridulator or not, I must
leave to future observers.

Lastly, with regard to the *Tinea*, I would call attention to
page 362 of the second volume of "Stainton's Manual," where
it is stated :—"The species of the genus *Glyphipteryx* have a
peculiar habit of slightly raising and then depressing their wings,
as though fanning themselves ; and *G. Thrasonella*, which is

generally abundant amongst rushes, may be often observed thus
employed in the months of June and July." If such be stridulators, how truly may it be said of man, "The gateway of sound
is impervious to the cry of that life which we crush beneath
our feet, and to the joyous myriads which sport in the sunbeam."

Lastly, butterflies and moths, in the larval and pupal stage,
emit sounds which have been referred to stridulation. The green-
and-brown caterpillars of the Death's Head Moth when tickled
hastily sway their head and first segment to and fro, emitting a
succession of crackling noises resembling the sound produced from
the winding up of a watch ; and those of *Langia Zeuzeroides*, a
stridulating Sphinx found in North India, emit a strong and
sharp " hiss," degenerating to a " squeak " when the caterpillar
becomes lethargic. The chrysalids of the Green Hair Streak, if
females (?), " chirp " or " creak," especially it is said when dry and
placed in the same receptacle ; and a similar sound has been heard
produced by the pupæ of the Death's Head Moth. Perhaps all
that can at present be stated regarding these notes is that they
appear uniformly to arise from muscular contractions arising from
touch, and seem to predict a stridulating imago. Another
sound-producing pupa is that of a large *Ornothoptera* butterfly,
which possesses the power of making a curious noise, like " Pha !
pha !" resembling aërial expiration. It makes it very loudly
when touched, and the noise is accompanied (perhaps produced)
by a sharp contraction of the abdominal segments.

The larvæ of some Nearctic Sphinx moths behave similarly.
That of *Smerinthus excæctus*, Lin., found on beech, when handled
or disturbed, emits a singing noise resembling the stridor of
Lema (*Crioceris ?*) *trilineata*, a relative of our Lily Beetle,
common in gardens in Canada. The larvæ of *Cressonia* (*Smerinthus*) *juglandis* (Grote), found on the hickories (*Carya alba* and
porcina), when shaken on the tree, give utterance to the notes
"Teep ! teep !" flexing the body sharply from side to side.

STRIDULATION OF THE COLEOPTERA.

The music of beetles is accomplished by vibration of the body
or elytra, and the stridulators are terrestrial or aquatic, indicating
musical groups in the various families of a class, pre-eminently

constituted by a rigid and compact skin, or dermis, for their
presence. The files, or limæ, also exhibit every degree of
perfection, from the pectinate structures found in the Dung and
Burying Beetles, arising from coalescence of transverse dermal
wrinkles, and presenting regular raised ridges of truncate triangular section to the rudimentary and sub-regular shagreening
seen in the Lily Beetles and Weevils. The musical organs in
our Palæarctic species are usually sexually common; and though
commonly employed to produce but a passive plaint of fear
resulting on touch, yet in some groups they acquire that spontaneity of action compatible with the attainment of the higher
economic expressions of anger, love, or rivalry, and in these kinds
they likewise postulate the existence of auditory organs. The
Dung Beetles (*Geotrupes*), a genus of Longhorns, and the Death
Watches obtain thus pre-eminence in their respective families.

In Coleoptera there are either one or two limæ developed at
points of similar friction, and the action of stridulation is commonly the respiratory protrusion and contraction of the abdominal
articulations, in respect to which the limæ are found on various
parts of the segmental surface or on the elytra or coxæ. In the
Longicornes, one Lamellicorn genus, and in one of the Elateridæ,
the music results from the action of the depressor and elevator
prothoracic muscles, and the single lima in the first is on the
mesonotum, in the second on the prosternum, and in the third on
the prosternal spine. It is surmised likewise in certain genera
that stridulation is effected by the action of the elevator and
depressor muscles of the hind legs, as in grasshoppers, suggestive
limæform structures existing in connection on the abdomen.

We also see that while the Longhorns present a uniform
method of stridulation, and have musical organs similarly placed,
differentiated by the number and fineness of their striæ, in the
Lamellicornes the action and limæ have considerable diversification; and the Ground and Water Beetles manifest considerable
latitude in the position of the limæ. As regards the periods of
music, in those kinds where the stridulation is voluntary, some
longhorns creak in sunshine and some of those with lamellated
antennæ at evening.

The Geodephaga, or Ground Beetles proper, include those
long-legged, flattish, voracious creatures that make us start

sometimes because they live darkling and are repulsively black, and sometimes because they sparkle with the gayest greens and violets of the sunlight, who, like raptorial aves and mammalia, scour waste land, ascend the tree trunk, or lie lurking beneath damp stones for unwary and weaker insects or chance refuse that may fall to their lot. Of these a few species are ascertained to stridulate, when seized or alarmed, by rubbing the abdomen against the elytra; others, again, seem to make a sound by rubbing the thighs against the borders of the elytra. The files, or limæ, are placed on the dorsal arc of the anal segment, or at the inflexed lateral margins of the elytra.

When primroses, anemones, and knots of violets carpet the oak wood, and the fig-wort ranunculus glistens on damp meadows, an elegant predaceous beetle may be started from beneath decaying leaves. This intrusion it resents by a "low, angry, hissing sound, distinctly audible at some distance," ascribed by the Rev. T. Marshall to the friction of the ventral arcs of the abdomen in the grooves formed by the inflexed margins of the elytra. The sound, indeed, may be imitated by rubbing the edge of a piece of stiff paper in the channel. Herr Westring remarks, regarding the ridge (lima) which runs parallel with the wing margin at the place a little distant from the tip where it is inflexed, that it is very subtilely scabrous, and Mr. F. Smith terms this coleopteron (*Cychrus rostratus*, L.), the most accomplished English musician.

Certain little metallic Ground Beetles are in the perfect state when the white water ranunculus begins to fleck the village pond with its stars, and pass the sunshine of the spring in hunting its muddy margin for such game as its fœtid banks supply. These little Nimrods—recognised by the rows of ocellated depressions on their brassy elytra—possess the power to utter a faint musical note, first put on record by Herr Westring in Sweden respecting a *Blethisa multipunctata*, found *semi-torpid* (?) in October at the root of a spindle-tree. Their stridulation is accomplished by rubbing the upper surface of the last or anal segment of the abdomen on the under side of the elytra. The musical limæ are easily found with a magnifying glass (Plate III., Fig. 4), and constitute two serrated and raised posteriorly diverging ridges (*l*), one at either side of the anal segment,

J

running from the membranous penultimate dorsal arc. The raised teeth in *Elaphrus uliginosus* (Fab.), and *cupreus*, (Dufr.), do not stand close together, being separated by intervals of a tooth. Their numbers do not appear to exceed ten or twelve. The teeth in *B. multipunctata* are similar, but *E. riparius* has them closer placed, and twice as numerous. At the bend of the wing-covers before the apex are placed the passive organs of stridulation, in the form of a raised, somewhat convex, and wrinkled callosity (*s*), wider anteriorly, and gradually attenuated backwards. The stridulation of a Ceylon ground beetle (*Cerapterus latipes*, Swed.), has been recorded by Mr. Thwaites. Many other exotics are, with little doubt, musical.

The Water Beetles (*Hydradephaga*) the summer idler by the silent pool sees rise to the surface, hang for a moment suspended with their head downwards, and then, having secured a silvery globule of air, dive seal-like down through the slimy abyss, at first sight, from an oval form common to aqueous life, seem more akin to shell-fish than insecta; and yet these amphibious creatures are really ferocious, voracious beings, differing in no essential from the ground beetles, distinguished alone by a slight modification of certain parts, sufficing to adapt them to their abode. Although they thus rise and sink at intervals during the day, for the sake of breathing atmospheric air, it corresponds to the period of sleep and dreams, their time of activity being the evening twilight or shadow of night, when they not uncommonly, leaving the pool, take wing, and traverse swiftly the air, to found new colonies and to populate new waters.

Pelobius Hermanni (Fab.), a smallish red Water Beetle, somewhat plentiful in stagnant ponds on the clayey district surrounding London, where it delights to grovel in the mud or cling to the weeds, utters a sharp noise on capture, when escaping the grasp of a Water Scorpion, or when fighting a rival, when it likewise exhibits some of that pugnacious temper characterising the stickleback fish. Mr. Douglas says that on this subject a friend writes as follows :—" Of the beetles popularly known as ' screech beetles,' I have several in my aquarium, which contains more than twenty gallons of water, and I have heard the peculiar noise emitted when two beetles were quarrelling for a piece of worm. I think the noise came from two at the bottom, but as several couples were quarrelling at the time, it might have proceeded from a couple nearer the top." Herr W. L. Schmidt, having taken a very rich catch of this beetle, detected the proportionately loud creaking was produced each time the insect raised the last segment of the abdomen and pushed it under the wing-covers, in and out. Applying himself to account for this motion, he was led to the discovery of a ridge (*lima*), commencing near the sutural edge of the elytra, at about one-third its length from the extremity, and from thence running obliquely to the apex, which as it approaches it dilates. On examining this ridge with a glass, he noticed closely placed, deep cross furrows, and these gave out a note similar to the cry of the beetle when scratched by a quill pen. The active organ in the creaking is an extremely sharp, thin, and prominent horny edge, that surrounds the last abdominal segment. Some other Water Beetles, as *Acilius sulcatus*, L., both male and female, and the male of *Colymbetes fuscus*, L., stridulate in the same fashion, producing a humming sound.

The Philhydrida present a transition from the aqueous to the terrestrial, include minute beetles, aquatic, or merely as their name denotes, fond of the vicinity of the calm and whirl of the stream and pond. As the mouths of these are deficient in a pair of palpi, their powers of seizure are less, and they have not the rapacity of the first two groups. The Danish coleopterist, Professor Schjödte, in noticing the habits of the minute species of *Bledius*, *Heterocerus*, and *Dyschirius*, says :— "The connection between these three genera is not of a systematic character, for they belong to widely different families, but they are closely connected by their habits, living together as they do on the shores of fresh and salt waters, where they excavate tunnels and galleries, which betray their presence on the surface by small heaps of earth, like diminutive molehills. The short elytras, *Bledius*, and small *Heteroceri*, are not seen about in the day-time, but come out of their habitations on warm summer evenings after sunset, flying in numbers near the surface of the water. Herr Erichson, having possibly reason to suppose the tiny kinds of *Heterocerus* produced an airy creak, submitted some to a magnifying glass, and discovered the existence of a peculiar arched ridge on each side of the first ventral arc of the abdomen, and a corresponding straight and sharp ridge on the inner side of the third pair of femora, which he interpreted as the constituents of an organ of sound. In this he is confirmed by Schjödte, who submitted this minute ridge to a strong side-light with a microscopic power of 50 to 100 diams.; the part having been previously detached and cleansed with solution of caustic potash, when on being adjusted so that the rays should strike along it, this ridge further proved to be striated inwardly, the cross bridges standing in relief from reason of their shadows, "regular, close, and minute, in proportion to the size of the insect," and therefore constituting a file as we find it in larger musical beetles, over which the ridge on the femora might play.* Herr Westring has mentioned two other micro-beetles of this group that utter a sound—namely, *Berosus luridus*, a coleopteron about the size of a grain of sand or pin's head, and the, in this country, rather scarce *Sperchus emarginatus*, which is scarcely its superior, thus ushering us to the very fount of those passions that fired a Troy and shook the pinnacles of heaven.

One large black fluvial species of this family, the great Water Beetle (*Hydrophilus piceus*, L.), also stridulates; the male when touched or held in the hand contracting the abdominal base within its elytra with a sound ("crumph") like that produced by crumpling a cabbage-leaf. The limæ are situated beneath the tip of the elytra, half-lunate in shape, wider inwardly, and shagreened—an appearance here arising from contiguous punctures; but in the female this structure is somewhat less marked and hairy.

The common *Necrophaga*, or Carrion Beetles, are a provision to remove the remnants of the feast of vulture and hyæna, in which they find able co-operation on the part of various flies. But Nature hastening in her phases, and imperious of speedy assimilation, commissions them to consign to it their thousand eggs, and then (the species being often, as the blue fly, viviparous) insect and grub work rapidly and simultaneously, while the tainted organism shrivels in the sun. The genus *Necrophorus* (Plate III., Fig 6), contains some kinds wholly black and some

---

* The arched limæ are bent back in *Heterocerus sericans* and *intermedius*, *Physites aureolus* and *Augytes hispidus*.

pleasantly coloured with orange; and these nocturnally bury
dead birds and mice in our gardens, for which task the sexes wing
at evening in pairs, guided by scent. This occupation, as with
fossorial bees, is mainly undertaken to form a nidus for their
ova, and it is performed mostly, as Gleditsh has long ago
recorded, by the male, who laboriously shovels out the earth
beneath the carcase with his head and thorax. On picking up a
belated Burying Beetle, fresh from his work, we find it invariably
exudes a musky aroma and produces a buzzing sound—provisions
that doubtless supplement colour in effecting companionship and
gregariousness. And in this case it is evident, from a protrusion
and contraction of the abdominal segments accompanying the
music, that when the breathing is augmented and accelerated,
as in the exertion of excavation, the beetle must be most predisposed to stridulation. The limæ (*l*) of these Burying Beetles
(Plate III., Figs. 6 and 7), are two somewhat posteriorly converging, rounded and slightly lenticular, parallel striated ridges,
situated centrally on the dilated and horny fifth dorsal arc.
They creak by moving these against a raised indurated ridge (*s*)
about ·86 Mm. long, situated at a slight distance from the
sinuated extremity of the elytron beneath; and this musical
movement is favoured by the membranous condition of the
sub-elytral dorsal arcs. The stridulation, which can in a measure
be reproduced in dry specimens from the cabinet, is strongest,
Dr. Landois fancies, during the protrusion of the abdomen.
The same writer enumerates 153 striæ in the limæ of *N. mor-
tuorum*, 126 to 140 in those of *N. vespillo*, and 133 only in
those of the large black *N. humator* (Fig. 7), and gives their
length and maximum breadth from ·23 × 1·95 Mm. to ·2 × 2·2
Mm.; they likewise converge about ·5 — ·46 Mm. posteriorly.
So that we have here ample reason for specific sound colour in
each kind. These beetles appear confined to Palæarctic and
Nearctic regions.

We are able in Lamellicornia to trace further the operation of the stimuli to the emotions of fear, love, and rivalry, and
the files or limæ show every degree of development, and are very
variously situated at the apex or base of the abdomen, on the
elytra or coxæ of the hinder legs, on the prothorax beneath.
The stercorarious groups of Dung Beetles exhibit pageants

of love and rivalry, or the music is maternal, since they stridulate when aggregating to feed, when chasing the female, or
when rolling the balls which contain their ova. But on
Palæarctic areas the intelligence of the large stercorarious
coleoptera is not in direct ratio to the stridulation, as the
Scarabæidæ appear certainly less musical but most gifted. Thus
while the female of our common Dung Beetles (*Geotrupes*) merely
grubs a vertical hole and deposits her ova in some dung she
drags to the bottom, the Sacred Beetles belonging to the
genera *Ateuchus* (Fab.), and their American allies, *Gymnopleurus*,
Illig., roll a mystic ball in company, pushing it backwards,
and then inter it, as the ancients imagined, with solemn rite, to
await a recreation of their kind. Our indigenous Lunar Beetle,
Copris lunaris, also works in couples, as may be ascertained
by examining its holes.

The stercorarious species of *Copris*, *Onthophagus*, and *Ateuchus*,
have the abdominal rings so anchylosed as to preclude segmental
motion; yet the Lunar Beetle is stated to produce a considerable
noise by the friction of the abdomen against the hinder margin
of the elytra; and Westring and Darwin go so far as to find
minute limæ on raised oblique ridges running along the sutural
margins from the elytral tip to a quarter their length, which
appear to rub on the sides of a posteriorly converging depression
at the centre of the last dorsal arc but one. Other Coprini
have the files transposed to the dorsal surface of the abdomen.
In the genera of another group the limæ are on the hind
coxæ. Darwin, quoting M. P. de la Brulerie, says : "The male
Ateuchus stridulates to encourage the female in rolling her ball,
and from distress when she is removed;" and in the sexes of
a Sacred Beetle, common in foul ravines behind Castellamare, I
found a prominent, though not well defined, linear lima on the
inferior surface of the hind coxæ near the femoral articulation,
which, during the rotatory action of ball-rolling, naturally
pressed against the inferior metathoracic surface, producing
thereby a just distinguishable sand-papery note.

The cow, camel, ass, and even the marsupial quadrupeds of
Australia, have all allotted satellite beetles to follow them on
their pastures, although these are quite independent as regards
their law of distribution. Coeval with the cow and third period

of geologists, our northern Lamellicorn associates of miry lanes are, despite their employment and livery of decay, creatures of superior instinct, and the creaking of the common species may at all times afford entertainment, from the windy vernal equinox when, ere the snow has vanished, the keen air of our southern heaths resounds to the boom of the bull-horned *Typhœus vulgaris*, until late in October, when the drone of the Watchman Beetle lends solemnity to the chilly evenings and blackberry rambles. Like all burrowing insects, both these kinds especially propagate on sandy soils.

One May afternoon, chancing to pass under the shadow of some stunted beeches in the grounds of Ochtertyre, Perthshire, at about 3 o'clock p.m., I espied a male of a small Dung Beetle (*Geotrupes vernalis*), there common in the woods, chasing its female and uttering at intervals a plaintive buzzing cry; while the latter, with feminine tact, Daphne-like, eluded his pursuit, and eventually took to burrowing in the loose carpet of feathery moss, under which they then both disappeared. This circumstance goes to prove a sound can be uttered by the male of this beetle as a call-note; but since the stridor is produced equally well by the female, we are led to the inquiry whether the stridulation may not supplement the gregariousness due to scent in these scavengers. For, as is known, when a Dung Beetle is picked up on the road and held in the hand extending its legs, it simulates death, and utters a creak intimating perception of fear; but if we enclose a number of these in a box, they, without fail, one and all, commence stridulating loudly, the female seeking to escape by burrowing downward, and the male by taking wing to the light.

The rounded, somewhat *f*-shaped limæ in the species of Geotrupes, were discovered by Herr Westring. They form sharpish ridges crossing the inferior coxal surface, converge posteriorly, and are furnished with parallel striæ (Plate III., Fig. 1, *l*). These are worked by the beetles in faint grooves (*s*) on the surface of the third ventral arc, and passed obliquely over a minute, thin, horny ridge at their posterior edge by protrusion and recession of the abdomen. The music resulting may be fairly reproduced with a quill, or by rubbing the abdomen against the limæ. The organs effecting it vary in the species.

Thus, Dr. Landois finds only eighty-four ridges in either limæ of the common Dung Beetle (*G. stercorarius*), each ·025 Mm. thick, ·36 Mm. broad, but 100 in those of the small vernal sort (*G. vernalis*), placed at a distance of ·02 Mm., and in the green (*G. sylvaticus*) 101, ·025 Mm. apart. These smaller species, where the limæ are finer, have a louder, higher, and clearer note than the first large violet Dung Beetle. In *Typhæus vulgaris* (Leach) the musical ridge is similar, but the filing very much finer—as fine, indeed, as in the British *Cerambyces*. Its leathery creak, therefore, high in pitch, is less intense, resembling the sound produced by drawing the finger over a hair-comb. The lima of this genus is first noticed, I believe, in the " Descent of Man." In some of the warty species of *Trox* limæ are found, as in the Lunar Beetle, according to Westring; and the *T. sabulosus*, a little rough coleopteron partial to sandy soil, makes a rattling stridulation with its tense elytra, so loud that, it is said, the late Mr. Frederick Smith, on capturing one, induced a gamekeeper to suppose he had found a mouse.

Passing from the swarthy stercorarious tribes to the bright array of Lamellicornes with larger antennal leaves, that spend the earlier stages of their life under vegetable soil, or in the heart of trees, until, issuing to light, they no longer sap the domain of Flora from beneath, but boom around the leaves inwove in the forest crown, hover at the balmy sap, or rob the flowers of their petals, we again come upon a familiar group—for such are the habits of the nocturnal Atlas Beetles of equatorial America, of the Goliath Beetles of equatorial Africa, and of the cosmopolitan birth of Rose Beetles, Cockchafers, or of strongly mandibulate Stag Beetles that resound in the cool shadows of fleeting summer.

The large horned males of the Neo-tropical *Dynastidæ*, or Atlas Beetles, all produce a shrill noise by the friction of the abdomen's superior surface on the elytral tips ; and, according to Darwin, in the Rhinoceros Beetle (*Oryctes nasicornis*, L.), their Palæarctic representative, the limæ are seated on the last abdominal segment but one. The females appear to want the musical organs. Spontaneous action of the nervous system to produce fear in the guise of spite or anger is well seen in one of the tropic Stag Beetle kind. " *Chiasognathus grantii*, of

South Chili," says Dr. Darwin, " is bold and pugnacious, and when threatened on any side, he faces round, opening his great jaws, and at the same time stridulating loudly." The *Melolonthidæ* or chafers add several instrumentalists to the insect choir. In many, the active organ of stridulation consists in two curved depressions, furnished with oblique irregular limæ, often little more than roughnesses, superiorly situated on the last dorsal are but one; and the rare mottled *Melolontha Fullo* executes an intense creaking by rubbing the abdomen against the elytra. The musical apparatus is also slightly marked in either sex of the common May Cockchafers, *Melolontha vulgaris*, and in their summer relative, *solstitialis*, as in *Anomala*, Köppe, and other genera.

But among the Melolonthidæ the stridulation of *Serica* is exceptional. Herr Westring being convinced the delicate *S. brunnea*, L., when struggling between his fingers at times produced a soft creaking, and tracing this to the thorax, was led to examine the inner side of the spoon-shaped under-part or prosternum, and in this singular situation he found a single elongate lima, which was evidently capable of being moved over the thin ridge on the metasternum's fore margin—both structures being blackened and indurated (Plate VI., Fig. 12, *l.*).

The common green chafers of Australasian forest clearings, *Lomaptera*, it is stated by Dr. Sharp, probably stridulate as the orthopterous *Pneumora*, by means of a somewhat raised space, thickly set with fine slightly-curved limaform lines on the second and third abdominal segments, from which a loud sound may be elicited by moving the inner surface of the femur covered with coarser and less regular lines backwards and forwards over their superficies. Lastly, *Euchirus longimanus*, a South American beetle, makes whilst moving a low hissing by the protrusion and contraction of the abdomen, due to a narrow rasp running along the sutural margin of each elytron; and the Ne-arctic genus *Gymnodus* emits a musky smell, and stridulates on seizure. In our Rose Beetle scarcely perceptible indications of limæ, as in the cockchafer, may be seen. The spontaneity of music we here see in some of the Lamellicorn is common to the stercorarious and phytophagous kinds.

I took, at the commencement of July, on a willow sapling

near Turin, where Stag Beetles were collected sucking, and
some Longicorns were licking sap, in company with butterflies
and insects of other orders, a stridulating male of the Skip
Jacks or *Elateridæ*. This individual (*Lacon murinus*, L.), while
I had it in captivity, would on seizure invariably lift its head,
like others of this group, when preparing for a spring; but,
instead of immediately afterwards depressing it to insert the
elastic pro-sternal spine in the meso-sternal groove, it first
nodded the head and thorax thrice with a movement and noise
resembling that of a Longicorn—"Whee! whee! whee!"—a sound,
I conceive, which originated in the rounded superior surfaces of
the twin claws that constitute the spine-point, which are faintly
striate, vibrating over the entrance of the meso-sternal groove.

The music of the Death-watch Beetles, by association with
the latter days of a British sovereign, has attained familiarity
with those even not professedly scientific. It was, we think,
one of those bright melancholy days, towards the latter part of
February, when the almond is already hoar with blossom, and
minute beetles float moat-like in the sunbeam, that King
William III. mounted Sir John Fenwick's sorrel pony, and rode into
the Home Park to see an excavation for a new canal, which was
to run in another longitudinal stripe, by the side of the existing
one, when just as he came to the head of the two canals opposite
the Ranger's Park pales, the sorrel pony trod on a molehill—
thrown up by one of those purblind creatures enjoying the
reviving tepid warmth—stumbled, and threw the monarch,
fracturing his collar-bone, an injury which subsequently caused
his death. The news of this sudden termination to the career
of the warrior king seems not to have broken altogether un-
looked for on his subjects; for we must regard it as ominous
that the early volumes of the "Philosophical Transactions" ushering in the accession of Queen Anne are occupied with long
disquisitions on the doings of the Death-watches by the savants
of the day, Allen, Derham, and Stackhouse. These are described as little beetles that ticked like a watch, and answered
one another from parts of the old wooden rooms, or in rotten
posts, each giving seven or eight cries, and then pausing for
the reply of its comrade, which is described as much smaller.

The way in which the noise was produced, all agree, was

that the insect "lifted up itself on its hinder legs, and somewhat extending, or rather inclining its neck, beat down its face," and Stackhouse declares the sedge of a chair where one was beating was depressed, "for about the compass of a silver penny." When the beetles were placed in the sun they attempted to fly. Rude magnified figures of the Death-watch Beetle are appended to the descriptions. These accounts seem to have provoked the muse of Swift, who probably had in mind the sophistry of one of these savants, "that he had known the noise to be heard by many when no mortality followed, and had taken two seven years since without any death following that year."

The story of the ticking of the Death-watches recorded in these now venerable volumes seems to be in the main very accurate, as it has been of late years served up and re-served, to the popular taste in entomological magazines with little variation or addition. Certain is it that the Death-watch Beetles exhibit the phenomena of love and rivalry by responding to each other's ticking, and by answering their notes approximated by a tap of the finger-nail. Male doubtless challenges male, but the female *attracts* the male. Darwin says that Mr. Doubleday twice or thrice observed a female (*A. tessellatum*), ticking, and in the course of an hour or two found her united with a male, and on one occasion surrounded by several males. This is the view taken by the contributors to the "Philosophical Transactions." Dereham, in the month of May kept two of these little creatures in a box in his study, a male and a female, and proved that it was the call of the female seeking a partner, for he not only got the couple to answer an imitation of their "beating," artificially produced with the finger-nail, the male very freely and the female rather reluctantly, but when the beating of the former at times got very eager, he actually observed them to pair.

These species of Anobium beat in any position, with a movement like that of the Pendulum Crane Fly, oscillating their bodies up and down. They are heard in May and July, and beat for a week or a fortnight previous to pairing. *A. tessellatum* beats seven to eleven strokes at uncertain intervals, and another kind, possibly *striatum*, six to eight, or *vice versa*.

The idea that the Death-watches actually strike the substance on which they rest with their jaws, head, or thorax, is, I believe, the only one rife in this country, but from this Herr Westring dissents, and considers the spectral ticking to be a species of stridulation, finding a lima in a little channel beneath the apex of the elytra, that may be rubbed by a central minutely elevated portion, of a terminal, indurated, apical, and dorsal plate of the abdomen, apparently. *A. pertinax* and the above two species all tick in the same fashion, and all alike pertinaciously simulate death on seizure, by drawing in their limbs; so that we have thus manifestation of the phenomena of love and rivalry in beetles of small dimensions, which also evince, in high degree, the perception of fear, by muscular contraction.

The dull, mottled, or diamond-sparkling pachydermatous Weevils (*Curculionidæ*) that dwell in porcine flocks on flowery spikes, cling tenaciously as glued on twigs, or move slowly and sloth-like on the branches of trees, seem to oppose a principle of obstinacy to the assaults of time and tide. At the gentlest touch they evince a perception of fear, becoming passive, or if loosed from their hold they find concealment by drawing in their legs with an iron clutch, and contracting their snout-like heads into a groove, when they assume the aspect of a round ball, shapeless, motionless, and inanimate, scarcely distinguishable from a neighbouring flower-bud, seed, or patch of grey lichen. The species are often highly vindictive, despite the small size of the mandibles, and their disadvantageous position at the extremity of the rostrum. Some stridulate from alarm, when shaken or submitted to a slight pressure, and others do so when recovering from their first swoon. The lima is placed on the anal, on the last segment but one of the abdomen, or beneath the apex of the elytra—the latter position is most frequent. They are questionably considered the oldest of coleoptera, but seem at least traceable to the fir-clad secondary period of geologists.

Dr. Lister, who has left an account of a trip to Paris undertaken in the reign of Queen Anne, was also a naturalist, and had remarked the creaking noise of *Cryptorhynchus lapathi*, L. a black and white weevil that crawls on willows and decayed alders. This coleopteron belongs to the number that place themselves as if dead when captured, folding in its feet and compressed rostrum close to the sternum. After he has remained in this position for some seconds, he begins to move on his legs, at which time, if held fast between the fingers, or in the closed hands, his sound becomes audible. Westring, who noticed a lifting movement of the abdomen during the production of this creaking, ascribes to it a lima, in the form

PEAK OF TENERIFFE, N. 70° W.
(An outlier of the Palæarctic Fauna.)

of an oval opaque surface on the fore part of the last dorsal arc but one, at either side, which, under a powerful magnifying-glass, appears scabrous or shagreened. The structure he also finds in the genus *Erirhinus* of Schönherr.

The stridulation of the genus of Weevils (*Acalles*, Schön.), was established by the late coleopterist, Mr. T. V. Wollaston, who informs us that while residing in a remote village in the north of Teneriffe, during a yachting voyage in the spring of 1859, his Portuguese servant brought him eleven specimens of, he believed, *Acalles argillosus* (Schön.), he had captured within the rotten stem of an Euphorbia (*Kleinia neriifolia*, D.C.), which, he narrates, he was about to throw away as worthless, when he was arrested by a loud, grating, almost chirping, noise, as of many creatures in concert; and in looking closer for the mysterious cause, he detected a truant beetle from which it was quite evident that a portion of the noise proceeded. On shaking the hollow stem, so as to arouse its inmates, and putting his ear alongside it, the plant appeared musical, as though enchanted. Mr. Wollaston kept some of these weevils in confinement, and discovered a constant source of amusement in making them sing. It was some time before he could satisfy himself, not only as to the *modus operandi* of this proceeding, but even as to the exact region of the body whence it emanated, for they would often stridulate lying on their sides, with their limbs closely retracted and their head applied to the sternum, and, in fact, whilst to all appearance perfectly passive and inanimate. At length his eye was arrested by a minute and rapid vibration of the apical segment of the abdomen. This solved the mystery. "The inner portion of the elytra (corresponding with the constricted apical region) against which the anal segment comes in contact at each of the pulsatory movements, is to the naked eye dull and sub-opaque; but when viewed beneath the microscope, this duller portion is coarsely shagreened, or sub-reticulate." In the gigantic *Acalles Neptunus* from the Salvages this lima presents the appearance of irregular elongate reticulations.

On Mr. Wollaston's arrival at Funchal from Teneriffe, in June, 1859, he exhibited his specimens (then in a lively state) of the *A. argillosus* to Mr. Bewicke, and requested him to listen attentively to the Madeiran species, whenever he chanced to meet with them, and he was afterwards assured by him that he had heard the music constantly in *Acalles dispar, nodiferus, terminalis,* and even in the minute *A. Wollastoni,* which is the smallest of all the Madeiran weevils then discovered. Mr. F. Smith, having read Mr. Wollaston's "amusing and instructive paper," says, "I felt a strong desire to ascertain, if possible, whether our British species possessed any amount of musical talent. During a few weeks' residence at Deal I had the good fortune to beat from the hedges considerable numbers of two small

indigenous species of Acalles. In the first instance I tried the
powers of single specimens by placing them in pill-boxes, which
I shook, to alarm the insects, and then applied them to my ear
for some time, without success; at length I distinctly heard
the notes of *A. roboris*—they were soft, gentle, and low."
The other sort, *A. misellus*, likewise proved to be musical, but
in a less degree. Mr. Smith also observed the motion of the
abdomen accompanying the music, and agreed with Mr. Wollaston as to the position of the musical lima in this genus.

Mr. T. V. Wollaston likewise remarks, "In another weevil, a
large and noble Plinthus, which seems to be peculiar to the
Canaries, the music was scarcely so loud, in proportion to the size
of the creature. The stridulating instrument, nevertheless, is perhaps somewhat better defined. It is entirely the same in position
and general character as the one that obtains in Acalles, except
that the sub-opaque portion of the inner tegument of the elytra
is, instead of being sub-recticular, strictly file-like, being made
up of a series of minute, close-set, regular, and parallel ridges,
similar to those on the mesonotum of the Longicorns. In this
insect, christened *P. musicus*, Mr. Wollaston not only observed
the stridulation frequently, but effected the noises artificially
by vibrating the terminal segment of the abdomen. He also
describes *P. velutinus* as a stridulator. There is a native species
of this genus found under stones on the Downs, but whether it
be musical or not I cannot say."

Some larvæ pertaining to the Curculionidæ are internal
feeders, while others, like the bright caterpillars of Lepidoptera,
live quite exposed, and even form network cocoons. One
indigenous weevil of musical capacity, the *Mononychus pseuda-
cori*, Fab., is cradled in the early stages of its existence in the
seed-pod of the yellow water iris, from which it emerges in
July. This insect, according to Dr. Darwin, Mr. F. Smith
once kept alive in confinement, and ascertained that both
male and female stridulate, and in an equal degree. During
the month of August I captured one of these stridulating
Rhynchophora near Turin. The music was, as in the former
case, produced by the coleopteron depressing its abdomen one-
sixteenth of an inch at the apex, and then withdrawing it
beneath the elytra; but although sharp and file-like, the sound

was audible only at a short distance from the ear. On examination it appeared the elytra were united, and posteriorly and inferiorly truncated for about one-third their length, where interiorly there was an ebon black triangular patch, wider posteriorly, that, showing longitudinal striation, evidently constituted a lima over which the lower edge of the latter superior arcs was adapted to sound.

It would thus seem the limæ in the weevils should be looked for beneath the elytral tips, although Herr Westring indeed finds them otherwise in *Cryptorhynchus lapathi*, L., and states that in some other small weevils there is an opaque transversely semicircular surface on the fore margin of the anal segment, but it does not appear that any of these last, save perchance *Nedyus Echii*, Fab., have claims to be considered musical. Thus, in these stridulating weevils we have probably the outliers of a musical group, which, like the creaking Longicornia, dwindle down to minute species, whose sounding striæ are scarcely capable of being discriminated beneath our most powerful lenses—one of these I have portrayed in its natural size on Plate III., Fig 2. Surely such little gongs, resounding within acorn-cup and heather-bell, present us with a marvel allied to the barking dog of the fairy tales enclosed in a nut-shell.

Emerging from the dim light of the Secondary ages in Europe, with their abundance of fish-lizards, bird-reptiles, birds with teeth, and occasionally strange insects, denoting a period of general metamorphosis, in the Tertiary we clearly see the giant Longicornia, inseparably connected with the hoary monarchs of the time-changed forest and its fallen mossed and lichened trunks, whose livery they wear, and which they bore into as larvæ, and sun themselves on when arrived at maturity. Like the leaf-devouring coleoptera they indicate at the present time a maximum as to size and number, as we advance towards the equator; on whose teeming growth they form a natural check by demolition, and stimulus by thinning out. The Longhorns are also beetles whose economy is not uninteresting or uninstructive, whether we regard the capability many possess of diffusing musky liquid perfumes of the odour of tea-roses or sweet-briar, the singular pruning and trimming the matron *Lamia* and *Prionus* confers with her mandibles to the twig she designs to oviposit on, the various trees some are allotted to, the various sunny flowers others frequent, their rich bold ornamentation, or cold licheny garb, their amours, music, pugnacity, and gregariousness. A very large number of Longicorns are musical. Many individuals of the groups Cerambycidæ, Kirby, and Lepturidæ, Leach, creak on seizure; some aspire to the phenomena of love and rivalry. I remember, as it were yesterday, watching how the apterous *Doricardion Pyrenæum* would climb the heated walls and trees in the Toulouse Botanic Garden illuminated by the hectic sunshine of April, and commencing to nod its prothorax, send forth a creaking, that on cessation would call forth quite a chorus in response from others of these beetles similarly perched in the neighbourhood. The Harlequin Beetle (*Acrocinus longimanus*) in the same fashion betrays its retreat by a rather loud noise, which it produces by the friction of the thorax. The *Prionidæ*, Leach, however, are erroneously noticed as stridulators.

Whether our present group of *Cerambycidæ*, which includes genera * which do not stridulate as the rest, is a natural group, or whether the musicians will have to stand out from among them when their biology is more fully investigated, posterity will decide. The sound colour of their stridor has been compared with various inarticulate expressions, such as the braying of an ass, and the squeaks of a mouse; or to others artificially produced, as the scraping of a fiddle-bow, or rasping of a saw. They all produce their sharp chirp (Plate V., Fig. 2ᴀ, 2ʙ) by rubbing the posterior saddle-shaped, inflexed, overlap of the prothorax (Fig. 2ʙ) on the front edge of the scutum of the mesothorax (*l*). Sometimes in the giants of the race the whole dorsal surface of the prescutum shows interferential colours, being covered with extremely close, deep, and minute transverse striæ, while in the smaller kinds the lima shrinks to a central elevation of triangular, semicircular, or lenticular outline. The musical striæ are invariably minute and numerous, causing the lima to appear glossy to the naked eye. In *Cerambyx heros* among the largest species of the European forests,

* *Callidium, Dorcasomus, Vesperus, Necydalis,* and *Molorchus* (excepting *M. major,* L.)

K

Landois enumerated as many as 238. They appear first to have been detected by Westring in Sweden.

Any one who cares to try the experiment, can easily test the musical capacity of a dead Longhorn by nodding the prothorax on the mesothorax; and for this purpose even dry and subsequently relaxed species are serviceable. That the respiration is intimately connected with the music, if it does not add to its intensity, I infer from the manner of the *Doricardions*, which only creak when the sunshine falls on them; and if the position of the large spiracles (2A spr.), lying beneath the prothorax, be remarked, it will be comprehended how, every time the prothorax nods up and down, these are exposed and recovered.

But the stridulation of these elegant beetles may also be studied in our walks in this country, northern and isolated though it be, as many musical kinds* are distributed through the wooded districts of England and Scotland, within their proper climatic and floral limits. The Longhorns in which a susceptibility for instrumental music is most often noticed are the Musk Beetles, found on the trunk of willows, and the yellow species of *Clytus*, occurring plentifully in England on the old posts in which they propagate. The musical organ of one of these is that shown on Plate V., Fig. 2A. Then, again, there is the Timber Man Beetle (Plate V., Fig. 2B), more abundant in Scotland, where it may be met with flying in the fragrant glades of the Braemar pine-forests, with its long antennæ streaming to the wind; or sometimes seated on felled logs of the sole indigenous fir, with these feelers timidly spread out like compasses, a habit which is said to have acquired it its Scandinavian nickname. Many more kinds are highly musical, such as the Poplar Beetle, and others of the genus *Saperda*, Fab.—a genus that includes species whose grubs, like those of certain weevils, occasionally form vegetable gall; and Mr. C. O. Waterhouse, who has had opportunities of listening to the creaking of the large black *Lamia textor*, Fab., a grim hermit of old and decayed willows, informed me it possesses by far the most powerful and effective instrument of our native Longhorns.

Besides the goat-like *Cerambycidæ*, many a slim, gaily-

* *Cerambyx*, L., *Aromia*, Serv., *Astynomus*, De Jean, *Lamia*, *Mesosa*, Meg., *Saperda*, Fab., *Clytus*, Fab.

coloured Lepturida peevishly stridulates as we draw it out of
some favourite and rank blossom, where, plunged head-fore-
most, it is idly toping the tepid nectaries, emitting an acrid,
waspish note, acute as a thimble scratched by a needle :
the little orange *Leptura livida*, Fab., for example, when we
snatch it from the composite head of a sea-pink on the Downs;
and Dr. Landois tells us he has watched some of these minuter
beetles nod as if creaking, yet their notes were too subtle for his
ears; which we are also given to infer from the fact that the
fine-spun musical organ in *Grammoptera ruficornis*, Fab., has a
lima only ·375 Mm. long, with 113 furrows—one-tenth of the
dimensions it attains in *C. heros*. Here a specially-constructed
stethoscope might be employed with advantage.

Among the small fauna of the *Phytophaga*, that appear as
by magic towards the close of summer to aid the defoliation
of certain trees and shrubs, on whose leafy hair, dank with
honey-dew, they depend, like lucid drops from the rainbow
borrowing lustre of the jewel casket, native musicians, despite
the gregariousness of the individuals, are seldom noticed. In
Germany there is a beetle of renown with the younger population, who call it " Musikant." Its grubs appear annually on
the white Persian lily, and soon disfigure its immaculate chalices
with excrement. These eventually become beetles of a coral
red (*Crioceris merdigera*, L.), which, when seized in the hand,
complain with a sharpish sound (Plate IV., Fig. 10). Others
of their congeners said to be musical are better known in this
island, such as the small yellow and black Asparagus Beetle,
(*C. asparagi*, L.), that makes its *début* on the reed-seeding
asparagus, the *C. melanopa*, L., and the metallic blue *C. cyanella*,
that holds gormandising parties on the dark alder-leaf. None
of these, though, perform very readily, and I am not able to add
any confirmation to the statements.

Westring ascribes the stridulation of these genera to the lowering and raising of the extremity of the abdomen, which works
the upper surface of the anal segment against the tips of the
elytra ; and in his second paper thus describes the lima—an apparently smooth, but really striated, elevated surface (Plate IV.,
Fig. 10, *l*), placed at the anterior edge of the anal segment,
which is either oval, transversely semi-circular, or semi-oval,

K 2

and is divided in some species by a central canal, which receives the sutural tips of the elytra into two somewhat triangular limæ, with the apex posterior. In many the clasp that works the lima, or limæ, is seen in an elongate callosity close to the sutural extremity of the elytra. The stridulation of the genus *Clythra*, Laich., was discovered by Mr. G. R. Crotch. Our *Clythra quadri-punctata*, L., an oblong beetle with orange, black-spotted elytra, is found disporting on tree-trunks near the nest of certain wood-ants, among whom it passes its youthful stages, creeping about in a portable case, like certain micro-Lepidoptera. In specimens of this insect kindly sent me by Dr. G. C. Champion, I found a lima in a glossy patch, somewhat triangular, and minutely striate, lying at the anterior edge of the hairy last dorsal arc of the abdomen, which was evidently sounded on the indurated, faintly-produced tip of either wing-case. The genus *Epilachna* also, I believe, stridulates, and in the same fashion.

Besides the South European genus *Pimelia*, and perhaps the Black Beetles (*Blaps*), which, Westring states, produce a sound by the friction of the abdomen against the wing-covers, there is the following mention of the creaking of Heteromera in "The Descent of Man," in which we are presented with an exception to the rule that both sexes sing. Dr. Darwin there says:—"Mr. Crotch has discovered that the males alone of two species of *Heliopathes* possess stridulating organs"; and that "he examined five males of *H. gibbus*, a small black lenticular coleopteron found on sand, and that in all these there was a well-developed rasp, partially divided into two, on the dorsal surface of the terminal abdominal segment; whilst in the same number of females there was not even a rudiment of the rasp, the membrane of this segment being transparent and much thinner than in the male. In the exotic (*H. cribratostriatus*), the male has a similar rasp, excepting that it is not partially divided into two portions, and the female is completely destitute of the organ." This heterogeneous group of beetles has been founded on the circumstance that the feet or tarsi of their hind legs have often a joint less— or rather, one marked and undeveloped—and it is this circumstance which confers a pace slow and solemn on our darkling churchyard kind and on that of his southern relatives that stalk among the tombs of Italy and of Egypt.

CHAPTER IV.

INSTRUMENTAL MUSIC CONSIDERED AS A MATERIAL AGENT IN REPRODUCTION AND DISTRIBUTION.

STRIDULATION OF THE ORTHOPTERA.

THE battles of Orthoptera indicate that prevailing angry and jealous temperament, which in Longicorn coleoptera we see take precedence of fear in determining a capacity for crepitation, if not in postulating a higher development of the organs of stridulation.

With regard to the three groups of leaping Orthoptera, the male is heard to stridulate by the vibration of its abridged fore-wings, termed from their office and coriaceous consistence hind-wing covers or semi-elytra. But this musical vibration is variously performed according to the group to which the musician belongs. Crickets and leaf-crickets (Plate III., Fig. 3) sing by raising their wing-covers more or less and rubbing them together, producing in this way a continuous and lively trill, such as the "Cree-cree" of the house-cricket and grinding of the Great Green Leaf-cricket. The grasshopper or locust raises its hind legs, doubles together the femur and tibia, and then moves the bent legs briskly up and down in various ways over its semi-elytra, which are previously slightly raised, producing thereby a broken harmonious rattle (Plate II., Fig. 3). That this is the case any one may convince themselves when a seasonable opportunity occurs of lying on the grass and watching these insects sing over the meadow lands, waste places, and swamps they enliven and populate.

The Greeks were not ignorant how the grasshoppers murmured; and when the Swedish Prime Minister wrote his "Memoires pour Servir," we find him fully acquainted with the mode of performance of grasshoppers and leaf-crickets, for he goes a step further, and says the friction cannot take place between the membranes of the leaf-cricket's wings (semi-elytra),

because they are so thin and fragile that they could not resist it ;
but there is every reason to think that the very quick movement
which the leaf-cricket gives to the veins, rubbing them on one
another, produces a kind of fluttering in the membrane, which
gives an augmentation to the sound.   But De Geer nevertheless
was ignorant of the microscopic aspect of the rubbing vein, as
was L. Frischs, who otherwise gives an accurate description of
the crepitation of the Vernal Field-cricket in his pretentious work
on " All the Insects of Germany," published at Berlin in 1766,
and in which he goes so far as to state that the shrilling vein,
which he correctly indicates, acts on the principle of a lock-spring,
so that the important discovery of the file or lima remained for
Goureau, a colonel in the French engineers, being first brought to
notice in his paper in the Transactions of the French Entomological Society for 1837.   And since then this interesting
mechanism, originally found to be present on the shrilling vein
of the crickets (Plate II., Fig. 8, l), and leaf-crickets (Fig. 5B, l)
and on the ridge at the inner side of the femora of the grasshoppers (Fig. 6), has come to be recognised as the source of
sound in insects of other orders.

These stridulating organs and their accessories characterise
the *males*, being alone perfect and efficient in male
grasshoppers and crickets, although in female leaf-crickets
they may be found exceptionally developed.   Their masculine
music, presenting a twofold stimulus, love and rivalry, fulfils a
twofold object, reproduction and dispersion ; and while the
variety and modulation in specific music at the pairing season,
paralleling or rivalling that of the rostrate cicadæ, confers on this
group precedence as instrumentalists, the ordinary challenge and
answers of the males indicate a phenomenon of rivalry cognate
with the dances in other orders, serving to band the species in
terrestrial migration with the mute females following attentively
in the track of their musical chanticleers.   Fear, elsewhere the
source of music, in this order is rarely, if ever, expressed by a
modulation of the notes.

The structure of wings in relation to their action as aërial
levers we have already discussed.   In investigating the music it
becomes necessary to consider the design evinced by the fore-wing
in a new and strange adaptation to an instrument of sound, a

subject both interesting and important, forming as it does a key
to the classification of the species and a method of distinguishing
the sexes. For in all Orthoptera we are led to recognise three
fields or discs in the fore-wing or semi-elytron, parted by constant
veins, namely, a marginal, M, an intermediate, I, and
an anal, A. (See Figure). Out of these areas, Nature,
in an ingenious hinging and folding by a method of
parallel veins—indicated in the diagram by plain and
dotted lines—forms at once a series of acoustic instruments, and an elytron or cover for the delicate
hind-wings. From these sutures it is evident we can
form an acoustic box in at least three ways. If
we bend down the marginal areas and cross the
wings, we have the section M of the elytron as we find
it in crickets. If we incline the marginal and intermediate, we shall have the section A presented by leaf-
crickets and grasshoppers. A transition form, M I A, is indicated in the exotic genus *Gryllacris*. We likewise notice when a
species of Orthoptera has finished its music, and is desirous to fly
or launch on a floating leap, it becomes not only necessary for it to
expand the organs of flight, but also to turn these areas on their
hinges, so as to restore the normal sustaining surface.

It appears, then, that all the tuneful species
of Saltatoria produce their notes by vibration of
the elytra, but they may also be observed lowering
their prothorax, in order to elevate their wing-covers
more or less previous to stridulating, and thereby
enclosing a portion of air above the body, which,
when these begin to vibrate during the friction
of the lima over its clasp, takes up these horizontal waves of sonorous vibration, and intensifies
them in ratio of its volume, as the sound of the
tuning-fork is increased when placed on its stand,
or the musical cylinder increases the sound generated by its teeth. It is at least probable that the
sonorous pulses are likewise transmitted to the large air-bladders
that distend the bodies of these insects, and that the notes thus
receive a further increment. We know at least this is a direct
sound accumulator, with regard to the males of the South African

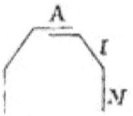

genus *Pneumora*, who have their musical organs removed from
the elytra to the abdomen. In genera of Orthoptera, as in
butterflies, the denticulations of the lima arise from abnormal
protrusion or constriction of the helix of the alar tracheæ or
veins; but in one group they present the appearance of a line of
acorn-shaped tubercles.

The subterranean species of cricket prolong their nympha
stage through the winter's cold, and many are heard only in
spring and summer; but the bulk of this tribe does not appear
till autumn on Palæarctic areas. Owing to their highly organised
instruments of music, with which they spontaneously express
their emotions, the periods of stridulation and repose of the
various species are rendered conspicuous. Some Leaf-crickets
crepitate by night; others are diurnal. As regards the crickets,
many genera are musical by day, but others, as the Mole-
crickets, are evening stridulators; and the House-cricket is
nocturnal. The grasshopper tribe is diurnal.

The conditions which excite specific music may be either
observed in the open air or experimented on in the vivarium, when
both methods will reveal love and rivalry as its chief phenomena,
accompanied with fear proceeding from hearing or sight, but
seldom from touch as heretofore. For if we imprison a male, his
music, desultory or coming in snatches responsive to artificial
tremors from clocks, bells, wheels, horse-hoofs, steam, or to the
chirp of birds, is rarely pronounced until another male or his
female be introduced, when the specific notes, with some exception, ring out intense and clear. Thus, if two males be confined,
they maintain incessant stridor, and nocturnal kinds prolong
their recital through diurnal hours, or *vice versa*; with the
crickets mortal combat invariably concludes this contest. If, on
the other hand, the two sexes be enclosed, whenever they approach the male starts into music; or if a male should bestow a
stealthy antennal touch on a female, his rival becomes musical;
it is also alleged that female Leaf-crickets, if free, will wing to
the stridor of their males. These characters may also be detected
in grasshoppers when confined.

Operating in Nature, such laws tend to band leaf-crickets and
grasshoppers in wandering societies, and crickets in associated
colonies, due to reiterated masculine challenges and the allure-

ment exerted over females and pupæ; although it would appear
in some groups of Acridiidæ as generally with the more inert
pupæ, that colour gradation is implicated in these phenomena,
eyes as well as cavities of hearing being well developed, especially in diurnæ, while jealousy is common. For this reason is
it likewise that orthopterous flocks and wind-borne clouds are
often so heterogeneous in character.

Male Saltatoria break into music when in proximity to a
female even differing in species; and a note or antennal caress
from an admiring male evokes a mixed chorus. But it is when
the season of reproduction and contest for the sex comes on that
grasshoppers especially are mutually incited to give their stridor
its utmost modulation. And this they effect by placing themselves at an angle with the auditory cavities of the female, and
shortening and jerking their femoral play; so that the notation
of the rival challenge differs both in character and gamut from
the new impassioned grating of the pairing note. Similar is the
stimulus that incites singing birds at the time of amour to pour
on the woodland enchanting strains, or inflorescence horticultur-
ally deprived of sexual character, to lavish sweeter colour and
perfume. Nor is the mere challenge note wholly intuitive, for
neither the alar nor femoral organs of stridulation in Orthoptera
attain efficiency with the assumption of the perfect state; and
the grasshoppers on emerging from the nympha may be observed
resting with a femur lowered, as if listening to the notes around
them, and learning their own proper and elaborate harmony.

LOCUSTINA (LEAF-CRICKETS).

No one who, towards the winter solstice, when days shorten
and grow chilly, wanders along our lanes, where berries pink and
black, hiding amid red and yellow leaves, announce the year's decline and completion of Nature's annual travail, can be ignorant
of the music of the leaf-crickets. The hedges, with dried grass-
stems and floating seeds of fantastic shape, no longer the abode
of variegated flowers melodious with winged forms, are become
the harbour of a new race. Little societies of the Cinereous
Cricket early in the noon begin a melancholy and puzzling
"crink-crink" that echoes from we know not where among the ivy-
tangle, and then as the clear penumbra of evening draws on the

red twilight, there arises all at once that weird and dreadful whistling of the great Green Leaf-cricket on the dusky top of the oaken spray, which in our younger days has so often induced us to quicken our steps homeward over the burnt heather, arising like the prophet at sunrise, to find this Gob Gobim vanished, or the solitary place of its hiding at most betrayed by a momentary chirp, and the feathered tribes again asserting their despondent supremacy of song.

In the musical species of saltatorial Orthoptera, where the males are uniformly pugnacious with cannibal females, the pre-eminence of this character seems to predispose higher capacity for stridulation ; for which we may seek parallel in the long-horned beetles, while a masculine differentiation of music indi-cates migratory banding of the species.   Thus, the tribe of Locus-tina, whose females are discriminated by sabre-shaped ovipositors serrated at the extremity, has males with alar musical organs highly organised, although, indeed, a few species are mute, and the females of others crepitate.   In Leaf-crickets, as in Crickets, stridulation results from sexual differential development of the semi-elytra, which in the male of the latter mostly correspond ; while in the former, the formation of the musical organ's two parts is usually accompanied with diverse development of the anal fields, and induration or expansion of their veins and membrane. In many species* a f shaped lima (Plate II., Fig. 5, B, l)—helically prominent on an inwardly-grooved outwardly-bowed transversal branch of the anal vein, and depressed beneath the left elytron's coriaceous elastic membrane—presents a serried row of obliquely parallel parallelopiped dentations (Fig. 4, c), small and soft at its extremities, elongate, black, and indurated at its centre ; apparently abnormal excrescences of the tracheal thread, or as Herr Vitis Graber affirms, modifications of the scaly membraneous folds that cover the wings and body.   These are fiddled obliquely over the rounded, raised, and incrassated inner basal edge of the right elytrons, near which is a tense and transparent iridescent talc-like spot, embossed superiorly on the anal field in fashion of a blue spectacle-glass, exteriorly to the rudiments of an un-indurated lima (sm, Fig. 5A).   As in musical insects generally,

* *Locusta* (*Phasgonura*, Westw.): *Thamnotrizon*, Fisch. ; *Decticus*, Ser. ; *Cholocœlus*.

reciprocal attenuation of the teeth is calculated to produce a musical sound or double gamut; while the raised mirror represents a bell in accumulating resulting vibrations. In the Saddled Leaf-crickets (*Ephippiger*, Fisch.), the elytra—abbreviated, dense, and scaly, and concealed beneath the prothorax—have an effective lima, situate in the males as previously on the left elytron's under surface; but in their *females* there is another on the glassy talc-like spot on the upper surface of the right. Numerous exotic modifications occur, and sometimes, as in the little autumnal blight-like *Meconema*, Ser., the musical organ is absent. The elytron also is subject to abbreviation, often presenting but the anal field; this takes place, too, even in one genus, *Decticus*, Ser.; and it appears the rule the longer the elytra in a species, the less unbroken its music and the richer in half-tones.

All species of Leaf-cricket are at once recognisable by the sabre-shaped ovipositor of their females; and their long antennæ. Their wing-covers, with talc-like spots in the males, are constructed like those of the cursorial kinds, to resemble in shape and colour the green and brown autumnal decay of the kingdom of Flora, where the majority preserve their delicate hues amid equinoctial damp and shade. Their thread-like antennæ seem designed to guard against the approach of friend or foe, and their leaping-legs have joints longer and more slim than either the grasshoppers or crickets, which causes many to be excellent leapers. Their elytra and ample wings act as gauzy parachutes as they spring, and either sustain or gently float them onwards. Like grasshoppers, the leaf-crickets live in wandering caravans, and while the males challenge and move imperceptibly forwards, they are heard and followed up by their mute females, who, like dames of romance, are subjected to life-long serenades. Their societies are nocturnal or diurnal, and the *shrill* of the former, affecting a light atmosphere, frequently presages a downpour of rain.

The larger kinds, whose wing-covers, broad and thick, resemble the glossy foliage of exotic shrubs and trees, come to us from abroad, and these are arboreal and nocturnal in habit, frequenting the thick shades of primeval forests. One, *Chlorocœlus Tanana*, discovered by Mr. Bates while exploring the river jungle of the Amazons, has a powerful wing drum of the type seen in our Great Green Leaf-cricket (Plate II., Fig. 5) ; and when a male was captured by that gentleman, and confined in a cage, its intermitting quirks of music, sharp and resonant—"Ta-na-na ! ta-na-na !"—could be heard echoing from one end of the village in which he was staying to the other. The notes of a West Indian kind, perhaps that termed by an old writer the Gully Bell (*Locusta camelifolia*), are described as slow and measured, resembling an articulation of the words, "Shock ! shock !" Of a third giant flying leaf (*Platyphyllum concavum*), we learn that the males during September and October band darkling among the tall summits of Canadian forests, and there, shrouded in the dusk of eventide, commence a screeching cadence of " Katy-did-she-did!" continued throughout the live-long night. Their mute females, unprovided with the noisy organ, can still on capture evince resentment by shuffling their elytra, thereby raising a low grating sound.

The shrill of the smaller species of *Phaneroptera* may be heard by the traveller almost everywhere within the warmer zones of the earth at the close of the day. In Northern Italy the antique and quaint little *Phaneroptera falcata* of Scopuli, with its knife-shaped wing-covers shorter than the projecting wings they conceal, remains throughout the day clinging on the sepia-green tufts of the acacia, where the similarity of its tint affords chromatic protection and harbour, so that the very brown summer freckles on the soft leaflets find a counterpart on the insect's thorax. And so they stay unobserved until about half-past five in the evening, when the slim-legged males, still spectre-like and motionless, kindle up in running serenade to a brisk momentary crepitation, uttered in the society of their cimetered females, which bears apt comparison with the catch heard on winding up a watch, " Crick ! crick !" They are at all hours easily provoked and sound-sensitive, becoming especially lively to the quickening stanzas of the guitar, and they never remain mute when the bells and crackers of a church festa invade their seclusion ; but if we seek to confine them in a gauze-covered box they without fail mope, and cease their music. The New World *P. curvicauda* is by preference nocturnal, according to Dr. Scudder. His diurnal music is " bzrwi " and lasts one-third of a second. But at night it becomes a repetition about eight times of a note resembling " tchw," uttered at the rate of five times in a second, and making each note one-half as long as the diurnal.

The musical vein in this genus is straight, attenuated exteriorly ; above it is flat, and, as might be inferred from the notes, but coarsely pectinated, showing in the common European species only about twenty denticulations. The mirror is oval. A slight rasping made by the females on closing their elytra may be attributed to a limaform structure on the marginal edge of the left anal field, probably a rudiment of the masculine organ of music. In the island of Antigua one of these leaf-crickets, in company with another of another genus, flies to light in the evenings.

The species of *Conocephalus* are as widely distributed as the previous, but they are of humble habit, and creep in among the dewy meadow grass, for which an elongate pisciform shape has admirably adapted them. The shrill whistling " vhree ! " of *C. mandibularis*, Rossi, lasting three to four minutes with pauses broken by " Wheat ! wheat ! " nightly resounds along the margin of damp herbage at the river-banks in northern Italy, with hollow conocephalian intonation, resembling the sound winding from an oaten reed or reverberating in a sea-conch. This music, that mingles with the wash of the Alpine torrents in August, arises about eight in the evening, and sometimes continues until dawn, but in bright moonlight it will suddenly subside and cease about midnight. The males are only too easily captured, and if encaged at a window as night comes on vigorously answer their comrades without. And it will then be found, that if the ear be approached the sound comes full and intense with bee-like symphony, but let the observer retire ever so little, the sound becomes deceptive with a dirling and hazy echo well calculated to afford protection to the performers. *C. robustus*, a congener found near the sea-shore in the southern parts of New England, emits a certain note at the rate of two-and-a-half times a second, which to an observer near at hand rises and falls rhythmically, and appears accompanied by a like loud droning noise. The musical vein in these species is straight, bowed downwards, fusiform and slightly *f* shaped ;

attenuated at either extremity, but more sharply so inwardly. Superiorly it is flat, with indurated black pectinations of a fineness that accords with the short aërial beats. During confinement we often notice the notes become more vehement in their production, and this is a sure prelude to a mortal combat, and is accompanied with the death of the musicians.

The Meadow Cricket (*Orchelimum vulgare*), heard by Dr. Scudder near Boston in America, commences with " ts," which it changes almost instantly into a trill of " zr"; at first there is a crescendo movement which reaches its volume in half a second; the trill is then sustained for a period varying from one to twenty seconds, and closes suddenly with " p." This strain is followed by a series of staccato notes sounding like "jip!"; they are one-eighth of a second in length, and are produced at one-half second intervals. The staccato notes and the trill alternate *ad libitum*. The night song differs from that of the day simply in its slower movement; the pitch of both is at B flat, two octaves above the middle C. The kind is exotic.

The genus *Locusta* of Burmeister, containing our familiar Green Leaf-crickets, is not recognised as extending much beyond Europe and the coasts of the central sea. The three or more kinds likewise vary according to locality, and approach each other very nearly in characters. One, *Locusta cantanas*, has shorter and broader elytra, and is chiefly confined to the mountain-ranges of Europe, where its notes, Yersin states, change with the time of observation. When the sun hangs on the horizon, its recitals are scarcely sustained during two or three seconds with similar intervals; later on they are prolonged. These hill-side sounds seem to come sharper and briefer than those of the next species I shall notice, and more run together, but its way of life is very similar. The commoner kind, the Great Green Leaf-cricket, *Locusta viridissima* of Linnæus, mingles its dirl pleasantly with our recollections of warm and dusty autumnal weather, and sweltering strolls and seats beside fragrant brushwood, sun-tinged vineyards, and cottage potherbs. In this country its chief resort is the potato fields, to whose new-world foliage, like the Death's Head Moth, it has habituated itself, prompted by some unseen harmony, and here over these noisome farm acres, during the months of July,

August, and September, its ravenous flocks silently steal like armed
warriors. Their music is properly nocturnal, and is thus extem-
porised. The male at about 8 p.m., as the marish mist rises in the
meadows, gains the summit of a leafy potato-stalk, or crawls
up the neighbouring hedgerow or low oak-bush, and hanging
there, head downwards, stretches out his fore-legs presenting
his auditory organs. He then, depressing the locking prothorax, unhooks with a click his elytra, and takes part in an
upspringing dithyrambic "Whitt! whitt! whitt!" or "Zie! zie!
zie!" if your ear be classic, resembling the spurts of a knife-
grinder, that subside ever and anon, and gather in a dull vitreous
resonance (Plate III., Fig. 3). If approached stealthily by a
female, he ceases a moment, and bestows an antennal caress.
This evening concert terminates about 11. During dull weather,
the hedge-encamping outposts of the drove continue their dizzy
ring during the day, or fling momentary responses to passing
vehicles. The ordinary musical snatches, lasting about four
minutes, with chirp-broken pauses of four seconds or longer, are
higher pitched than those of the *Dectici,* and the auditory
organs that receive them appear more acute, for while the green
minstrel in captivity becomes vocal with the hissing kettle or
domestic knife-cleaning, the Wart-biter responds to jingling
harness; and the reason of this is, the file of the second is long
and coarse, harmonious with the sunlight, that of our nocturnal
species broad and fine; and for this reason, crepitating together
in an apartment, the two extemporise pleasant part-playing of
the Christy-Minstrel order. The females are attracted by the
music, and approach, or even fly to, their males in captivity.
The lima carries about ninety-three dentations, of which only
some thirty-two are effective. The chirping of *Locusta caudata,*
a species of Eastern Europe, smaller and more graceful in
appearance, is described as shriller and weaker.

The plaintive cloth-snipping "Sprink! sprink!" of the
Cinereous Cricket (*Thamnotrizon cinereus,* Linn.), commencing in
slant sunlight among lank sunburnt grass, beneath bushes where
it is protected by its umber tints, greets the mild autumnal nights
from September to November. This scintillating madrigal, when
the insect is unexcited, presents a distinctly-repeated note (*Se!*)
with pauses recurring about every five seconds, for the intervals

perceptibly decrease after dusk, and increase towards morning. In solitary captivity in a box a male will respond throughout the hours of shadow to the chipping of freestone, strokes from a village clock, champing bit, or twitter of a startled finch; but if two be enclosed, they will be found to sustain a running cymbal-clash day and night, accelerated to as many as twelve chirps a minute—"Kre! kre! kre!" Often, too, we may thus hear them, angrily responding head to head, alcoved beneath a weedy bush or ivied hedge-bank, demonstrating a most marvellous temper and elasticity, in the one hundred and twenty incessantly-scraping indentations on the circular lima of their squamose and cup-shaped wing-cymbals. The female squats among them as they chirp, and is herself virtually without wings or elytra. This species is one of the best of the indigenous leaf-crickets to keep in a box, from the low tone and pleasing nature of its note; and it will thrive well and sing on till the first frosts, if only supplied with a few blades of grass. The similar music of the Apterous Thamnotrizon of Fabricius is perceptible from a considerable distance, and consists of from six to nine quickly-consecutive strokes broken by short pauses.

The diurnal species of the genus *Decticus* (Ser.) frequent the high gramineæ of all temperate regions, and crepitate in sunshine. Who has not heard of the *Pah-puk-keena*, the patronymic of the storm-fool, "shrill and ceaseless," during the summer heat, in the corn-fields of Gitche Gumee? The lima of the Wart-biter (*Decticus verrucivorus*), the European corn-frequenter, has from seventy-one to ninety dentations, and its music, a harsh knife-whetting, possesses a soft hazy resonance like murmuring steam, predisposing to summer slumber and reverie. These dreamy snatches continue from four to five minutes, with intervals broken by "Screet! screet!" a sound like a stray sheep bell, or one canary-like and low. And then a fitful part-playing from two of the males in a barley-field has quite the zest of a band of Christy Minstrels; or should a male bestow an antennal caress on a female among the cockles and poppies, his rivals immediately wander about crepitating with a sweeter plaint, for their jealousy is excessive. As regards pairing, Roesel informs us that when a female is attracted by a male, previous to coupling, she approaches from behind, and informs him of her proximity by strokes of the antennæ; he then ceases his music, throws back his feelers to determine whether it be friend or foe, and breaks forth into soft twittering notes, at the same time drawing in his hind legs so as not to hinder her advance. This species, termed by the Swedish peasants the Wart-biter, is normally green, with invisible clytral spots which the sun photographs to brown, while the verdant hue is gradually bleached off the elytra, and changed to brown and purple on the body. It eats gramineous shoots, docks, and plantains, or bread, indiscriminately, which it touches first with its palpi. After feeding, it moves the oral appendages as if ruminating.* In Southern Europe, the Wart-biter gives place to *Decticus albifrons*, Fab., another large and conspicuous species, there common on reeds in marine marshes during the month of August, whose note Yersin compares to that of the Saddled Leaf-crickets.

The smaller European *Dectici*, the same writer remarks, repeat their notes more slowly, and they thus seem to have connection with those of the crickets. Their mirror is tolerably circular. A common and distributed species, *Decticus griseus*, has a whirring note " cri," that in many places begins to be heard in tall rank grass at the commencement of July. The smaller short-clytraed *D. brachypterus*, partial to damp soils, where heath, moss, and low tangle mark unreclaimed or virgin land, jerks its elytra from four to seven times with a sound of " Wirr ! wirr ! wirr ! " or as Yersin gives it " Rice ! " and awaits response. If two males be captured and enclosed together, they approach in the receptacle head to head, or side to side, and maintain a ceaseless stridor ; the male likewise crepitates when he encounters his female, or when she meets him, and continues his music while in her proximity. The stridulation of the very much alike *brevipennis*, Charp., Yersin tells us, is first heard in the sunshine on grassy spots in the Vaudois valleys, at the end of June or beginning of July. At this period the sound of the males, of which the timbre is perfectly characteristic, is sustained but a few minutes. The maturer note heard later on is indistinctly trilled and sustained for an indefinite time, resembling much a distant hum, " éééé."

---

* The abortive lima of the female Wart-biter is shown on Plate VI.

L

Among tall grass and bushes, at Nice, occurs another species Yersin terms *Decticus sepium*. Its music is described as a short "cré," repeated at intervals of uncertain length during the day. The males of *Decticus bicolor*, Philip, are stated to have a music shrilly and continuous.

In the Saddled Leaf-crickets, or Cymbal Players, Herr Graber finds the teeth have the form of triangular prisms, instead of being, as in the majority of the leaf-crickets and crickets, lamellar, elliptical, trapezoidal, or crescent-shaped. The common Saddled Leaf-cricket of the Vine (*Ephippigera vitium*) has especially thick cup-shaped elytra (*e*), in which is set a circular smooth mirror, somewhat obliquely situated, and also a shrill-vein of uniform thickness, with from ninety to one hundred dentations. The sexes crepitate among bushes on elevations near vineyards in South Europe from August until October, where the male chirps twice and his female once in the sunshine. A

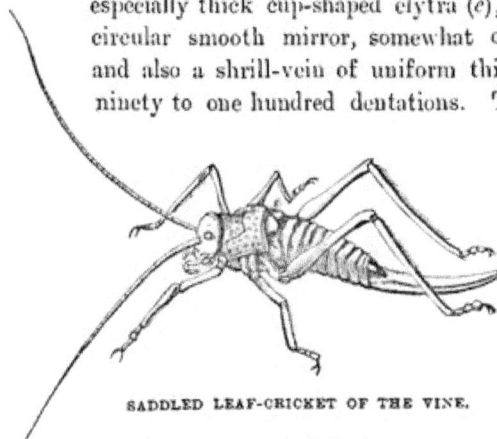

SADDLED LEAF-CRICKET OF THE VINE.

kindred species, *Ephippigera terrestris*, Yersin found at Frejus and Grass, in the wheat-fields after the harvest, haunting the borders of unfrequented paths, where the ferruginous colour of the soil afforded it mimetic protection. Its diurnal music much resembled that of the Cinereous Cricket, being short, plaintive, somewhat intense, and repeated at considerable intervals. Another species or variety of Saddled Leaf-cricket (*E. provincialis*), noticed by the same observer, holds the covert of thick bushes and hedges, or may be observed climbing the vines and young trees, and here during the day it emits a music resembling that of the Great Green Leaf-cricket, or Wart-biter —an acute "zig-zig-zig," repeated indefinitely and without interruption.

Orphania denticauda, Charp., a smaller green cricket with cup-shaped elytra that Yersin used to meet with, in his

Vaudois Alps, stridulates in the sun, when walking over the grass, emitting a music intermediate in measure and timbre between that of a little *Decticus* and the Great Green Leaf-cricket—" Zie, zie, zie." On approach the male intimates alarm, and softens his tone, hardly moving the elytra; and when closing them he emits a sharp sound. The long-winged *Xyphidium fuscum*, Fab., as he stridulates ascends and descends the reed-stems which he frequents, moving the elongate antennæ with a slow oscillatory movement. His music consists in an indefinite series of notes shorter than those of the Great Green Leaf-cricket, run together, and so feeble as to be audible only to an observer close at hand. The teeth of the shrill-vein number about thirty, and are scarcely ·08 Mm. broad. Some leaf-crickets produce a stridulatory sound during flight, for we find it stated, in Harris's " Insects of New England," of *Phylloptera oblongifolia*, which obtains the perfect state in September and October, that it frequents trees, and when it flies makes a whizzing noise somewhat like that of a weaver's shuttle. The notes of the male, though grating, are compara-tively feeble. The anal fields of the elytra in the males of *Odontura* and *Xyphidium* Vitis Graber notices are identical as to venation; while in *Odontura Fischeri* and *albovittata* the mirror is small and rough, and the shrill-vein simply bowed instead of S-shaped, denoting here a low state of development in these organs. The same authority states that the mirror in the leaf-crickets is formed from the space enclosed by four intersecting veinlets, these he terms the vena specularis anterior, posterior, interna, and externa. He considers certain tubercles on the dorsal surface of the second and third abdominal segments in the genera *Gryllacris* and *Dienacrida* postulate stridulation.

The file and mirror are sometimes noticed in fossil species. Dr. Scudder discovered a lima in his *Xenoneura antiquorum*, from the Devonian of Lancaster, New Brunswick; I have myself detected it in *Corydalis* (?) *Brogniarti*, from the Carboniferous of Coalbrookdale, Shropshire—a wing fragment figured in many geological text-books, that is closely allied by wing venation with the *Gryllacri*, Heer, from Saarbrucken mines; and also with *Lithomantis carbonarius*, Woodward, from Scottish coal-fields. These gigantic winged insects of Sigillaria and Calamites-

stalks, with elytral expanse often exceeding six inches, approximate the present leaf-crickets, although absence of saltatorial legs leaves the matter still contested. The mirror
sexually discriminative, is likewise traceable in *Locusta speciosa*
and *L. amada*, Hag., from the Oolitic Solenhofen ; extinct types
in which an attempt at restoration I was once kindly per-
mitted to undertake at the British Museum, seemed to reveal
a complex and oblique elytral venation, like that of the aerial
migratory species of locust, *Pachytylus*, Fieb., united with a
sabre-shaped locustinal ovipositor in their females; while the
thorax and hind legs affiliated them to the genus *Decticus* of
Serville. This circumstance has led me to surmise that these
leaf-crickets in ancient times had much greater powers of dissemination than now possessed by their family. Lastly, there is a
cast of an elytron of a large species of true *Decticus*, about twice
the size of our present Wart-biter, in the national collection,
from the Purbeck of Swanage, with about as clear indications
of the usual lima as can be drawn from a plaster impress.

ACRIDIIDÆ (LOCUSTS AND GRASSHOPPERS).

The Akrides, or Locusts, " produce their song by the hind legs
with which they leap," says Aristotle, writing over 2,000 years
ago. And in this group (Plate II., Fig. 3) we find them fashioned
so as to answer both the purpose of leaping and that of music.
For here, as in other saltatorial insects, they are longer than the
two fore pair, with the femoral joint consequently raised and forming an acute angle with the tibia. Then the thicker crank-shaped
thigh or femur is characteristically large, tough, and powerful,
and along its inner and hinder part runs a furrow in which the
shin or tibia is placed previous to leaping, and in which it is carried
likewise when producing the music. In front of this groove runs
the hard keel or ridge that plays the active part in the stridor
(Plate II., Fig. 6, *l*). The passive organ is seen in a certain vein
which is raised from the elytral surface at either side (Plate II.,
fig. 3, *y* or *x*) ; or in one singular instance it is transposed to the
abdomen.

The timbre of the notes of the grasshoppers, according to
Yersin, is less adapted than that of the cricket to be attained
and notated by human musical instruments, having most in

common with the percussional sounds of a child's rattle, or the frictional notes of scraping cardboard, wood, or metal; a reason, by the way, why their perceptive organs should widely differ from ours. The strokes of the femora when prolonged are slow, producing more basal notes; the shorter are always rapid, effecting those higher in pitch. They are almost invariably diurnal musicians, and the greater portion are tuneful only in the glow of the sun. The notes of the male, when in the presence of a female, diminish in intensity, or change completely in rhythm.

Certain males of the larger locust kinds are noticed to produce notes by slightly raising their elytra roof-shaped, and bending the tibia of the hind leg beneath the femora, where they are lodged in the furrows designed to receive them. After this they either draw both legs briskly up and down through a small angle (Plate II., Fig. 3, *a' b'*), or scrape first on one elytron and then on the other, using the right and left thighs alternately. Thus Yersin observes, that when *Pachytylus nigrofasciatus* walks on the sand it keeps the legs somewhat removed from the body; but that from time to time it raises one or the other, and passes it with considerable play and rapidity over the elytron, bringing forth a brief indistinct note. Another locust (*Paracinema bisignatum*) effects an analogous movement, but employs the legs simultaneously. In these and other species of *Pachytylus* the raised keel on the inner side of the groove of the hind legs appears smooth. It is scraped during the to-and-fro action of the hind leg on a vein intercalated between the internomediate and externomediate, which is bowed out at either side of the roof-shaped elytra. This is minutely serrated to form a lima, here constituting the passive organ of music (Plate II., Fig. 3, *x*). The wing-covers, over which the resulting vibrations spread and accumulate by the conduction of the veins, are tough, and must be considered as forming a delicate musical box, where the tremors induce and lead sonoral pulses. These poured on the ambient ether produce the feeble trill of the locust.

Herr Vitis Graber relates of *Stetheophyma grossum*, L., that when a number of individuals of this species were springing around him in the high grass, as they leapt the peculiar noise of the males came forth. This sound resembled most the chirp elicited on closing the elytra of a leaf-cricket; it was short and

tolerably high—" tzi, tzzi." These notes had their origin in an
apparatus similar to that just noticed. About 4 mm. from the
wing-base the vena subexternomedia in the male is somewhat
elevated, and the ordinary scales which cover it increase in size
and thickness, passing into a lima such as we find in the
Saddled Leaf-crickets. This character extends to about 10 mm.
from the wing-base, where the vein disperses in a fine network.
In respect to a Nearctic species, Dr. Scudder says: "Some of
the veins in the centre of the wing of the male *Arcyptera lineata*
(one of which he indicates) have a rasp-like surface, upon which
the hind thighs are scraped up and down."

The hyaline-winged *Epacromia thalassia*, Charp., according
to Yersin, is partial to the plains in Northern Italy, where it
haunts the dust and stones. Here the males assemble together,
place themselves side by side, move over one another, advance
and retire, as if in sport. When one retires he vibrates a single
leg, so as to produce for less than half a second a dull and feeble
sound. He then takes some steps forward and vibrates the other,
and so continues till he meets another individual, when he
prolongs his strokes. "Once I saw a male vibrate both legs at
a time. He was alone, with the head applied to the ground,
and the abdomen with the posterior portion of the body raised at
an angle of 45°." To observe the stridulation of the ink-lined
Parapleurus typus, Fischer, it is necessary to choose a sunny
day, and place oneself in the midst of the continental meadows
where they abound. When two or three are gathered together
they commence movements with the legs, some long, others short
and rapid. The latter only are accompanied with a sound, which
is distinct at a short distance, and lasts not more than half a
second. Yersin informs us that the males of *Œdipoda fasciata*,
Fab., also move their hind legs, and perhaps produce a sound,
which may be said likewise of those of the pink-flushed *Caloptenus italicus*, Lin.

But the larger grasshoppers with coloured wings, where the
attraction of colour appears to supplant that of sound, as, for
example, those of the genus *Œdipoda*, seem to be mostly incapable
of much music. These certainly have the elytral vein previously
alluded to raised, and presenting, in common with its fellows, a
structure resembling obliquely strung beads, but I have never

heard the males stridulate, nor do I know of any record of their
so doing. If, nevertheless, the ordinary music be virtually absent,
many of these brightly-coloured beauties as they pass through
the air emit a whizzing sound on the wing, resembling somewhat
that of the firework termed a squib, and usually ascribed to
the thickening and serration of the costal wing-veins, which, at
least as regards their extremity, appear actually to chafe as they
dart, beneath those at the anal portion of the covering elytra.
This, however, is not exclusively characteristic of these paragons,
for we find it in some duller sorts, and even in those possessing
ordinary organs of terrestrial music. In exemplification of this
we may notice the Stridulating Locust (*Pachytylus stridulus*),
occurring, according to Roesel and Fischer, not rarely in the
woods and on the mountain-slopes of Central Europe during
August and September, the male or female of which, when
started in some solitary wood clearing, or from the embers of the
charcoal-burner, leaps, emitting a crackling sound resembling
the stirring rattle of a night watchman or holiday chasseur of
hares, and as they do this, expanding the sheen of their floating
brick-red under-wings, they are rendered doubly charming and
attractive. The males of the common blue-winged variety of
Œdipoda fasciata, Siebold, frequenting the opener sunny banks
and fields, when disturbed, similarly leap high, with a sharp,
rustling, autumnal sound, flashing back as they go the translucent glow of their gay under-wings; and we may again witness
this music in some Neartic species of the genus, as, for
example, in the common yellow-winged autumnal Rattling
Locust of Canada (*Œdipoda sulphurea*). Either of these species,
whether from natural gregariousness or a sense of local protection, is singularly partial to circumscribed spots, to which the
latter, on being driven away, as invariably returns again.
Another brilliant American locust, termed *Locusta corallina* by
Harris, and appearing in New England as early as the middle of
April or first of May, is similarly described as making a loud
noise in flying. Among the smaller European grasshoppers, the
males of *Stenobothrus melanopterus, miniatus, viridulus*, and *variegatus*, Sulz., emit a brisk, scintillating sound as they leap, and
those of the Asiatic Migratory Locust (*Pachytylus migratorius*) do
so when on the point of settling. Further than the apparent

attractive design of this sound, it is as difficult to say whether
production be in measure subservient to the will or no, as
to specify the emotional principles, if any, it is capable of
expressing.

Locusts, contrary to the general notion, have some diversity
of habit; while the migratory kinds live in flocks, and move over
the pasturing grounds, or rising, waft through the breezy air
to distances proportionate to their capability of flight, Fischer
speaks of one European species (*Acridium tartaricum*), a hermit
from its kith and kin, that lives on trees, and has not only
a song deceptively like that of a bird, but also a mimicry of
habits, for, he mentions, if any one approaches the perch on
which it is sitting, it at once flies off to another at a short
distance. This large species is not rarely met with in Italy
among the mountain gorges.

The males of our little autumnal grasshoppers sing in the
same fashion as the larger kinds; ascending some elevated perch
on a grass-stem, they lower their prothorax, thereby unlocking
their wing-covers, which they raise pent-like to uncover the
organs of audition at the base of the abdomen, then, doubling
up their hind legs, they pass these backwards and forwards,
usually simultaneously, over the exterior surfaces of the wing-
covers, winnowing them briskly through a small angle, and
producing a blithe note, which varies with the speed, pressure,
and length of the stroke indulged in (Plate III., Fig. 8). You
may convince yourself it is so, according to Fischer, by watching
these minstrels in the open air, or otherwise by the experiment
of imitating the motion of the femora, in a living or recently-
killed insect. But the song of these merry species only resembles that of the former as regards mode of production, for
by aid of the magnifying-glass and microscope we soon discover
the musical organ is transposed to the keel on the hind femora
(Plate II., Fig. 6, *l*), which in the males bear at their lower end,
for two-thirds of their length, a row of acorn-shaped, or lanceolate, tubercles (Fig. 7), as invariably absent in the mute female.
These, by the action of the femora, are played over the inner
branch of the chief or scapular vein of the elytra (Plate II., Fig.
3, *y*), which in *Stenobothrus*, Fisch., and *Gomphocerus*, Thumb.,
is raised on the surface; in the latter genus these tubercles are

more numerous, and the song comparatively more powerful. The wing-covers of the males are adapted as sounding-boards to receive the music by their tough membranes netted rectangularly; these in the females become extended, thinner, and the meshes branch or stretch into V's, Y's, and natural hexagons, with the evident object of strengthening the attenuated structure, and thus too, the V's and Y's are observed invariably to point outwardly.

In the existing rage for cheap music, when flashing lights, impassioned notes, and sweet warblings greet the man of business homeward wending, and drive far into the sorrows of the night, it is scarcely to be wondered refrains so full of small peaceful harmonies as those complaining notes, that each autumn echo beneath the blithe ring of the mowers, should continue a study for poet and musician. And it is thus we not only hear of them blending in the luxuriant tide of song on Transatlantic pianos, but what is more generally feasible, find them adapted to rhythmic notations by admiring frequenters of the green banks of the Rhine and Alpine glaciers, where they possibly lend much to the charms of the scenery. The sweetest minstrels of these classic spots, according to Fischer, are *Stenobothrus miniatus*, heard all over Europe, and high up among the purple gentians in the rarefied air of the Alpine snow-line, where his liquid metallic trill has often drawn attention; *S. variabilis*, that has enchanted the trained ears of three German savans; and *Stetheophyma variegatum*, whose stilly song Yersin, in his Vaudois valleys, fancied to vary with the hours of the clock.

Nor are the pastorals of our insular troubadours to be despised. How often do the young in years, who listlessly recline in zephyry hay-fields, take lovers' walks or meditative strolls, receive brisk overtures, which haunt the mind and whisper back the cheerful voices of seasons that have flown! Nor in themselves do these songs of emulation want that charm of variety, the invariable outcome of a busy life. The stridulation of our most powerful violoncellist, the *Stenobothrus viridulus* of Linnæus, with plain brown wing-cases, slashed with shady green above, as it arises in the parching heats of summer from the patches of damp, rank grass in marsh-lands, has a delightfully cool and refreshing sound, resembling the fitful chafing of the silicious reeds, or vapoury sound of escaping steam, "hisss!" You first hear the saucy rattle of its fine-wrought musical-box, lasting from ten to fifteen seconds, when another more distant performer springs on the alert, and another and another; and as you proceed over the green oozy tufts in quest of game, the scared and agile males leap high from under your feet, with a sudden rustle like the whispering eddy catching the brown beech-leaves in autumn, yet sharp as the ring of a whinstone or flint. These largeish green grasshoppers appear in the perfect state over the plains and on the mountain-sides of Central and Northern Europe at the end of June, and in August they pair. When this all-important season approaches, the males gather round their females on the warm and sunny banks in coteries of four or more, and challenge. Then one or other of their number advances to the sex, discriminated by her larger size and more portly ways, and, placing himself alongside, or with his head directed toward the centre of her body, sings briskly on one or both legs, producing a harsh grating call-note, like a snapped watch-spring or dancing-hoop, "Tit! tit! tit!" This jarring discord he now and then diversifies with a strange, asinine kicking out of his hinder legs; and while it proceeds, the complacent female will lower one or other hind leg horizontally, so as to expose her proximate auditory cavity, and thus listen, sunk in cold apathy, for hours to the ardours of her serenader, who varies his stridor to *piano* or *forte*, in harmony with the passing gleams of light and shadow. This state of things goes on day after day, and but rarely in the afternoon a pair may be observed accoupled. Not then, however, does the beau cease his music; he briskly prattles or snarls to every interloper who dares approach his seclusion, and, with true grasshopper spirit, will continue snatches of recitative, even when the throes of death press upon him, and the golden meads of Proserpine are all his perspective. In regard to the instrument of stridulation in this species, Dr. Landois says the row of tubercles on the femur of the leaping legs in the male amount to one hundred and nineteen, and this is doubtless tolerably correct; I myself have found an average of one hundred and ten in some British species I examined beneath the microscope. The stridor is thought chiefly to arise during their downward stroke on the scapular vein.

The common little Red-legged Grasshopper (*Stenobothrus rufipes*) is generally of a fuscous or yellowish colour, with red tibiæ, but some examples, accredited a variation of its female, which I have frequently taken in Lombardy, have their wing-covers above slashed with deep green, and this circumstance, taken in conjunction with a similarity of thoracic configuration, an important feature in grasshopper charms, has induced the author of the "Orthoptera Europea" to consider this but a variety of the sort just noticed. But when we turn to the subjects of distribution and biology, a discrepancy of kinds becomes more pronounced. While the former damp-lurking grasshopper finds a southern limit of dispersion among the Alpine rivulets, the latter vagrant of the arid pebbles boldly descends into the plains, and penetrates to Sicily, regions in which it often assumes a swarthy hue, with legs wholly red and coralline. It appears in the perfect state, too, earlier in the year, being noticed from May to September, and its pedestrianism and music form quite a distinctive character. On dry sandy banks in the month of July I have frequently reclined on the herbage and watched the male quietly sitting on a grass-stem or squatting on the ground, with his right hind leg lowered to listen, and then, as if prepossessed with idea, or feeling a sense of loneliness, suddenly raising and moving both femora from six to eleven times forwards over the wing-covers, with pauses of about seven seconds, fling forth long trilled defiances, "Rete-tetee!" If not answered he will run, or take a flying leap, and challenge again. These grasshoppers are incessantly in motion, with their red-tipped abdomen kindling like a glowing coal in the misty sunshine; and when one male meets with another he dashes at him and bites him, and the two then adjourn to the nearest available grass-stalks, or fallen twigs, and respond sharply for a few seconds, the provocation and reply blending in a guttural "Ritta! ritta!" But a female, even if in a different dress, is greeted with a distinct recognition, "Thirph! thirph!" There must be nevertheless a slight latitude allowed as regards the notation of these slender notes, which to our gross ears will ever vary with the attention bestowed upon them. My first studies were made on the sand-dunes covered with scant grass that form the French sea-board, and which in summer-time are all resounding with these populous cries; and here I had inserted the ordinary challenge in my pocket-book as "wrree." Yersin also similarly notates it "vrrriiii," and he says it lasts rarely more than seven or eight seconds, and consists of from one to three notes; but on again hearing the same music trilling fifteen to twenty seconds in the hay-fields of Piedmont, I sat down, and after some cogitation, decided on the present enunciation as more expressive. The various sounds are all due to the vibration produced by about seventy tubercular teeth, very uniform as to size and distance of insertion along the ridge of either hind femur, and their distinction and modulation is due to the play of the hind legs.

The *Stenobothrus vagans* inhabits Central Europe. Yersin says it is abundant in the Valais from Fouly to Sion, where it may be commonly heard on the stones at the wayside. It emits four notes a second, which are clear and more or less separated, "ééé," or "rrréééé," and this music lasts from two to five seconds. Dry stony margins of hilly fields are the common resort of the species, which resembles greatly the Variable Grasshopper, save in its being less downy, and having the lateral keels on the thorax less cross-shaped. Another grass-hopper with more northern distribution, the *Stenobothrus apricarius*, is heard on the Alps from August to October. It seems an unusually powerful minstrel, since it emits ninety notes in fourteen or fifteen seconds, which gradually augment in fulness, "E-tin! e-tin! e-tin!" the two sounds appearing to arise from the backward and forward movements of the femora being made with different pressure. It frequents similar spots to the former.

The male of *Stenobothrus melanopterus*, with transparent wings strongly bowed outward in front, and dirty umber veins, does not appear recorded as being found in this country, although it is by no means unfrequent on the opposite French coast, and is distributed over the greater part of Europe. Its powerful notes with a timbre "sssin," we glean from Yersin's papers in the bulletins of the Vaudoise Society, are followed by another graver, resulting from a simple elongate stroke of the legs which press on the elytra during their upward movement, "trrraa," or the two notes combined, "ssssin, trrraa." Some-

times when walking over the ground only "ssssin" is heard.
Dr. Fischer finds it common in parts of Germany in the fields
and meadows, where the clear stridor of the male may be
perceived at some paces. It is caused, he says, by the hind legs
being drawn upwards, and then from five to six times briskly
rubbed on the wing-covers; and although different rhythms
sometimes arise during the upward movement, the music is
commonly " Rtsch-ssssss! rtsch-ssssss!"—a Hebraic concate-
nation of consonants an Englishman may interpret or translate,
I presume, " Ritisch-sisisisisis!" Grasshopper language inclines
to the penetrating *i* or sorrowful *e*. But then, why on comparing the two notations are the sounds reversed in either case?
We can hardly suppose the dwellers in the Alpine fastnesses, and
along the banks of the arrowy Rhone to have so accommodated
themselves to the neighbourhood of Fribourg and the Rosskopf
mountain.   Herr Graber gives the teeth on the femur of this species as one hundred and forty, and attributes the loudness of their
music in Austria to their number and wideness of interspacing.

The Variable Grasshopper (*Stenobothrus variabilis*), Fieber
is one of the commonest and most distributed of European
kinds, as well as the most changeable. Generally it is grey,
fuscous, reddish, or of a dull, greenish hue above, with the
elytra covered often with black speckles. The extremity of
the abdomen is not rarely lacquered red, especially in the male.
The varieties, too, occur on the same spot, and couple, as far
as I can learn, indiscriminately, the negro with the brown, and
green with the fuscous, a fact perhaps owing to the fine ears
of grasshoppers, and a likeness in specific music. October
ushers in their season of reproduction, and at this period the
morose males, lit by the chill sunshine of the waning year,
collect on dry banks, often in numbers, and form groups round
their apathetic and coy females, nestling to bask on the bare
patches of the warmth-retaining soil; or they depend around
like a forbidding fruit on low shrubs in the vicinity, a seat of
vantage derived from a power of agile flight.   Inspired by
their ardours arises their seething stridor in fierce unison, now
dragging as the salt waves when heard receding along an unfrequented shore, and now presenting the illusion of a bubbling
source, or the harmonious warbling of nightingales. Fitfully

this angry turmoil is heard to surge and drive along the ground, as the prattling of a hasty shower, or the wind caught in the aspen. This is a sure token a male is advancing to one of the females: he does so delicately or persuasively, with a pre-meditative feint at stridulation, vibrating the hind legs for a time, and eventually raising a faint sound, an overture never-theless understood, and immediately chorused from all sides. The would-be suitor patiently receives the general ovation with a listening leg wakefully lowered, and then sharply gives his response by jerking the right femur forwards on the wing-covers, with a sound of "Tirp! tirp!" the proper call-note. He also, as the fit takes him, leaps on the female, bites her, and moves one of his legs from one to five times forwards, when the note runs to a species of cackle. The males also dash at one another; and when wearied with their solicitations, or annoyed at a bite, the portly female leaps off, one or the other follows on her bounds, and endeavours to arrest her with a twang on his violin. When united, the note is softly mitigated to a low squeak, " Weha ! "

Regarding the notation of the ordinary challenge-note of the Variable Grasshopper some confusion prevails. Fischer says that the male, when stridulating, rubs the femora simultaneously against the elytra, and gives out a powerful song, now and again rising with a metallic accent, and then decreasing. This sound may be imitated with the breath ; but those males who have the anterior area of their wing-covers less dilated stridu-late more gently. Yersin merely gives the notation he before attributes to the Red-legged Grasshopper, and adds, the stridor consists of from one to three notes. He likewise notices that when the female is present the legs vibrate before the note is heard, and then it arises shorter and softer. He next proceeds to give his studies of the music of two varieties, which in his time were considered distinct species. The greenish sport Char-pentier, in his " Horæ Entomologicæ," terms *mollis*, he says has a music composed of from twenty to thirty notes, with in-creasing intensity, the former lasting less than half a second, and more acute than the latter, which besides are much slower, so that it even happens that the eight or ten concluding sounds attain each nearly a second's duration. As regards other melanic deviations with uniform black elytra and sandy ochreous sides, occurring by the sea-shore and on mountain-slopes, he finds them to have a single note, lasting scarcely half or the third of a second, and not repeated till after a repose of double and triple that length, sometimes much longer; so that we may commonly reckon only one note from two grasshoppers in two seconds. For my own part, in the hay-fields of Italy, where during the month of August I first learnt to distinguish the Variable Grasshopper from its fellows by its habit of extending its wing-covers when leaping, I, by dint of watching, came to perceive that it meted its stridor to some eight slow and distinct double strokes, "Wheeh! wheeh! wheeh!" &c.; but on returning to the Surrey Downs in the late autumn of the same year, some straggling individuals lingering on in sheltered spots along the blackberry-hedges were only venting themselves in two short drowsy notes, "Whirr! whirr!" In the pairing season, beneath the Calais ramparts, I have remarked this music becomes considerably accelerated, and that it is prolonged to ten or eleven simultaneous strokes of the femora, lasting a quarter of a minute. The teeth on the femora producing the music are fewer than those enumerated previously for the grasshoppers with brisker action, numbering at most about seventy.

The various grasshoppers now noticed belong to a group where the keels on the pronotum of the thorax approach in an hour-glass form. In another division these ridges, distinguished often by their white or lighter tint anteriorly to the wings, approach somewhat, but less markedly. One of these, the common though badly named *Stenobothrus pratorum*, the beginner will perhaps earliest recognise by its thick cord-like antennæ and wing-covers coarsely netted in rectangles, resembling in shape the blade of a penknife. The head is a little prominent, and the hind wings when expanded are found to be often quite rudimentary and unserviceable for sustention in the air. As other green insects, this grass-stained grasshopper frequents shade, and lurks on heaths and among bush and scrub. Its prattle with that of the kind last noticed, belongs to the category where the snatches of music are composed of a succession of brief notes, repeated at perceptible intervals. If we trace the shadow thrown by a tree, according to Yersin, from

sun to shade, we find the individuals decrease the fulness of their
notes, but prolong their length and number, the strokes of the
hind legs attaining to twelve or fourteen in the dusk and the
stridor lasting from four to five seconds. So at morning the
male commences a single prolonged note, which only relapses
into the ordinary music when the sun appears and he has reposed
a little in its beams. At midsummer, in the Alps, the snatches
barely exceed a second, while during September they become
lengthened out to two. Generally we notice from seven to
eight notes, " Rééé! " or " Rrééé! " which increase in intensity ;
these last for about two seconds, with intervals of about three.
The male still continues its recitative until late in October over
heaths and on woodland greensward, where the sunbeams penetrate ; and here I have myself come upon him meting out a slow
series of four sounding and emphatic strokes, "Thiph! thiph!
thiph! thiph!" curtained by the deep umbrage and silent
gloom cast by afternoon shadow. Sundown even does not
invariably quench his chronic loquacity from out the damp
fungi and mouldering vegetation. The musical tubercles on the
hind legs of this wiry grasshopper I have portrayed, after Dr.
Landois, on Plate II., Fig. 7. In those examples I have myself
examined they were invariably less pointed, and in places were
apt to be wanting, irregular, or laterally supplemented. They
are rather wide apart as to insertion, and vary as to number
even in the same species ; Dr. Landois having enumerated in
one male ninety-three on the left femur and only eighty-five on
the right. The stridor is thought to arise from their impinging
on the elytral vein only when the hind legs are thrust downwards.

Stenobothrus lineatus, Panz., another species not uncommon
in fields among rank grass, is at first sight liable to be confounded
with this grasshopper ; but after a while one may learn to distinguish it by its delicate wing-covers and brighter hue, as also
by a white streak along the front margin of the abbreviated
wing-cases of the sluggish female—a characteristic that has
given rise to its trivial name. Both kinds are found in this
country, and have very similar distribution over the plains and
mountains of Europe ; though if any one on meeting with
lineatus in an idle hour should care to lounge on the grass and
observe its laconic ways, he would at once be struck with its

alien nature.   It is the type of indolence, often playing but on
one leg, or alternately moving both four times forwards and back
wards with pauses of about nine seconds.   Sometimes, however,
the overtures are prolonged; for, Yersin tells us, one note of
the male is always fuller and of a different timbre from the rest,
" In ! in ! in ! in ! in !"   These two notes which remind us of
the Piedmontese negative, together last about a second, and
the insect is said to repeat them each twenty times without
taking repose.   When in the presence of a female he kindles to
yet greater ardour.   Then both femora simultaneously spring to
action and move concisely and with sufficient rapidity to raise a
feeble note recalling that in " in," which continues for about a
minute.   But such tardy enthusiasm soon flickers and expires;
for if the female should chance to abscond or an observer to
approach, the music afterwards is recommenced in the ordinary
dilatory fashion, first with one leg and then with the other.
As the male walks and eats the grass-leaves, he also emits
a sound scarcely audible, of less than half a second's duration,
with one or the other femur, as if he found a satisfaction in
gastronomy.   Dr. Fischer in his work notates the music of this
grasshopper with the German breathing " Sch ! sch !—sch !
sch !" in place of the nasal sound adopted by Yersin, but he
gives no further particulars.   Herr Vitis Graber enumerates
the femoral musical tubercles at two hundred, and finds them
transverse elliptical callosities, giving the little row somewhat
the appearance of the wing-tiles of the crickets and leaf-crickets.
To their number he attributes the loudness of the music.

  Stenobothrus dorsatus, Lett., another prattler of this group
indigenous to Northern and Central Europe and likewise occurring in this country, commences its stridulation by drawing both
femora five times swiftly downwards, according to Fischer, with a
sharp sound, " Rrrt ! rrrt ! rrrt !" they then move up and down
to a gentler tune with somewhat this rhythm, " - - - - - UU, UU !"
&c.   Yersin says the male takes four simultaneous strokes of
the posterior legs, emitting as many short notes, " rrééé ; "
these are followed by another which is different in character,
more acute and prolonged, produced by a rapid alternating
movement of the legs, " tzin."   The music does not last more
than a second or a second and a half, and commonly the individual

M

repeats it a certain number of times, as many as twenty-five,
with intervals of about a second. When the male encounters a
female, even, it is said, should she be of a different species, he
suddenly stops, directs to her his antennæ, and approaches as
near as possible. This insect and canine formality being com-
pleted, he then commences to stridulate almost insensibly, and
will continue his snatches for as many as a hundred times.
The female the while usually remains immovable; but should she,
on the contrary, move off, she is then followed by the male, who,
if he lose sight of her, sounds the note "tzin," and seeks about
until he meets with other males or a female. The former partake
in his uneasiness, reply with the same note, and join in the chase;
but if disconcerted, and one should resume the ordinary music,
the others flag and similarly reply.

If this species has a superficial resemblance to our common
large green grasshopper, another (*Stenobothrus stigmaticus*,
Rambur) has decidedly the form of *S. pratorum*, with the colour-
pattern of *S. lineatus*. In the cabinet it may be discriminated
by a slight difference in venation, it is said, and by the structure of the antennæ, though to a more superficial observer,
perhaps, the green slash down the upper side of the femora and
the distinguishable black spots on the elytra will afford the
most tangible points. In its hardy habits and music, however,
it is quite distinct; and in Piedmont, where I have myself
listened to the males stridulating, its dashing "Whir, whir-he-
wee" constitutes one of the chief charms of the alluvial
meadows. This singular challenge, which I have even caused
an over-heedy individual to reply to in the hay-fields by whistling
in imitation, usually lasts some fifteen seconds, and is produced by
some four femoral strokes; but, as with other grasshoppers, during
its reproductive season in September, when engaged in soliciting
his female, a shorter and harsher pairing note is adopted, produced by a grating stroke from one or both hind legs—a sound
nevertheless as carefully attuned to vibrate through the female
chords of sense, as the brisker notes that are thrown out to
exasperate his fellows.

In a few grasshoppers we find the keels on the pronotum
parallel. *Stenobothrus elegans*, one of these kinds, rare in this
country, has a note first heard in June, which lasts from half a

second to a second, and this is repeated as many as five times
at intervals of a second. The species disappears in August. The
music of another (*Stenobothrus declivus*, Bris.) is heard only in
South and Central Europe. It is described as very similar, low
and intense, and commences about the same period of the year.
Herr Vitis Graber gives the teeth in the lima of *Stenobothrus
petraeus*, Brisout—a grasshopper I am unacquainted with—as
numbering over one hundred and thirty.

Opsomala brachypterus, Ocskay, which is common on the
Alps and Jura, has an acute note approaching that of the leaf-
crickets, and lasting less than half a second; a performance
nevertheless that is most feeble, owing to the shortness of the
elytra—"Ree." The music of *Stethophyma variegatus*, Sulzer,
likewise found on the Alps, normally consists of five notes, two
grave and short, lasting less than a second, one longer and
acuter, and then two others resembling the first—"Drrii, drrii,
iiiiiiiiiiii, drrii, drrii." In the morning, when the male commences his stridulation, only the grave note is heard. The
teeth in the lima of *Stauronotus flavicosta*, Fisch., Vitis Graber
gives as sixty-one.

The *Gomphoceri* are discriminated from other grasshoppers
by possessing a clubbed termination to the antennæ, most
marked in the females, a character which also distinguishes
moths from diurnal butterflies. The species vibrate their
femora some time before the sound of stridulation is caught.
Gomphocerus sibiricus, an interesting species, is found, according
to Yersin, on the mountains near Vevey, at 1,500 mètres above
the sea-level. The movement of the legs is short and rapid.
producing a note—"Tré, tré, tré, tré"—which is repeated two
hundred times at the rate of five to seven sounds a second. The
overture lasts from half a minute to a minute, but the last notes
of the series are often emitted lower, as if the grasshopper was
exhausted—"Ri, ri, ri." Fischer says the music of this species
is lisping, and notates it as "Ts! ts!" *Gomphocerus rufus*,
L., will produce a note resembling "Tiph! tiph!" the legs
oscillating at least ten seconds and ten times before the sound
is caught; while the swift vibration of the femora causes the
music produced by twenty-four strokes to have a duration of only
fifteen seconds. When soliciting a female, his legs are jerked

forward, " Wuf ! wuf !" The effeminacy of this grasshopper
has come under the observation of Fischer, who relates: " I have
sometimes seen the male stretch out his fore feet on the ground
with a most ridiculous gesture, sway his body backwards and
forwards, erect and wave his antennæ, and employ all his art of
flattery." Yersin says the note is rapidly trilled and silvery,
lasting from three to four seconds. The teeth in the femoral
limæ Herr Graber gives as one hundred and fifty. The male of
the minute and agile *Gomphocerus bigulatus*, L., I have observed
singing on bare sand-flats at Calais, during July, and later on
near Largs on the Clyde. The powerful femora move swiftly
as many as from eleven to twenty-one times on the elytra, and
at the fourth double stroke the sound is borne on the ear; it
then increases, and again lapses away, leaving a floating echo in
the air. The teeth on the ridges of the femora number from
one hundred and thirty to one hundred and forty. Largest at
the lower end, they decrease in size and become more thickly
placed toward the upper ; so that when they strike on the scapular
vein, the double note may be compared to running up and down
the octave, but the illusion of the stridulation increasing and
diminishing in intensity is more probably due to a variation in
the celerity of the movement of the femora. Yersin says of
this species that it emits from nine to twelve notes of about a
second's duration, and that these are slower and stronger at the
conclusion—" Vrrrééé." The legs vibrate imperceptibly, and
press on the elytra only during their forward movement.

Dr. Scudder gives us the stridulation of some of the Neartic
grasshoppers adapted to musical notation. One (*Stenobothrus
curtipennis*) produces about six notes per second, and continues
them from one and a half to two and a half seconds ; another
(*S. melanopleurus*) emits from nine to twelve notes in about
three seconds. The music of the latter when in the shade is
slower.

The South African genus *Pneumora* forms a third group of
the order, distinguished in their mode of stridulation from the
preceding. The males of these pass the inner side of the posterior
femora over a little semicircular row of oblong horny teeth (*f*)
placed on the skin of the abdomen, which in this sex is inflated
like a bladder, a circumstance that has earned the species the

name of Blazops from the Dutch colonists of the Cape. Unlike the rest of the order likewise, they are said to stridulate at night.

From the instances now cited, it will be sufficiently manifest that the music of the grasshoppers is due to the rivalry of the males, and that it fulfils the end of collecting and distributing, as well as of propagating, the races. As a love-note, supplanting the vocal music of the vertebrata at a period of pairing previous to the disappearance of the species, the employment of this mechanical stridor is also conspicuous.

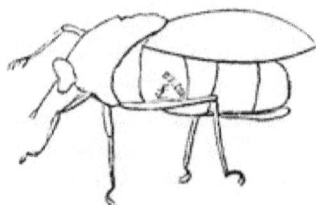

PNEUMORA.

The operation of the music in collecting and distributing the species will be inferred while listening to the nimble males singing and responding over the meadows in the sunshine, or analytically, on confining them in a well-ventilated receptacle, when the males will be seen spitefully approaching one another and chirping, repeating their stridor in snatches at midnight. Often when the masculine sex are thus wrangling, a little attention will be rewarded by bringing to view the newly-emerged females and pupæ, winding in among the grass-stems, and sluggishly nearing or following the performers, guided by and attending to the repartees. As there is a manifest tendency in orthopterous species to amalgamate during migration, due to indiscriminate jealousy, that allows a loud stridulator to collect all the others in a knot around him, the various means that diversify and appropriate the acridiideous music to the species, seen in the variation of the veins played on, or the graduated development and different position of the limæ, must be all-important; and it is significant that, although male grasshoppers of divers species may often be observed on grass-blades engaged in musical contests, or a solitary individual energetically responding to the ring of a leaf-cricket, each is able to recognise his proper and often very variably coloured partner by sight. And on the other hand I have observed two male *Stenobothri*, differing as to colour and species, when contending, draw two reluctant females, that approached together, but who were at once discriminated by their mates, who, after a murmur of recognition, leapt from their perches, and each bit the head of the female wearing his livery.

On Plate III., Fig. 8, are portrayed two males of the larger green grasshopper thus engaged in a musical contest on a grass-stem, and tracked by the larger female. In this country none of this family seem in any way injurious, and if we accept the green-coloured kinds, which as green leaves, seek dusky shade and moisture, where, with other green insects, they conserve their delicate charm, the relish of this family is wanting in an island with an equable temperature, for these are pre-eminently lovers of Nature parched and dry, manœuvring their armies on plains scorched by summer suns, led on by the chanticleer males quarrelling over the best pasture and winds of heaven, " that return again according to their circuits."

SONG OF THE GRASSHOPPERS.
From Landois' " Ton und Stimmapparate der Insecten."

GRYLLIDÆ (CRICKETS).

In this dull-coloured subterranean group of Orthoptera, the masculine charm that gathers together the locust legions to devastate grassy plains, and flocks of leaf-crickets to defoliate forest verdure, manifests itself as a social character, and the rival music of the males and subjection of the females originates associated burrows, where the sexes live solitary in angry neighbourhood. By their music they also express perception of rivalry and love, with a general notion of fear communicated by the auditory nerve, which finds musical expression. Thus the males of the field-crickets are diurnal, stridulating in sunshine. The music of the European kinds, *G. campestris, melas,* or *sylvestris,* is a brisk " Cree-cree !" loudest in the first. The sound increases when the males are in proximity, and is mitigated and broken by short sharp notes at equal intervals during the presence of the female; it ceases at the approach of an observer. The rhythm of the nocturnal house-cricket varies in pitch, and it has also an alarm-note, said to be understood by its kind. *Gryllotalpa* and *Œcanthus* are musical towards evening.

In *Locustina*, *Gryllidæ*, and *Phasmidæ* (?) stridulation results
from mutual friction of the elytra, which generally assume
masculine differential development in their membranes, with
tumerosity and induration of certain veins. Among the crickets
it is the male alone that sings; and their elytra folded, as in Fig. M,
page 151, have the whole membrane more or less tense and
glassy to act in the capacity of sounding-boards, and impress
the vibrations caused by the file on the included air. The file
does not vary much in regard to position; if we examine with
a lens the wing-covers of the House Cricket (*Gryllus domesticus*),
or of the Field Cricket (*G. campestris*), or of the Mole Cricket
(*Gryllotalpa vulgaris*), we shall find it well developed (Plate VI.,
Fig. 10) beneath either wing-cover, on the vein separating the
intermediate field from the anal, which is strong, horny, or chitineous, sunk in the membrane above, and prominent at the
under side, in the form of an italic *f*, and somewhat parabolic
(Plate II., Fig. 8, *l*). Now, as this vein constitutes a much more
elongate file than is found in the *Locustina*, its teeth are consequently more numerous; Dr. Landois estimates the number in
the field-cricket from one hundred and thirty-one to one hundred
and thirty-eight; in the house-cricket he finds about one hundred
and thirteen, and in the mole-cricket eighty to eighty-two. Their
shape also is different; for in the leaf-crickets they are formed
like fiddle-bridges, and here they are solid triangles. (Compare
Fig. 4c with Figs. 4A, 4B.) This vein, then, being the bow,
shrill-vein, or file, where is the clasp over which it sounds? If
we next examine the part above, where the file of either elytron
scrapes when the wing-covers are rubbed together, we shall not
fail to find this in a branch the notched vein gives off towards
the base of the wing-cover, which is smooth and prominent, and
along which the file works a little obliquely (Fig. 8, *s*).

"The field-cricket," says Colonel Goureau, "is common
in the province of Gex, where the warm and sandy soil is very
favourable to its increase. The larva is produced from an egg of
a dirty-whitish colour at the end of July, and these inhabit a
little hole scooped in the soil. At the entrance of this they conceal themselves. At this period of their lives they are sometimes met with in the evening during twilight, collected together
in great numbers, and crossing roads and footways, leaping like

toads.] These young insects then pass the winter in these holes, protected by a stone which, frequently covers them. About the 10th of March, the crickets reappear at the mouths of their cells, which they then open and bore, and shape very elegantly with their strong jaws, toothed like the shears of a lobster's claws. Of such herbs as grow before the mouths of their burrows they eat indiscriminately; and on a little platform, which they make just by, they drop their dung, and never in the day-time seem to stir more than two or three inches from home. Sitting at the entrance of their caverns, they chirp all night as well as all day, from the middle of the month of May to the middle of July; and in hot weather, when they are most vigorous, they make the hills echo, and in the still hours of darkness may be heard to a considerable distance. In the beginning of the season their notes are more faint and inward, but become louder as the summer advances, and so die away again by degrees. In August their holes begin to be obliterated.

"When they quit the covering of the pupa in April, they are white and soft, and incapable of producing sounds; soon, however, their colour deepens, their elytra become firm and sonorous, 'and they stridulate. The male alone possesses the power of stridulation; he makes use of it to attract and please the female. Placing himself at the entrance of his habitation, he begins by stretching out his legs, placing his breast against the ground, at the same time slightly elevating the abdomen; in this attitude he raises his wing-covers, and rubs them briskly against each other, incessantly repeating his song, which is loud, sharp, short, and monotonous. When a female, attracted by his music, approaches, he advances towards her, touches her with his antennæ, and modifies his accents; his song becomes softer and less loud, and is interrupted by a short, sharp sound, occurring at frequent intervals of equal length. The crickets then take several little turns about the habitation of the male, from which they do not go far. He precedes his mate, walking with short steps, if I may be allowed the expression, *en rampant.*

"When at liberty these crickets are very timid, and they are not easily surprised whilst engaged in singing, or in the execution of the other functions of their lives. On the least noise they are immediately silent, and run into their holes; and one is surprised, in passing through a country abounding in these insects, to hear the songs cease as you advance. But if you confine a male and female in a box, they soon become familiar, and an opportunity is afforded of observing their amours, and listening to their song. It is a good plan to shut up two males and one female, for the jealousy between the former makes them redouble their ardour. They at first keep at some distance, and call the female with loud songs; when they meet they fight, seizing each other with their strong jaws. Mostly one of them falls a victim, and is devoured."

Whether the house-cricket be indigenous and degenerate, or whether it be an imported insect, is a question; for, although it is accused of living in garden walls, cracks in the earth, and hedge-banks during the summer, in the autumn, as far as my experience goes, it does not survive a removal from the fire. Its powers of burrowing are small, and if not enfeebled by domestication, it is probable that it is its nature to seek some such ready concealment as a crack or fissure. At all events, the history of its introduction into houses forms an interesting problem that awaits solution. The male cricket does not arrive at the perfect state, and begin its song, till the latter portion of the year, and sings during the autumn and through the winter months, thus surviving the oviposition of the female and fresh brood that emerges in October.

"Tender insects," says Gilbert White, "that live abroad, either enjoy only the short period of one summer, or else doze away the cold uncomfortable months in profound slumbers; but these, residing as it were in a torrid zone, are always alert and merry; a good Christmas fire is to them like the heats of the dog-days. Though they are frequently heard by day, yet is their natural time of motion only in the night. As soon as it grows dusk, the chirping increases, and they come running forth, and are from the size of a flea to that of their full stature. As one should suppose, from the burning atmosphere which they inhabit, they are a thirsty race, and show a great propensity for liquids, being found frequently drowned in pans of water, milk, broth, or the like.

" In the summer we have observed them to fly when it became dusk out of the windows and over the neighbouring roofs.

This feat of activity accounts for the sudden manner in which they often leave their haunts, as it does for the method by which they come to houses where they were not known before. It is remarkable that many sorts of insects seem never to use their wings but when they have a mind to shift their quarters and settle new colonies. When in the air, they move *rolatu undoso*, in waves or curves, like woodpeckers, opening and shutting their wings at every stroke, and so are always rising or sinking. When they increase to a degree, as they did once in the house where I am now writing, they become noisome pests, flying into the candles, and dashing into people's faces."

When the male sings he elevates the wing-covers so as to form an angle with the body, and then rubs them against each other by a horizontal and very brisk motion. The music is capable of considerable modulation, and as in the case of the Leaf-crickets, the chirps come quicker and angrily when male meets male, especially if a female, the object of all dispute, be present. For though from a distance the note falls on the ear with a monotonous sound of "Cree-cree!" if the observer will station himself near the hearth when these insects are briskly stridulating on a frosty morning, at which time the air is especially sound-transparent, and they most noisy, two notes may be distinguished—a loud "Awhit! awhit! awhit!" uttered in unison with the ticks of the clock, and a lower occasional "Wee, wee!" uttered hastily, I presume, when Greek meets Greek. There is also an alarm-note, "Chreek!" at which it is said all scamper to their retreats. If the wanton sounds of the male should allure a female, she intimates her presence with her long feelers, and the two engage in antennal discourse; but tempers are quick in crickets, and the crannies where they lurk become sad dens of cannibalism, since the males, females, and immature pupæ alike bite and devour one another, although their rapacity is probably mitigated by the provision of music and their proverbial nimbleness.

The third indigenous *Gryllus*, the Little Wood-cricket, has the musical mechanism as before, but differs in that the right wing-case has a firmer consistence than the left; so that whereas the former species, in singing, rub the elytra together with either uppermost, and even change them during stridulation, this cricket is thought to be capable of producing its notes only when the right is below, the exact converse to the rule that holds with the leaf-crickets. The cry, "Ru!" or "Rrruu!" Yersin says, is feeble, scarcely lasting half or quarter of a second. The wood-cricket, rare in this country, sings in the autumn; but being destitute of the burrowing hand-like fore-tarsi of the mole-crickets, and excavating jaws of other *Grylli*, Nature has destined it to a life beneath the fallen leaves of the forest. Colonel Goureau says, " Besides these two species another is found in the province of Gex, the Wood Cricket (*Gryllus sylvestris*), which does not appear at the same time as the field-cricket. Its larvæ are seen in the spring, and the perfect insect from near the end of August to the beginning of winter. Some individuals would appear to survive this rigorous season, as they have been found under stones in the month of February. I have not observed that it inhabits a burrow; I have always found it under stones, or on grass under trees at the foot of mountains. Placed in a box with a female, the actions of this insect resemble those of the field-cricket. The male approaches the female frequently, extends his legs, places his breast against the ground, and elevates the abdomen. In this position he raises the elytra, and rubs them with rapidity against each other. A feeble mono-tonous noise results, very different from the short, sharp sound produced by the field-cricket on a like occasion." Fischer thus describes its song :—"The male, hid beneath leaves, adjusts itself like others of the genus, with outstretched legs, and for many minutes emits a rhythm in this fashion : '- ∪ - - - - - - - ∪ - ∪ - ∪ - - - ∪ - -.' This, although the insect is small, is heard at a distance of some yards, and that not only in the evening, but also during the afternoon. Some males I had caged in my room were accustomed to retire beneath a dry leaf provided them for the purpose, and there sing. The stridulation is compara-tively more sonorous than that of the leaf-crickets of the genera *Decticus* and *Locusta*. *Gryllus melas* (Charp.), a black species intermediate in size between the field-cricket and wood-cricket, is exceedingly plentiful in the meadows near Turin. It has the habits of the first, is equally lively, and fond of basking in the light and movement of the sun, and its music, heard during July and August, is a very similar 'Cree-cree.'"

"While," says Gilbert White, "the field-cricket rejoices amid the glowing heat of the kitchen hearth or oven, the Mole Cricket haunts moist meadows, and frequents the sides of ponds and banks of streams, performing all its functions in a swampy, wet soil. With a pair of fore-feet curiously adapted to the purpose, it burrows and works underground like the mole, raising a ridge as it proceeds, but seldom throwing up hillocks. In fine weather, about the middle of April, and just at the close of the day, they begin to solace themselves with a low, dull, jarring note, continued for a long time without interruption, and not unlike the chattering of the fern-owl, or goatsucker, but more inward." Another observer thinks the song more shrill, but softer, than the croak of the frogs; and it has been compared to the note of the tree-frog, and to that of the Corn-crake. Latreille thinks it soft and pleasing. Yersin calls it a single grave but feeble note, "Rié, rié," which, when the cricket is seized in the fingers, is changed to acute and short cries of "Ié, ié, ié," indicating perception of fear, the cricket, at the same time, ejecting an offensive liquid from the anus. The ordinary stridulation can be heard at a distance of from one hundred to one hundred and fifty feet. About the middle of May the female cricket lays her eggs in a subterranean chamber shaped like a bottle with a curved neck, neatly smoothed and rounded, about the size of a moderate snuff-box, with many caverns and winding passages communicating.

The mole-cricket common in the United States (*Gryllotalpa borealis*, Burm.), according to Dr. Scudder, commences its daily chirp between two and four p.m., but stridulates more actively at dusk. It is only about half the size of the European species, yet can be heard at a distance of five rods. Its chirp—a guttural infantile trill, "Gru," or "Green," sounding exceedingly like the croak of the toads at the spawning-season—lasts for two or three minutes at a time, and is pitched at two octaves above the middle C. The separate notes are usually repeated at the rate of one hundred and thirty to one hundred and thirty-five a minute, but when many individuals are singing together their rate of utterance is increased to a hundred and fifty a minute. The sharp, querulous plaint of the crickets, resembling the first few strokes of a professional violinist, thus differ greatly from the soft trill of the mole-cricket, corresponding to the fuller play of the fiddle-bow; and this is not alone due to a difference in the method of stridulation, but appears, as far as tone is concerned, directly owing to the formation of the elytra, as in the one we find the membrane hard and tense, giving rise to large glassy areas similar to those which exist in the leaf-crickets, while in the other the membrane is softer and less elastic, presenting fewer dilated fields.

Since the sonorous areas in this tribe are not here confined to the anal field, but the whole of the wing-covers, and flattened central disc more especially, partake of their glassy nature, it naturally follows, among species provided with such effective acoustic boxes, individuals should occur that rival the Locustina alike in power and modulation of tone. And if we also consider the tenseness of the wing-covers in the various species, the size of the teeth in the wing-file, their shape, and variable number, we shall see ample cause why the song of each should possess an individual character, and contribute to an extensive scale of cricket-music. Some, indeed, are truly powerful musicians, as the *Œcanthi*, tiny Continental crickets, not, I believe, occurring in this country, which are remarkable on account of the great angle to which the males are accustomed to raise their elytra when stridulating; and this, Herr Fischer considers, has the effect of a sounding-board, carrying their trill, at sunset, to a distance truly wonderful for such weak and tender insects. The larvæ of *O. pellucens* emerge in June, and the perfect insects may be found beneath the leaves of bramble and other bushes during August and September. In appearance they might be easily mistaken for the Lace-wing Flies, on account of their narrow, fragile wings, which confer a very fly-like look; and it has even been said they are in the habit of fluttering about flowers. Their song, which has a rhythm, "ᑌ ᑌ ᑌ ᑌ ‑ ᑌ ᑌ ᑌ ᑌ," ceases on the approach of an observer, showing that, in spite of their diminutive size, their perception of sound is yet keen.

The loudest of all European crickets is the bold and savage *Brachytypus megacephalus*, whose male is heard at the middle of April, on sandy flats near the sea in the Val de Noto, in Sicily. It begins at the entrance of its burrow about four in the after-

noon, and its music is not interrupted like that of the common
field-cricket, but more sustained, sonorous, and clear. The
notes are repeated every second, and can be heard distinctly
at the distance of a mile. In appearance, it has the look of
a gigantic house-cricket. The *Gryllus pipens*, a loud-singing
Andalusian cricket, warbles melodiously as a thrush at twilight
and daybreak among the lower hills of Arragon and Catalonia,
his bird-like gush cheating the fowler in pursuit of game. But
others, again, are thought not to be capable of singing, as the
kinds of *Nya*, of which one, a small digging species akin to the
mole-crickets, inhabits the sandy river-banks in Southern Europe,
or the members of the genus *Trigonidium*, that extend to the
Mauritius and Java. One of the latter, erroneously indicated by
the Rev. J. A. Marshall as confined to Corsica and the neighbouring islands, is described by him as a beautiful little cricket,
shining black, with red hind femora.

Dr. Scudder has made some observations on the crickets in the
neighbourhood of Boston, New England. There the Spotted
Cricket (*Nemobius vittatus*) appears simultaneously with the
Black Cricket (*Gryllus niger*). The chirping of the two insects
is very similar, "Crrri;" but that of the former may be better
expressed by " R-r-r-u," with the French pronunciation. One of
these insects was once observed while singing to its mate. At
first the music was mild, and frequently broken; afterwards it
grew impetuous, forcible, and more prolonged; then it decreased
in volume and extent, until it became quite soft and feeble.
At this point the male began to approach the female, uttering
a series of twittering chirps; the female ran away, and the male,
after a short chase, returned to its old haunt, singing with the
same vigour, but with frequent pauses.

GRESSORIA.

The leaping genera have their respective prototypes in the
walking and running kinds, with which they present more or
less perfect continuity, but the latter generally want a musical
apparatus, or have it poorly developed. Mr. Wood Mason has,
however, noticed the probable existence of such an organ in a
female of the *Phasmidæ*, or Stick Insects (*Pterinoxylus difformis* of
Serville), in connection with which he mentions a mirror, or

talc-like spot; and the same gentleman, in a subsequent communication to the Entomological Society, considers certain of the *Mantidæ*,* or Leaf Insects, stridulate by rubbing the abdomen against the lamellar expansions at the anterior margins of the wing-covers, which, when the latter are in repose, are inflexed beneath it. These extensions are in both sexes converted along a greater or lesser portion of their length into highly indurated, erect, and obtuse teeth, which, being furnished with lateral setæ, appear morphologically identical with the microscopically small, blunt serratures seen in a Malayan kind. He also stated he was informed a certain Indian Leaf Insect, when tormented, kept making a "hissing noise," without obvious movement of the alar organs—a notion previously, though seemingly incorrectly, advanced in respect of the common European Preying Mantis, by Colonel Goureau, who says that when this species is alarmed on a tree, it places itself in an attitude of defence, and rubs the sides of the abdomen against the interior borders of the wings and elytra, so as to produce a noise like that of pieces of parchment rubbed together.

CURSORIA.

Individuals of a West Indian cockroach, nicknamed the Drummer, and identified lately as *Panchlora Maderæ*, F., are frequently mentioned in books of the eighteenth century, as responding from wooden structures in houses or in trading vessels, with a reiterated drumming or knocking. This nocturnal sound, likened by the Rev. J. A. Marshall to the "chur" of a very distant nightjar, is said not to be produced by the species when in confinement, or when they find themselves observed, fear then serving as a check on its emission.

The stridulation of spiders and crabs is similar in production, and appears to fulfil the same ends as in insects. The first crepitate from fear or anger. Westring artificially discovered that several poisonous black spiders decorated with blood-red spots,† found in the vineyards of Southern Europe and elsewhere, have the power of making a very feeble sound, while their females are mute. This they effect by means of a serrated

---

* *Empusidæ.*

† *Theridon (Asagena,* Lund) *serratipes, quadri-punctatum, guttatum.*

ridge at the base of the abdomen, which is rubbed against the hard hinder part of the thorax. Regarding the crepitation of the *Mygale stridulans* of Assam, Mr. S. E. Peal relates: " The noise is both peculiar and loud; it resembles that made by pouring out small shot on to a plate from a height of a few inches, or, better still, by drawing the back of a knife along the edge of a strong comb. The stridulation is very distinct, and has a ring about it which I do not notice in the Orthoptera, wherein it more closely resembles a whistling sound. It is now some six years since I first heard it, and under the following circumstances : Some Assamese were cutting out an old bamboo-clump, the ground under which was dry and full of decayed roots, and of holes. White ants had made a nest there, and I had collected several ' queens.' While attending to these, with my back to the clump, at a distance of some four or five feet, I suddenly heard this peculiar noise, and turning, saw the man who was hoeing the mound making futile blows with his hoe at a huge black spider that kept up this curious sound; but the ground fortunately being uneven, none of the blows took effect, and I soon secured the prize. On reaching the bungalow, I undid the cloth in which it had hastily been secured beneath an inverted tea-sieve. On stirring the cloth the spider ran out, whereupon my cat pounced forward; but the spider, instead of retreating, ran round and round inside its prison, following the movements of the cat and stridulating louder than ever. When thus roused the spider usually rested on the four posterior legs, raising the other four and shaking them in the air, with the thorax thrown up almost at right angles to the abdomen, and the cheliceræ in rapid motion." The sound-producing apparatus in this Mygale has been found to consist of a comb composed of a number of highly elastic and indurated club-shaped chitineous rods arranged close together comb-like on the inner face of the basal joint of the palpi, and of a *clasp*, formed by an irregular row of sharp erect spines on the outer surface of the penultimate joint of the cheliceræ, and equally well developed in the two sexes.

Two large scorpions allied to *S. afer*, obtained by Mr. Wood Mason from some Hindustani conjurors, when fixed face to face on a light metal table, and goaded into fury, commenced to beat the air with their palpi, and simultaneously to emit sounds resembling those produced by scraping a piece of silk-woven fabric, or a stiff tooth-brush, with one's finger-nails. The musical apparatus of these and other scorpions, from which notes can be elicited in dry and alcoholic specimens by friction of the parts, is stated to be duplicate. The *clasp*, thickly beset with stout, conical, curved and sharp spinules, is situated on a slightly raised oval area, of lighter coloration than the surrounding chitine, and placed outwardly, at the base of the basal joint of the palp-fingers ; while the *lima*, crowdedly studded with minute tubercles, shaped like the tops of mushrooms, is similarly situated on the produced inner face of the corresponding joint of the first pair of legs.

Crustacea crepitate when alarmed, &c. Mr. Saville Kent found that a small *Spheroma*, a species of the Isopodous order, which he kept in a glass jar, made a sharp "tapping" sound, produced three or four times consecutively with intervals of about one second's duration ; but he failed to determine the cause, as on being approached this little creature always eluded notice by passing to the opposite side of his stalk of seaweed. Among the Decapodous crustaceans, the genus *Alpheus* and allies produce "clicking" noises beneath the water, by a sudden extension of the terminal joint of their larger claw, familiar to those who have searched for animals on coral-reefs, or dredged in tropical seas. One collected by Mr. Kent in Guernsey, *Alpheus ruber*, emitted a "snapping" noise audible at a considerable distance, that at once betrayed its lurking-place. The large sea crayfish, or thorny lobster of the London and Paris fish-markets, is stated to emit, on handling, a "shrill squeaking" sound by rubbing together the spinous abdominal segments. The organs of stridulation in Crustacea, according to Mr. Wood Mason, are paired as in Arachnida and Insecta. In some the scrapers are on the carapace, and the rasps on a pair of appendages, as in both sexes of *Matuta* ; or the structures are transposed, as in the males of *Macrophthalmus* and allies, in which the scraper is formed by a sharp-edged lamellar projection on the meropodite of each cheliped, and the rasp is the crenulated infra-orbital margin. In other species, the *lima* and *clasp* are on different parts of the same appendages, as in male *Ocypoda* ; or on two pairs of appendages, as in those of *Platyonchus bipsululosus*.

N

A TABLE OF GENERA CONTAINING INSECTS THAT STRIDULATE.

COLEOPTERA, OR BEETLES.

Rule :—Both sexes have well-developed files, and stridulate.

GENERA.	POSITION OF FILES.	POSITION OF CLASPS.	MECHANISM: BY WHOM INDICATED OR DESCRIBED.	BY WHOM HEARD TO STRIDULATE.
GEODEPHAGA.				
Megacephala ?	...	...	...	Saîtes à Buffon.
Euryprospus ?	...	...	...	Saîtes à Buffon.
Oxychella ?	...	...	...	Saîtes à Buffon.
Cuclicus ...	...	...	...	♂'s stridulate?
Cychrus, Fab. ...	In marginal grooves of elytra (in which they rub)	Edges of inferior plates of abdomen	Rev. T. Marshall, "Ent. Mag.," 1881, vol. i., p. 213	T. Marshall, F. Smith, &c.
Elaphrus, Fab. ...	On the last segment of the abdomen (which they rub against)	A projection beneath elytra	Westring, "Naturhist. Tids.," Kröyer, Bind. 2, 1846-49	G. R. Crotch?
Methisa, Bonelli...	On the last segment of the abdomen (which they rub against)	A projection beneath elytra	Westring, "Naturhist. Tids.," Kröyer, Bind. 2, 1846-49	Thorell (quot. by Westring).
Cerapteras...	...	...	Thwaites	"Trans. Ent. Soc.," s. 2, v. 2, proc. p. 2, 1852.
HYDRADEPHAGA.				
Pœlobius, Schön	At the extremity of the elytra, beneath (over which they rub)	The last segment of the abdomen	W. L. Schmidt, "Stett. Ent. Zeit.," 1840, T. 1, pp. 10-12.	See "Ent. Month. Mag.," vol. viii., 1872, p. 69; "Entomol.," vol. xi., p. 255.
Dytiscus ? Geof. ...	...	...	A. L. Dufour	
Colymbetes, Clairv.	At the extremity of the elytra, beneath (over which they rub)	The last segment of the abdomen	...	A. G. Larker, "Ent. Mag.," vol. xii., 1879, p. 21.
Acilius, Leach ...	At the extremity of the elytra, beneath (over which they rub)	The last segment of the abdomen	...	A. G. Larker, "Ent. Mag.," vol. xii., 1879, p. 21.
PHILHYDRIDA.				
Heterocerus, Bosc, ?	At the sides of abdomen, on first ventral segment (which they rub with)	The hind femora	Erichson, "Natur, und Ins Deutschl.," 3, 539; Schiödte, trans. "An. and Mag. Nat. Hist.," 3rd series, vol. xx., p. 37, 1867	—
Physites ? ...				
Augytes ? ...				
Spercheus, Fab. ...	At the extremity of the elytra, beneath (over which they rub)	The last segment of the abdomen	Westring, "Göthcborg's Kongl. Vet. und Vit.," iv., Haf. s. 47	Leprieur (quot. by Westring).

A TABLE OF GENERA CONTAINING INSECTS THAT STRIDULATE—continued.

COLEOPTERA, OR BEETLES—continued.

GENERA.	POSITION OF FILES.	POSITION OF CLASPS.	MECHANISM: BY WHOM AND WHERE INDICATED OR DESCRIBED.	BY WHOM HEARD TO STRIDULATE.
PHILHYDRIDA.				
Berosus, Germ. ...	At the extremity of the elytra, beneath (over which they rub)	The last segment of the abdomen ...	Westring, "Götheborg's Kongl. Vet. und Vit., iv., Haf. s. 47	Fkeberg (quot. by Westring).
Hydrophilus, Geoffr.	At the extremity of the elytra, beneath (over which they rub)	The last segment of the abdomen ...	A. H. Swinton ...	(Mihi) H. piceus L.
NECROPHAGA.				
Necrophorus, Fab.	On fifth segment of abdomen (which they rub against)	Hind edges of elytra	Landois, Goureau ...	Stridulate readily on seizure.
Lethrustes ...				F. Darwin, "Des. of Man," vol. i., chap. x.
LAMELLICORNIA.				
Chiasognathus ...	At extremity of abdomen (which they rub against)	Elytra ...		Darwin, "Des. of Man," vol. i., chap. x.
Passalus, Fab. ...	On inferior surface of hind coxæ?			Moquerys, "Soc. Ent. de France," 1814; Séance du 28th Sept. M. P. de la Brulerie, quot. by Darwin. Westring.
Ateuchus, Web.& Fab.	On thickening at base of outer margin of elytra (which they rub against)?	Metathorax?		Stridulate readily on seizure?
Gymnopleurus ...				
Copris, Geof.		Metathorax?	Darwin, "Descent of Man," vol. i., chap. x.	
Other Coprini ...	On dorsal surface of abdomen		Lecomte, "Int. to Ent.," pp.101, 113	
Geotrupes, Lat. ...	On coxa of hind legs (which they rub over)	Ridge on the third segment of the abdomen	Landois, "Zeitschr. für Wis. Zoologie," xvii., Bd.	Stridulate readily on seizure.
Typhœus, Leach...	On coxa of hind legs (which they rub over)	Ridge on the third segment of the abdomen	Darwin, "Descent of Man," vol. i., chap. x.	Stridulate readily on seizure.
Trox, Fab. ...	In marginal grooves of elytra (in which they rub) ...	The edges of the abdomen?	Westring ...	
Oryctes, Illig.	On last segment but one of abdomen (which they rub on) ...	The elytra	Lecomte, "Int. to Ent.," pp. 101, 113, quot. by Darwin	Kirby, F. Smith.

A TABLE OF GENERA CONTAINING INSECTS THAT STRIDULATE—continued.

COLEOPTERA, OR BEETLES—continued.

GENERA	POSITION OF FILES	POSITION OF CLASPS	MECHANISM: By whom described and where indicated	By whom heard to stridulate
LAMELLICORNIA.				
Scarabæus, Lat.; Dynastes, MacLeay	On last segment but one of abdomen (which they rub on)	The elytra		M. Lacordaire, quot. "Nat. Library," vol. iii. p. 206.
Euchirus	At sutural margin of elytra (beneath which they rub)	The abdomen	Darwin, "Descent of Man," vol. i. chap. x.... Wood-Mason, "Trans. Ent. Soc.," 1878, p. liv.	
Pintelidæ				P. H. Gosse, "Canad. Naturalist," p. 272.
Gymnoitus				
Series, MacLeay(Linna single)	On prothorax beneath (which they move over)	Sternum of metathorax	Westring, "Naturhist. Tids. Kreyer," Bind 2. 1846-19	Westring.
Melolontha, &c.	In curved groove at either side of last segment but one of abdomen (which they rub over)...	The tip of the elytra	Roesel, "Insect Belustigung," T. 4. p. 298; E. Blanchard, "Metamor. Mœurs of Inst. des Insectes," Paris, 1868	Roesel cites M. Fullo as a stridulator.
Lomaptera?	At side of second abdominal segment (which they scrape with)	The hind femora?	Dr. Sharp, "Ent. Month. Mag.," vol. xi. p. 126	
Cetonia, Fab.	At either side of last segment but one of abdomen (which they rub over?	The tip of the elytra?		
MALACODERMATA.				
Anobium, Fab.	Beneath the apices of the elytra (on which they rub)	The abdomen?	Westring "Göteberg's Kong. Vet. och Vitter. Hand.," iv. Heft 8. 6	Often heard in old timber, called Death Watches.
HETEROMERA.				
Pimelia? Moluris	On second segment of abdomen		Latreille, quot. by Westring "Oliv. Entomol.," 1. Pref. ix. Westring, "Göteberg's Kong. Vet och Vitter Hand." N. Haf. s. 17	
Blaps? Fab.				
Heliopathes	Exception, male alone can creak. In males, on last segment of the abdomen (which they rub beneath). In female, absent.	The elytra	G. R. Crotch; Darwin, "Des. of Man"	

A TABLE OF GENERA CONTAINING INSECTS THAT STRIDULATE—*continued.*
COLEOPTERA, OR BEETLES—*continued.*

GENERA.	POSITION OF FILES.	POSITION OF CLASPS.	MECHANISM: By whom and where indicated or described.	BY WHOM HEARD TO STRIDULATE.
RYNCHOPHORA. (Tibia single.)				
Plinthus	Beneath apices of elytra (on which they rub) ...	Last abdominal segment, or *vice versa?*	T. V. Wollaston, "An. and Mag. Nat. Hist.," vol. iv., p. 11, 1860	T. V. Wollaston and Bewicke.
Acalles	Beneath apices of elytra (on which they rub) ...	Last abdominal segment, or *vice versa?*	F. Smith (verb.) ...	See F. Smith, "Zoologist," Oct., 1865, p. 7747. F. Smith—Darwin, "Des. of Man," vol. i., chap. x.; Lister, "Ray Hist. Ins.," sup. p. 383 (C. lapathi).
Mononychus ...	... ...	...	...	
Cryptorhynchus ...	Beneath apices of elytra (on which they rub) ...	Last abdominal segment, or *vice versa?*	Westring, "Naturhist. Tids. Kröyer," Bind. 2.	Westring, Wollaston.
Ceutorhynchus ...	Beneath apices of elytra (on which they rub) ...	Last abdominal segment, or *vice versa?*	Westring, "Naturhist. Tids. Kröyer," Bind. 2.	
Erirhinus? ...	Beneath apices of elytra (on which they rub) ...	Last abdominal segment, or *vice versa?*	Westring, "Naturhist. Tids. Kröyer," Bind. 2.	Aleen, "Edin. Month. Mag.," Nov., 1864, p. 130.
Scolytus ...	... ...	...	Kröyer, "Naturhist. Tids. Kröyer," Bind. 2.	
LONGICORNIA. (Tibia single.) Fam. Cerambycidae ...	... ...	Hinder edge of prothorax ...	...	Stridulate readily on seizure.
Cerambyx, Linn. ...	On mesothorax (over which they rub) ...	Hinder edge of prothorax ...	Landois, "Zeits für Wis. Zoologic," Bd. xvii.	
Astynomus, de Jean ...	On mesothorax (over which they rub) ...	Hinder edge of prothorax ...		
Clytus, Fab. ...	On mesothorax (over which they rub) ...	Hinder edge of prothorax ...		
Aromia, Ser. ...	On mesothorax (over which they rub) ...	Hinder edge of prothorax ...		
Deucalion ...	On mesothorax (over which they rub) ...	Hinder edge of prothorax ...	Wollaston, "Ins. Mad.," p. 452	Smiles à Buffon, "Coleop.," T. viii., p. 11, C. O. Waterhouse (verb.).
Lamia	On mesothorax (over which they rub) ...	Hinder edge of prothorax ...		
Monochamus, Megerle				
Saperda, Fab.				

A TABLE OF GENERA CONTAINING INSECTS THAT STRIDULATE—continued.

COLEOPTERA, OR BEETLES—concluded.

GENERA.	POSITION OF FILES.	POSITION OF CLASPS.	MECHANISM: By whom and where indicated or described.	By whom heard to stridulate.
LONGICORNIA. (Lina single.) Acrocinus ...	...	...	...	"Anal. des Scien. Natur.," xxi. p. 180.
Dorcadion, Dalman				
Fam. Lepturidæ ...		...		
Grammoptera ?	...		Landois, "Zeitschr. für wis. Zoologie," xvii. Bd. ...	
Leptura ...				
Pachyta, Megerle ...				
PHYTOPHAGA. (Lina often single.) Megalopus ?	...	...		
Donacia ? ...	At the side of ridges on metathorax (that lock the) ...	Sutural margin of elytra?	Westring, Liebe, "Der Zoologische Garten," Frankfurt, 1871; Haller ...	Some of this genus stridulate readily.
Crioceris ...				
Hispa ? ...	...	...		
Epilachna ?	...			(Mihi.)
Clythra, Latch ...	On last segment of abdomen (which they rub beneath) ...	The elytra	G. R. Crotch; Darwin, "Des. of Man," vol. i., chap. x.	

A TABLE OF GENERA CONTAINING INSECTS THAT STRIDULATE—continued.

ORTHOPTERA.

Rule:—Male alone has a well-developed file, and stridulates; female is mute.

GENERA.	POSITION OF FILES.	POSITION OF CLASPS.	MECHANISM: By whom and where indicated or described.	BY WHOM HEARD TO STRIDULATE.
BLATTIDÆ. Blatta	...	...	...	Kirby & Spence, "Lettr.," xxiv., B. gigantea. *Special Reference.*
GRYLLIDÆ. Gryllus	Beneath either elytron, on a projecting vein crossing the intermediate field, only developed on the right elytron in C. sylvestris (which they rub on)	The opposite elytron, over a branchlet of same vein	Gourean, "Annal. de la Soc. Ent. de France," 1837, p. 31; Frisch, "Beschr. v. Ins. in Deutschland, 1766; White, "Nat. Hist. of Selborne," vol. ii., 1825, p. 262, &c.	Most of the males stridulate when placed in the sun, subjected to heat, or when two of a species are enclosed together.
Œcanthus	...	...	Fischer, "Orthoptera Europæa Landois, "Zeit. für Wissenschaft Zool., 1818; Yersin, "Bull. Soc. Vaudoise," 1856, p. 108; Scudder, "American Nat.," vol. ii., 1868; ditto, vol. v., p. 97; "Zool. Record," 1867, p. 460.	
Brachytypus	...	...		
Gryllotalpa, Latr.	...	...	J. G. Wood. "Ins. at Home," p. 211, 242; Gourean, "Annal. de la Soc. Ent. de France," 1837, p. 31; Gilbert White, "Nat. Hist. of Selborne," &c.	May be heard at evening.
			Papers on the Stridulation of Locustidæ.	*Special Reference.*
LOCUSTIDÆ. Odontura, Ramb.	Beneath the left elytron, on a vein crossing the anal field	The rounded edge at base of opposite elytron	Described by Gourean in 1835-37, Newport 1839, Goldfuss 1843, Siebold 1844, Westring 1845. De Geer, "Mém. pour Servir." vol. iii., p. 423.	These insects may be heard to sing in a state of freedom, or will readily perform if two males, or a male and female, be enclosed together.

A TABLE OF GENERA CONTAINING INSECTS THAT STRIDULATE—continued.

ORTHOPTERA—continued.

Genera.	Position of Files.	Position of Class.	Mechanism: By whom and where indicated or described.	By whom heard to stridulate.
Locustidæ.				
Xiphidium, Serv.	...	...		—
Locusta	... ...	...	Goureau, "Annal. de la Soc. Ent. de France," 1837.	
Phaneroptera ...	Female also emits a sound when closing its elytra, due to a rough vein ...	...	Fischer, "Orthoptera Europea," Lip., 1853	Scudder, "American Nat." vol. ii., 1868 : Scudder, "Proc. Bost. Soc. Nat. Hist.," vol. xi., 1868.
Orchelimum ...	... ...		Fischer, "Wiener Entomol. Monatschr.," b. i., sft., p. 360	
Thamnotrizon, Fisch.	... ...	...	Newport, "Tod's Cyclopæd. of Anat. and Phys.," vol. ii., p. 925.	
Decticus, Serv. ...	...	...	Landois, "Zeit. für Wiss. Zool.," Leip., 1818, b. xvii.	
Orphania, Charp.	...	...	Harris, "Ins. of New Eng-land," 1842	Yersin, "Bul. Soc. Van-doise," T. iv., p. 108.
Phylloptera	...	...	Guilding, "Trans. Linn. Soc.," vol. xv., p. 154.	
			Siebold, "Erich. Archiv. für Naturges," 1844, s. 52-81.	
Campsocleis	...		Fischer, "Orthoptera Europea,"	Fischer, "Orthoptera Europea."
Platycleis ...	...		V. Graber, "Verhandl. der K. K. Zool.-bot. Gesell. in Wien," b. xxi., 1871, pp. 1057-1102.	
			V. Graber, "Mittheil. des Na-turwissenschaft. Ver. für Steiermark," 1874, pp. 32-46.	
			V. Graber, "Zeitschr. für Wissenschaft Zool.," b. 22, 1872, p. 100.	

A TABLE OF GENERA CONTAINING INSECTS THAT STRIDULATE—continued.

ORTHOPTERA—continued.

GENERA	POSITION OF FILES	POSITION OF CLASPS	MECHANISM: By whom and where indicated or described.	BY WHOM HEARD TO STRIDULATE.
LOCUSTIDÆ				
Chlorocoelus	...	...	...	Bates, "Nat. on Riv. Amazons" (the hardly shown); Harris, "Insects of New England," p. 128.
Platyphyllum	Female also emits a sound when captured, shutting its elytra. Darwin, "Des. of Man," vol. i., chap. x., p. 356	...	Yersin, "Bull. Soc. Vaudoise," t. iv., 1856, p. 108; Ditto, ditto, pp. 63-76; Darwin, "Des. of Man," vol. i., chap. x.; Scudder, "American Nat.," vol. ii., 1868.	
	*Exception to rule—Both sexes sing. In male, beneath left elytron, as before (rubbed over)			
Ephippiger	In female, above on right elytron (rubbed beneath)	Right elytron	Goureau, "Annl. de la Soc. Ent. de France," 1837	
Phylloptera	Makes a sound when flying	Left elytron		Harris, "Insects of New England," p. 110.
Gryllacris ?	At side of second and third dorsal arcs of abdomen	...		V. (Graber, "Mittheil. der Naturwissenschaft Vereines für Steiermark," Graz, 1874, pp. 32-46.
Dicuacrida ?	...	...	*Papers and Reference.*	*Special Reference.*
ACRIDIODEA				
Pachytylus, Fab.	On a vein on the elytron (over which they rub)	A ridge on the femora of hind legs	...	De Geer, "Mem. pour Servir," V. (Graber, "Zeitschr. für Wissenschaft Zool.," b. 22, 1872, pp. 121-125; Scudder, "American Nat.," vol. ii., 1868.
Stethophyma (Grossum, L.)	...	...	Siebold, "Erich. Archiv. für Naturges," 1844, s. 52-81	
Arcyptera	...	...		
Epacromia	...	...		
Parapleurus	...	...		
Œdipoda	...	...		

A TABLE OF GENERA CONTAINING INSECTS THAT STRIDULATE—continued.
ORTHOPTERA—concluded.

GENERA.	POSITION OF FILES.	POSITION OF CLASPS.	MECHANISM: BY WHOM AND WHERE IS INDICATED OR DESCRIBED.	BY WHOM HEARD TO STRIDULATE.
ACRIDIIDÆ.				
Stenobothrus, Fisch.	On a ridge on the femora of hind legs (which they rub over)	The scapular vein	Gourean, "An. de la Soc. Ent. de France, 1857. Gibb, "Canadian Nat. and Geologist," vol. iv., p. 121. v. Graber, "Verhand. der K. K. Zool.-bot. Gesell. in Wien," b. xxi., 1871, s. 1097-1102. Yersin, "Bull. Soc. Vaudoise," t. iv., 1856, p. 108.	Gourean, "An. de la Soc. Ent. de France."
Gomphocerus	On a ridge on the femora of hind legs (which they rub over)	The scapular vein	v. Graber, "Zeitschr. für Wissenschaft Zool.," b. 22, 1872, pp. 121-125	
Opsomala		The scapular vein	Scudder, "American Nat.," vol. ii., 1868	
Pneumora (Species that stridulate in flight.)	On inflated abdomen of male (rubbed by the)	Hind femora	Westring, "Naturhist. Tids.," Bd. "i., 1844-45, s. 5?	Burmeister, "Handb. Ent.," i. p. 512.
Œdipoda				
Stenobothrus nigropterus				
Stenobothrus miniatus				
" viridulus				
" variegatus				
Pachytylus stridulus				
" migratorius				
Mantis?				Gourean.
PHASMIDÆ.				
Pterinoxylus			Wood-Mason, "Trans. Ent. Soc.," p. 4, p. xxix., 1877	

A TABLE OF GENERA CONTAINING INSECTS THAT STRIDULATE—continued.

HEMIPTERA (HETEROPTERA).

Rule:—In the Heteroptera both sexes stridulate.

GENERA.	POSITION OF FILES.	POSITION OF CLASPS.	MECHANISM: BY WHOM AND WHERE INDICATED OR DESCRIBED.	BY WHOM HEARD TO STRIDULATE.
Pachycoris ...	On ventral arcs of abdominal segments?	...	Westring, "Götheborg's Kong. Vet. och Vitter," iv., Haf., s. 47 ...	...
Scutellera ...	...	...		
Stiretrus ...	...	...		
Optomus ...	...	...		
Cocloglossa	...	...		
Arctocoris ...	...	...		
Isacasta ...	...	...		
Halys ...	...	...		
Reduvius (pupa and imago)	In groove beneath prosternum (in which they rub) ...	The rostrum or sucker	Westring, "Naturalist. Tids. Kröyer," Blind. I., 1811-45, s. 5.	Hay, Goureau.
Coranus ...	...	...	O. M. Reuter, "Mittheil. de Schweiz. Ent. Gesells," iv. p. 139, and "Ent. Month. Mag." vol. xi, p. 137.	De Geer, iii., 289.
Pirates ...	...	...		Prof. Westwood, "Modern Class of Ins."
Harpactor (imago)	At upper lateral angles of mesothorax (over which they rub)	...	A. H. Swinton, "Ent. Month. Mag." vol. xv., p. 117.	
Naucoris? ...	...	The prothorax		L. Frischs, "Beschr. v. Ins. Deutsch," 1766, "Ent. Month. Mag." vol. xiv., pp. 253.
Nepa? ...	...	...		
Corixa? ...	On centre of mesothorax (over which they rub) ...	The prothorax?		R. Ball, "Report of the British Assoc.," 1845.
Notonecta? ...	...	...		

A TABLE OF GENERA CONTAINING INSECTS THAT STRIDULATE—continued.

HEMIPTERA HOMOPTERA (MUSIC VOCAL).

GENERA.	POSITION OF FILES.	POSITION OF CLASPS.	MECHANISM: BY WHOM AND WHERE INDICATED OR DESCRIBED.	BY WHOM HEARD TO STRIDULATE.
Various genera of Cicada ...	Convex vesicular membranes at the base of the abdomen, exposed or covered ...	Which they vibrate by means of a muscle	Réaumur, "Mém. pour Servir," v., t. xvi.	Sing during the day when held in the hand or tickled with a straw.
Fulgora? ...	...			

Rule:—Both sexes stridulate.

HYMENOPTERA.

Mutilla	Upper surface of third segment of abdomen (which they rub beneath)	The second ...	Goureau, "Annal. de la Soc. Ent. de France," 1837, p. 66; "Naturalist, Thos. Kröger," h. i. 1844; Darwin's View, "Des. of Man," vol. i., chap. x., p. 364, is incorrect	Stridulate readily on seizure.
Myrmica ...	Anterior constriction of abdomen (which they rub in) ...	Second knot of pedicle	A. H. Swinton, "Ent. Mon. Mag.," vol. xiv., p. 157	M. ruginodis ♂♀, placed beneath a glass.
Pepsis? ...	...	...	...	"Drury Illustrations," vol. ii., plate xxxix., fig. 4, a "clicking noise" when flying.

Rule:—

NEUROPTERA.

Atropa? ...	...	...	...	Has been discussed since the time of Derham, 1700.
Termes? ...	...	...	...	"They make a hiss," Smeathman, &c., "Phil. Trans.," 1781, p. 48.

A TABLE OF GENERA CONTAINING INSECTS THAT STRIDULATE—continued.

LEPIDOPTERA.

Rule:—Both sexes stridulate.

GENERA.	POSITION OF FILES.	POSITION OF CLASPS.	MECHANISM: BY WHOM AND WHERE INDICATED OR DESCRIBED.	BY WHOM HEARD TO STRIDULATE.
RHOPALOCERA (Butterflies).				
Ageronia	On costal vein of hind wing (over which they rub)	The anal vein of fore-wing	Doubleday, "Proc. Ent. Soc.," March 3rd, 1845, p. 123; "Ent. Mon. Mag.," vol. xiii. p. 207; Bigg-Wither, "Pioneering in South Brazil," p. 306.	Darwin, Wallace, click-flying.
Fanfea	...	...	Dr. Fritz Muller, "Trans. Ent. Soc.," 1878, p. 211.	—
Eupetychias	...	...	Dr. Fritz Muller, "Trans. Ent. Soc.," 1878, p. 211.	Greene, Hewitson (V. Io.), "Trans. Ent. Soc.," makes a noise like sand-paper," 1852, vol. ii. New Series; "Proc.," p. xcviii. 1856. vol. iv. N.S.; "Proc.," p. ii.
Vanessa	At base of anal vein of fore-wing (which they rub over)	The costal vein of hind wing	"Ent. Mon. Mag.," vol. xiii. p. 169.	—
Theola ?	At base of anal vein of fore-wing (which they rub over)	The costal vein of hind wing	A. H. Swinton, "Ent. Mon. Mag.," vol. xiv., p. 209	—
HETEROCERA (Moths).				
Hecastesia	As in the Cicada	...	...	Prof. Westwood, "Gen. Diurnl. Lep.;" H. Thyridon.
Hasiana	...	...	...	H. Postica "gives out sounds resembling those of a Lamia, for minutes together."—Walker, "Trans. Ent. Soc.," 1867, p. iii. p. 237.
Langia	...	...	...	Col. Gott., "Ent. Mon. Mag.," vol. xiv., p. 116 (L. zenzeroides).

A TABLE OF GENERA CONTAINING INSECTS THAT STRIDULATE—concluded.

LEPIDOPTERA—concluded.

GENERA.	POSITION OF FILES.	POSITION OF CLAWS.	MECHANISM: By whom and where indicated or described.	BY WHOM HEARD TO STRIDULATE.
Acherontia	On first joint of palpi (which they scratch)	On the proboscis	Landois, "Zeitschr. für Wis. Zoologie," xvii., Bd.	Squeak on seizure.
Sesia	...	...	...	G. Gibb, "Canad. Nat. and Geo.," vol. iv., p. 121 (S. pelasgus). "Gen. Diurn. Lepid." Prof. Westwood,
Glaucopis	As in the Cicadæ	...	...	
Halias (Silver-lines)	On hind wing vein?	Which the elbow on inner margin of fore-wing scrapes over	...	Various notices, ♂ squeaks flying.
Callimorpha, Phragmatobia, Arctia (Euprepia, Chelonia)	On metathoracic vesicle, on epi-sternum (rubbed by)	Hind femur?	Mihi; "Archiv. für Natur-gesch.," 1864, p. 375, &c.	E. matronula, (Czerney, "Yahresh., 1859-60, p. 215; C. Pudica, "Ann. de la Soc. Ent. do France," t.i.
Setina	On metathoracic vesicle, on epi-sternum (rubbed by)	Hind femur?	...	Guenée, "Ann. de la Soc. Ent. de France," 4me Ser., 1861.
Lithosia	On metathoracic vesicle, on epi-sternum (rubbed by)	Hind femur?	"Bullet. de la Soc. Ent.," 1859, p. 43	Erichson, "Arch. fur Na-turges," 1860-61, p. 381.
Miltochrista	On metathoracic vesicle, on epi-sternum (rubbed by)	Hind femur?	...	"Ent. Mon. Mag."
Gnophria, Nudaria	On metathoracic vesicle, on epi-sternum (rubbed by)	Hind femur?	...	
Dicrorampha	Along anterior margin of hind wings (rubbed by)	Inner margin of fore?	...	

DIPTERA?

Rule :—Male stridulates ?

CHAPTER V.

WING BEATING AND VOCAL MUSIC CONSIDERED AS A MATERIAL AGENT
IN REPRODUCTION AND DISTRIBUTION.

" O to watch the grape of Lemnos
    Swelling out its purple skin,
When the merry little chirpings
    Of the Tettiges begin !
For the Lemnian ripens early,
    And I watch the juicy fig,
Till at last I pick and eat it,
    When it hangeth soft and big."

Aristophanes (Trans.).

A LITTLE familiarity with the dissection of insects serves to show us that their system of respiration, as compared with the typical organs of the vertebrate animals, may be denominated complex. To examine this mechanism, the subject larva or perfect insect may be either killed, cut open longitudinally, and then placed beneath a vessel of water, or, after being prepared, it may be allowed to dry pinned out on a flat board. Then by manipulating with a needle-point, we may learn that the air-pipes or tracheæ, composed of a spiral elastic thread lined by a silvery membrane, pass in two main-tubes along either side of the body (Plate II., Fig. 1, t), and at each segment or division communicate with one of an external row of breathing-slits or *spiracles* (Plate II., Fig. 3). From the same points they send out a series of branches that replenish the internal air-bladders insects possess in common with all flying and swimming life (Plate II., Fig. 1, $t', t', t' \ldots$), or disperse in the muscular valves of the circulating vessels, and penetrate the ovaries, legs, and all parts of the organisation. To the anterior rings of the body the external breathing-pores are often wanting, and here in lieu, offsets of the tracheæ ramify in the antennæ, or wings; the latter perform

the office of alar expansion in the new-born insect on quitting the
swathing pupa mask ; or afterwards, where the hind-wings plait
and fold in repose, they inflate and spread them balloon-like
previous to taking flight.   It is evident, therefore, this similar
lateral and corresponding disposition of the respiratory organs in
each segment, typical of insects and spiders, excludes simple
laryngeal voice, where the lungs are the bellows, and the vocal
cords, palate, and teeth the facile notes of the flute ; so insects
can neither cry, sing, nor speak, and the term vocal can here only
be technically retained in treating of any spiracular sounds.

If we may suppose the instrumental music of Articulata had
its origin in muscular contractions when under the stimulus of
emotion, spiracular music may have taken origin in a wing
movement performed under similar conditions.   And while in
some insects these murmurs are in degree alar in character, in
others their production becomes more evidently dependent on
the internal muscular system, and more evidently produced by
the action of the tracheal : a specialisation which confers new
powers in the expression or communication of the emotions.
But before entering on the elucidation of this higher feature of
vocal music, it will be well to entertain clear views of the earlier
wing movements and their import.

Certain moths produce a sound by simple wing-percussion,
that appears to promote the intercourse of the sexes.   In proof
of this we may turn to the broad-wing silk-spinners, or to their
representatives among those with looping caterpillars, and observe the sonorous beating of wings that accompanies the brisk
gyrations of the male subsequent to its scenting a female.   This
is well exhibited by the common White Mulberry Silkworm
Moth.   The female also, I believe, thus assembles its suitors ;
for it is not unfrequently the well-attuned ear of the country fly-
catcher catches this soft fluttering sound, when out with his gauze
net on the dewy meadows, at the hush and solitude of nightfall,
when some bulky cynosure announces the term of her siesta by
climbing the flowery herbage, winnowing her echoing wings.
Then, if a male of one of these beating moths be confined in an
apartment, its period of flight will be intimated by its trilling at
intervals, with a sound not unlike that of a policeman's rattle,
which begins low and increases in loudness.   The grey-suited

lethargic Cockney Moth (*Biston Hirtaria*), of seasonable appearance in the London parks, when imprisoned in a box, will thus invariably respond for a minute at evening to the successive mural tremors arising from a passing rumble of wheels in the street below. Then if two stout-bodied moths, especially if male and female of the same species, be enclosed in chip-boxes and placed in too near proximity, they, as is well known, I imagine, to every patron of the lamp-posts and sugar-pots, beat alternately and in quick succession, despoiling themselves of their pleasant nap. Now the fact of these insects raising their emotions in alternation, or consecutively to certain sounds, seems to establish this as a provision for sexual communication, and as an intimation of jealousy ; while the ardency of its utterance when the opposite sexes are approximated, indicates a love-call that reproduces in miniature the strutting, wing-drumming, and rustling of the males of the turkey and grouse at the pairing time.

The popular science of acoustics has not failed to direct its researches to the investigation of the atmospheric sounds of insects, which it aspires to gauge both in pitch and as to the component wing-beats ; and for this purpose three instruments are available—the wheel of Savart, the syren, and the graphic cylinder. The wheel of Savart is simply an ordinary toothed wheel, which, when scraped over a simple piece of cardboard, gives a whizzing sound that varies in height with the celerity of revolution, the stroke of each tooth being considered to impart a simple vibration to the said cardboard. In the case of the syren, an invention ascribed to Cagniard Latour, a Frenchman, and so called because it can be made to sing beneath water— air is forced by a pair of bellows through holes in a flat and horizontal plate, which are covered and opened by similar orifices revolving in a circular disc; so that the faster the motion the quicker the jets of air escape, each puff being considered as forming a simple atmospheric vibration. Lastly, the graphic method is an ordinary revolving cylinder covered with soot, which, when applied to a vibrating object, writes the beats in wavy scratches on its surface, the celerity of the turning handle being employed to decipher the resulting superscription.

Even if we suppose the correctness of the indications afforded by these instruments, which must be in itself a matter

o

open to doubt, there yet remains somewhat to be desired, as all
experimenters take for granted the sounds emitted by insects
in flight are simply due to the wings beating the air at an
almost incredible speed; and to prove this, the would-be measurer
of movement or sound fixes his subject on a pin, when the
limber-vans, urged and vexed by impatience at captivity, make
unwonted exertions, and thus cease to manifest their imperturb-
able momentum of flight. Then, in the two cases where
sound is to be measured, we have to trust to the fineness of the
operator's ear, who compares mentally his buzzing insect with
the whirr of the cardboard or sing of the syren. On the other
hand, in measuring the wing-beats, the blackened cylinder must
chiefly fail, from a difficulty in the adjustment of mechanism to
the nervous motions of an animated object.

Turning to the physiological character of these sounds, we
find the males of various birds produce on the wing a strange
winnowing or atmospheric concussion, arising from a modification
in the form of certain of their wing or tail feathers, which are
pointed or clubbed, and during the pairing-time are thus
rendered subservient to the emotions in the production of these
singular calls. An example of this kind of music is afforded
us by the male of the common Snipe, which now and again
startles the wanderer in the fenland by a sudden and rapid fall
from the zenith, accompanied with a mysterious drumming.*
But in respect to insects—although among the aërial butterfly
and moth kind some are observed to possess various and fantastic wing outline—I am aware of no instance in which musical
modulations have been ascertained to arise from this peculiarity.
On the contrary, it is remarkable that bees and flies which
have the precedence as hummers and buzzers have uniformly
triangular wings, and that they give out their varied and
various sounds rather in proportion to the size or content of
their body than in the ratio of their wing area. A remarkable
phenomenon likewise presides over this music — namely, it
is subject to atmospheric perturbations. Gnats, species so
sonorous and vindictive during summer heats, when the damp

---

* Various instances of this music are given in "The Descent of Man."
Distinct from laryngeal voice, it seems to correspond to the instrumental music
of Insecta.

of autumn comes on may not unfrequently be observed flying
about our apartments quite noiselessly; but I will not pretend
to decide whether this may be owing directly to the weather, or
whether it proceeds indirectly from its chilling influence on the
slender palpitating frame.

That these aërial notes of various colour do not take origin
in flight may be easily shown in these orders; for if we capture
a fly or bee traversing our apartment with loud sound, we notice,
although the deep resonance of locomotion has ceased, it continues its song when retained in the hand. And if we then
begin and methodically clip the wings from the tip down to
their roots, it will become further evident that, although the
sound during this operation has risen in pitch, its emission continues in greater or less intensity as long as the thoracic muscles
remain in action. We have thus a fundamental music independent of wing-motion to account for, and that this arises
from increased respiration, induced by muscular action at the
tracheal spiracles, has been more or less satisfactorily shown by
Chabrier and Burmeister in the case of flies, and by John
Hunter as regards bees.

Chabrier informs us that in the Bluebottle the hinder
thoracic spiracles (metathoracic) are closed by (two) little scaly
lips, and if these be carefully removed with a fine needle the
buzz of the insect is scarcely audible during flight. Burmeister advanced further. Having removed all movable external parts from the common Drone Fly of our flower-beds,
which still continued its peculiar notes, he also became convinced
the sound arose at these same metathoracic spiracles, which
he proceeded to dissect. He then discovered their edges to be
furnished internally with a fringe of parallel membranous plates,
horizontally overlapping and decreasing in size towards either
extremity. In this insect there are fifteen such on either edge
(See Plate V., Fig. 6, *l*). Dr. Landois follows Burmeister. By
cutting the thorax in two longitudinally, and removing the
longitudinal and oblique wing-muscles so as to expose the mouth
of a posterior spiracle, we may with aid of magnifying power
discover these serrated laminae, comparable to a curry-comb or
Pan's-pipe, present in a majority of flies. In the Hover Flies,
where they are most conspicuous, their position and appearance

are invariably the same as that seen in the Drone Flies. But in
the Blue-bottle, agreeably to Chabrier's experiment, we find them
placed on the spiracular lips, nine on one and sixteen on the other
(Plate V., Fig. 3, *l*), and in the large bee-like *Tachina grossa*,
that sits on flower clusters at autumn, they are placed on the
single lip that shuts the spiracle. Behind these spiracles,
situated posteriorly and symmetrically in the metathorax, lie two
capacious air-vesicles bounded by the alar muscles, and thus
calculated to increase their power of respiration when these
muscles are vibrated.

Regarding bees, John Hunter had previously recorded in the
Philosophical Transactions for 1792 a similar conclusion he
had arrived at in respect of the hum of the Hive Bees. "They
produce," he says, "a noise independent of their wings; for if a
bee is smeared all over with honey so as to make the wings stick
together, it will be found to make a noise which is shrill and
peevish. To ascertain this further, I held a bee by the legs
with a pair of pincers, and observed that it then made a
peevish noise, although the wings were perfectly still. I then
cut the wings off, and found it made the same noise. I
examined it in water, but then it did not produce the noise till
it was very much teased, and then it made the same kind of
noise; and I could observe the water, or rather the surface of
contact of the water with the air at the mouth of an air-hole
(metathoracic spiracle?) at the root of the wing, vibrating."
Dr. Landois considers the spiracular notes of bees due to the
large metathoracic spiracles (Plate V., Fig. 4, B), and that the abdominal ones may assist in their production. In the Humble Bee,
according to this author, the spiracles (*s*) appear to want the
laminæ seen in flies, but their inner lips are chitineous, sharp-
edged, and immovable, suited to vibration; and the lower one (*b*)
forms internally a cup-shaped cavity, covered by the valve that
opens and closes the spiracle (*a*).

Whether or not, then, these Dipterous spiracular laminæ are
influential in producing their buzz—and there may be those who
are inclined to compare them with antennal extensions and
consider them an osmeterium—it cannot but be evident it
is from respiration rather than wing-beat the colour of the
aërial notes of insects arises—a physiological view which, if

generally correct, will render unnecessary the received physical notion enunciated to satisfy the present indication of the phonometer termed the syren, that requires the insect wing to strike the air with enormous rapidity; for here we have adequate mechanism to produce short wave lengths and high notes, independent of wing acceleration.

But certain species in these diaphanous-winged orders afford proof of the existence of a voluntary spiracular music. Thus the majority of the ornamental tribe of Hover Flies when captured, emit acute high notes in the gamut of a singing tea-kettle, slightly vibrating the shut wings, or retaining them in perfect repose. The Drone Flies, the Wasp Flies, with the smaller *Syrilla pipiens*, readily so perform; and among the bees we recognise these notes in the Sand Wasps, the Bumbles, and others. These, as the former, produce it with the wings closed over the back, either motionless or vibrating slightly to the thoracic muscles, the potency of whose play may be experienced as a species of electric shock by applying a pencil to the part. In bees a respiratory movement, an elongation and contraction of the abdomen, invariably accompanies this music, due to the chitineous plates covering the pedicled abdomen sliding one within the other telescopically. We may observe this singular action of pumping in air in both bees and beetles when inflating previous to flight, the abdominal distention being allowed for by intersegmental membrane; and, as we have seen likewise, this is often the movement of stridulation in the same orders.

Having treated of the existence and manner of production, we now come to the range of expression and import of vocal music. The British Diptera most noted for their humming or buzzing notes are the Gnats or Musquitoes, the two classes of Cattle Flies, the Mottled Clegs or Forest Flies, the vernal, flower-visiting Bee Flies, the road-settling Hornet Flies (*Asilus*, Lin.), the many genera of sparkling Hover Flies, the unpretentious Stinging Flies (*Stomorys*, Fab.), too often mistaken for their House Fly prototype, the species of *Tachina*, and the swarms of carcase and other domestic, refuse-born kinds, that haunt our apartments and back courts. Yet although we may at times observe the note of a fly who impels his companion on the breakfast cloth takes unwonted shrillness, or that incensed gnats

dance with peevish ring, that some flies rise to the sunbeam with
a gay tune, and others meet in air with jealous cries, nevertheless, owing to the bluntness of human sense and slow process
of thought, the delicate modulations, as the purport of this
widely-spread music, as a rule elude our senses. But we are not
without proof that in those species where it assumes spontaneousness, it is as capable as instrumental music of vindicating and
interpreting certain stimuli, as fear, love, and rivalry.

The high spiracular note of the *Syrphidæ* and Æolian murmur
of the *Muscidæ* breaks forth in its varied compass when one is
held in the hand, or, driven by fate, falls into the meshes of the
autumnal spider hordes ; and species of Hover Fly thus greet
a predaceous insect as readily as a cat hails a dog. I remember
one morning when sauntering along the highland glen immor-
talised in "The Queen's Wake," catching a *whining* cry from
the sunny side of the road, and on proceeding to the spot
detecting an orange-belted fly wrestling on a buttercup-head with
a small Hymenoptera. Boxing them together under the net,
the fly crouched down at the bottom of the receptacle, con-
tinuing its distressful note, until its aggressor, after crawling
about for some seconds, darting down, seized it beneath and
decapitated it.

A marked employment of this spiracular music in courtship
came under the notice of Col. Goureau. "On the 9th of July
I saw two Wasp Flies (*Chrysotoxum arcuatum*) sitting the one on
a fir branch, and the other on the leaf of a neighbouring beech.
Both were uttering a shrill and plaintive note. They flew away,
and returning, settled nearer to each other, and recommenced
their song. They repeated the manœuvre again and again, at
times as they took wing, meeting in the air, and appearing to
seize each other, or falling precipitately towards the ground,
and as I watched them utter their complaining cry, I noticed
a slight movement take place in the wings, which increased as
the sound gained in intensity, but ceased each time the insect
took flight, and was succeeded by the deep hum."

We are here introduced to a species of dance accompanying
the notes, and originating in the same stimuli, and no less do
spiracular murmurs of rivalry prompt aërial gambols. During
hot, cloudless July weather, there appears on the surface of wells

and inky ditches in the neighbourhood of London, the little, slender, long-legged *Dolichopus nobilitatus*, a kind of strolling player, who extemporises in the sunlight a round of tournament and mock amour. As in Racine's tragedies, a female opens the jousts. You first notice her tinsel-green body and immaculate wings as she sips marginal moisture or proceeds over the scummy water searching for a tasteful infusion, a feat performed by an oar-like movement of the legs. Some other flies and spiders, be it observed, similarly traverse the surface.

But not long does she remain thus unmolested, for of a sudden a glassy buzz announces an arrival of larger males, with black-tipped wings and portentous anal claspers, one of whom immediately begins court by horizontally opening and closing his wings, and as she makes off thus ludicrously dogs her retiring footsteps. His wings now open and close quicker and quicker, and commence to vibrate when expanded, till, having gained a certain proximity, taking several quick straight or circular darts within the compass of a few inches, he lightly floats buzzing around her head, in the style of Bacchus greeting Ariadne as depicted by Gaspar Poussin, and with an agility that would postulate some special adaptation of wing muscle.

The fugitive female resents this familiarity by adjourning to a proximate puddle, and now the male, with his eyes glowing like emeralds, occupies the deserted arena. At such a crisis, or during the mock courtship, should a rival alight in the vicinity, a sharp tilting ensues. Jerking in the air, these little atoms dash in one another's faces like game-cocks, while angry *spiracular* (?) tinklings resound in the vaporous herbage. This is the summons for other males, and the flirtation usually concludes in a vicious affray all round. The two phases of this suggestive little tale are depicted in Plate V., Fig. 1.

The horizontal wing movement here noticed, a phenomenon common to a group of little flies, and incessantly performed by some that crawl over plant leaves, *Scioptera*, is thought by Dr. Landois, in the case of the aquatic genus *Stratiomys*, to postulate some organ of instrumental music, though after all its production may be, I fancy, spiracular. "The Water Flies," he says, "make a noise of a cracking, crackling colour, which, in treating of the stridulation of insects, cannot be omitted. It is produced by the

articulation at the root of the wings, and when the wing is very
slowly moved up and down each time there originates a sound,
which clearly results.from the wing-roots jerking on the prothorax during their descent, a second note succeeding the first
when the wings have attained their lowest posture. The crackling sounds follow quick on one another, and originate the
stridulation, soon to be drowned in the notes of the vibrating
wing." In addition to the instances preferred, I have heard
the primitive spiracular notes of the Hover Flies given out by
a showy northern species, *Sericomya borealis* (Plate IV., Fig. 7),
that frequents brambles in the West Highlands, as it alighted on
rotten stumps with closed wings, to fulfil the maternal duty of
oviposition : others emit them, I believe, when simply sunning
themselves, a joyous sound harmonising with the country air.

In Bees, especially when apterous, the filiform character of
the antennæ allows the development of a language of touch, and
there are also instrumental performers. But with the aërial
species no medium is more suitable for intercommunication than
that various and varying hum that falls so pleasantly on the ear
whether overshadowed by the palm, the vine, or the sallow.
Among northern Hymenoptera most noticeable for their summer
notes are the genera Sirex, Bembex, Saropoda, Dasypoda, Vespa,
Crabro, Apis, Bombus, &c.

The modulation this music is susceptible of, expressive of
fear, anger, or pleasure, did not escape the attention of John
Hunter. Hive Bees may, he says, be said to have a voice.
" They are certainly capable of forming several sounds. They
give a sound when flying which they can vary according to
circumstances. One accustomed to bees can immediately tell
when a bee makes an attack by the sound, which is a very
different noise from that of the wings when coming home of a
fine evening loaded with farina or honey ; it is then a soft
contented noise. They may also be seen at the door of their
hive, with the belly rather raised, and moving their wings,
making a noise."* The male Hive Bee or drone, and the
females, queens and workers, can also severally be distinguished
by the timbre of their music.

Then, as we have already seen, the primitive spiracular note

* *Philosoph. Trans.*, Vol. LXXXII., p. 182 (1792).

is produced in the bee kind from a sense of fear; and in proof
of this we may hold a musical species by the wings, glue them
together, clip or remove them, stop the spiracles, or plunge the
body in water, and the subject of the operation continues its
piteous or angry whining; but a very similar sound is given
out by many individuals when at liberty, and then it is a
chronic wayward note of economic interpretation that springs
from exuberance of pleasure, or it is a manifestation of delight
in maternal labour.   Thus the long-bodied queen of the hive
produces a clicking with her wings folded in repose, a kind of ring
or toot of a small trumpet that John Hunter once found in the
lower A of the treble, and the object of this sound is to electrify
the workers, according to Huber, or to call the hive to swarm,
according to Kirby.   When wandering in the Western Highlands
I have likewise repeatedly noticed worker Humble Bees voluntarily
producing a sharp impatient note, when, in essaying the wayside
chalices of bramble, wild-rose, tall fox-glove, or other phane-
rogam, they chanced to settle on some nectary unpilfered, or
more palatable than usual.   The first kind I observed producing
this sound was the little red-tipped bee of Northern Europe
and Canada, *Bombus lapponicus* (Plate IV., Fig. 5), with its
baskets laden with bee-bread, meandering over a wild-rose fence
on the Duke of Argyll's estate at Roseneath; here, too, it
often flies in, and suavely murmurs over the opulent blossoms
of the conservatories.   Afterwards I repeatedly observed this
highland music in Argyleshire from the yellow-banded, white-
tipped *B. lucorum*, Lin., a worker-bee kindly determined by the
late Frederick Smith, who then informed me its true female is
the large vernal *Apis terrestris* of Linnæus, the commonest of
our Bumbles, who takes as its paramour the scientific male with
the above designation.*   A somewhat similar plaintive note

---

* The following is Mr. Smith's note on this curious subject:—
"British Museum, Nov. 8, 1877.
"My dear Sir,—I return the drawing of *Bombus lapponicus*; the sketch is
not very good, but there can be no doubt about the species. The specimen in
the box is the worker of *Bombus lucorum*, the female of that species is the *Apis
terrestris* of Linnæus. On the Continent many would call the bee in the box
B. terrestris, but they have no male for that insect, as they do not unite the true
male, *B. lucorum*, to it. But both of the types are in the Linnæan cabinet, and
I have taken the sexes in coitu and also out of their nests repeatedly.—Believe
me, yours sincerely,   "FREDK. SMITH."

uttered by the female of a Mason Bee (*Anthophora parietina*)
is recorded as having brought its offended males around an
audacious captor.

But most frequently this music is observed in another group
of the bee kind, where the first segment of the thorax is
narrowed in front into the form of a knot or joint, and the first
ring of the abdomen, and sometimes even the second, is
narrowed to an elongated pedicle, so that the hind body is, as it
were, hung by a stalk or thread. The females of the species of
Sphex, or Sand Wasp, agree with this description; and those
who on arenaceous soils watch these tunnel their holes in the
shifting bank, and successively close them with some insect
carcase to nourish their future progeny, repeatedly observe them
uttering their Æolian trill, with their wings folded in repose or imperceptibly vibrating, producing spiracular notes which have been
described as lying somewhere between the music of the common
little dipteron, *Syrilla pipiens*, and the stridor of a small locust.*
Their wasp-like relatives, *Pelopeus*, one black kind of which is
omnipresent on the roads in Northern Italy, hum a similar
brief bee tune when scraping up mud for their danby clay-
nests beneath the ledges and eaves; a proceeding which doubtless afforded Virgil the eccentric idea of bees gathering ballast
to steer in the wind's eye. The music in this case approaches
the whine of a Bumble in colour.

The wing movement in most insects is accompanied with a
sound. Hawk Moths that hover over flowers hum in flight.
A buzz is more characteristic of the night-fliers, whose tattoo
is so often heard at twilight on the window-pane, or during a
blind career over the ceiling of apartments that look out on a
garden. That these sounds are in part spiracular is more than
conjectural, as may be shown by clipping the wings of various
moths, when the hum or buzz becomes higher and higher, till
it resembles that of the *Culices*, but continues as the wing-roots
are capable of motion.

Many beetles have a sonorous flight, drone, and boom; and

* Goureau, *Ann. de la Soc. Ent. de France*, b. vi., p. 397. There is a
description of the way the Wasp Ichneumons (*Sphex floripennis, albisecta,* &c.)
tunnel in sandy soil and store insect carcases for their young, in the "Naturhist.
Tidsskrift. And. Rak.," p. 34.

as their spiracles (Plate V., Fig. 4, A) have horny lips (*l*) with a wavy or fluted construction, these may influence the production of music after the manner of a Pan's-pipe. The inner movable lips, which open and close the mouths of the tracheæ, appear similar, if not identical, with those of bees.

Of the Coleoptera, the Lamellicorns are the greatest fliers and aërial hummers. The boom of the Chafers and Stag Beetles has certainly the appearance of a tacitly understood evening call to feed on the foliage, as one of a calm night hears them take flight successively in one direction. The same may be observed in the stercoraceous beetles when at early dusk or fresh morning these bulky insects cross our path in the same line. But then also this head-to-tail flight is found in the nectar-feeding butterflies and moths, and, if independent of aërial agency, it appears implicated with scent, for it is here we find the antennæ extended, lamellated, or pectinated. Longhorns and Tiger Beetles also produce sounds on the wing, which in a water beetle (*Acilius*), have been ascribed by Mr. Rye to Aluke. With respect to the remaining orders which contain insects capable of flight, certain Dragon-flies, especially the large species of *Libellulina*, give out a melancholy sound on the wing, which Dr. Landois, who attributes its production to the spiracles, thinks they learnt from listening to the sighing of the reeds. And, turning from these to the Homoptera and Bugs, we have numerous records of the *Cicada Anglica* flying in warm ferny nooks with a rustling like a Dragon-fly; and concerning individuals of *Coreus marginatus*, we learn from Kirby[*] that when "flying, especially when hovering together in a sunny sheltered spot, they emit a hum as loud as that of the Hive Bee." The flying grasshoppers and locusts traverse the air with a sound that has been compared to the notes of an Æolian lyre, patter of rain-drops, or wind in the shrouds.

The Homoptera in the genera of Cicadidæ stand at the head of the vocal musicians; for these not only express considerable range of sensation in their music, but possess highly specialised organs for its production. In their genera we shall find, as Carus has already shown, that the central portion of the body is

---

[*] " Introduction," Lettr. XXIV., p. 485.

mainly occupied by a large centrally-constricted sub-œsophagal
air-bladder (Plate III., Fig. 5A, M,M), presenting an instance of the
coalescing of the metathoracic and anterior abdominal vesicles,
which extends from the anterior of the metathorax to the
termination of the second abdominal segment in the female,
sixth in the male; anteriorly, this bladder is closed by two symmetrical chitineous partitions forming the posterior termination
of the mesothorax; posteriorly, it is lined by an oblique
membranous diaphragm inclined from below forwards, and
above covering the viscera; it is also divided longitudinally
into two symmetrical parts by a membranous partition (d).
In this bladder are situated the organs of music and hearing
placed respectively in the first and second abdominal segments,
which are abnormal and constricted, and here they form local
boundaries, apparently but indurations and attenuations in the
texture of its membranous lining, with which they seem continuous. The organs of music usually take development in the
male sex, and are only indicated and impotent in the females.
They consist in two outwardly convex parchment-like membranes (T, Figs. 5A, 5B), situated at either side of the dorsal arcs
of the first abdominal ring, that exhibit transverse chitineous
indurations, calculated to counteract atmospheric pressure, and
disposed so as to retain the symmetry of the structures. Among
phanerotympanums, *Tibicen hæmatodes*, Latr. (Plate VI., Fig. 4),
has six long oar-shaped ridges on its tymbals, and nine short
central callosities, bent forward and directed in a line from
within outwards; but in *Tettigonia orni*, Fabr., where the
musical organs are slightly covered posteriorly by a projection of
the second dorsal arc, there are only five long and three short
ones wider interspaced. Among the cryptotympanous kind, in
Tettigonia plebeja, Rossi, where the tymbals are totally covered
superiorly by a prolongation of the second dorsal arc (c), there
are but three imperfectly indurated ridges, as shown in Fig. 5A,
Plate III.

Beneath this fluted structure the membrane has a lenticular
thickening, and at its anterior angle is a semi-pellucid spot (p),
where is inserted the tendon (t) of a special muscle that performs
the function of motor. To form a firm point for the insertion of
the opposite extremity of these large muscles, which run upwards

to either tymbal interiorly along the sides of the ventral integument of the first segment, the first two abdominal rings coalesce, so as to leave two cavities closed beneath by the metathoracic lamellar projections, or operculæ (*a*, Fig. 5B) ; and the sternum of the first is angular, forming with the circular ventral ring of the second two semi-lunar orifices (Plate III., Fig. 5A, *m, m* ; Plate VI., Fig. 5). To avoid further prolixity, as these parts have been so frequently the theme of general description, and since the tymbals, muscles, tendons, and other parts may be recognised by any inquirer even in desiccated Cicadæ, it will be merely necessary to add that in *hæmatodes* the connection between the metathoracic spiracle (*s*, Fig. 4, Plate VI.) and motive muscles is exceptional. This orifice enters the vesicle by a short tracheal pipe, which after debouching is continued by its branches across the vacuum ; these meet the adjacent muscle at its centre, penetrate and surround it, and, ascending, encircle the tendon.

The regularity of the series of curved chitineous ridges that cross the tymbals of phanerotympanous Cicadæ might suggest that these *Homoptera* produce their music by friction, and that they should be classed with stridulators. That this is not the case may be either deduced from the fact that in cryptotympanous species these are rudimentary, and only semi-indurated ; or it may be practically tested in the phanerotympanums by dissevering the wings, when the superior surface of the tymbals will be bared, and their sonorous action manifested as free collapse and rebound of the tense vibratile membrane from below upwards, produced and regulated in fulness and tone, by the varying tensions of the tendons of their twin and special motive muscular fascicles. In what measure the resultant sound is due to the tymbal's action on the air external, or on that enclosed in the long central air-bladders, is doubtful (M, M) ; but since it is similarly produced by muscular vibration of the integument, and is accompanied by respiratory abdominal movements, it appears in direct connection with the vocal music exemplified in *Hymenoptera* and *Diptera*. This latter action of the abdomen, ruling the respiration, and effected by special elevator and depressor muscles, has specific variation, and appears to determine the rhythm of the notes.

Thus the males of *Hæmatodes* when they commence a fitful

series of overtures, which they repeat at intervals, raise the
abdomen to an angle with their thorax, and retain it
thus, protruding and contracting its segments until exhausted,
when they lower it, and the sound simultaneously ceases.
The music resulting, which has duration of from a quarter
to half a minute, is a species of trumpeting or whistling;
two or three croaks, and then a continuous dirl resembling
a clock running down, or escape of steam, "Pip! pip!
pece . . . . .!" I have also noticed one male vibrating
or beating his closed wings in unison with the notes. The
males of *T. orni* vocalise with no perceptible abdominal motion,
and with slight thoracic tremor, and their descant has the
colour of stridulation, a metallic or vitreous tone resembling the
sound of rubbing rosin on a violin bow, filing of a blacksmith,
scraping of a slate, " Chip! chip! chip!" or mellowed by a
sylvan echo, " Derde! derde! derde!" But when commencing
this recital, or when alarmed, they deal out their notes without
blending them, "Tip! tip! tip!" Their ordinary overtures,
with pauses or stresses on a note, are drawn out to twenty
minutes, but frequently last only from two to ten, while the
harshness of the sounds composing them is often manifested in
the contrast presented by the stridor of a *Locusta viridissima*
from a willow, which then falls on the ear softly as a trickling
spring; plunged beneath water, these sharp notes become a
frog-like croak.

Plebeja as it sings crawls slowly backwards or forwards along
the sunny bough. The males incessantly move their abdomens
vertically up and down from ten to twenty seconds, a motion that
lapses into a tremor ; and these timed series of recitals, a quick
rattle like the shaking of coppers, severally shut with a harp-like
refrain of five seconds' duration, " Whee—whay!" This note,
resembling the breathing of a fly in a spider's toils, and slightly
marked also in the pauses of *orni*, appears due to the aërial
vibrations retained, echoed, or prolonged by the drum-covers.

While the males of *Plebeja* and *Hæmatodes* in the north of
Italy sing on the summit of brushwood at an elevation varying
from ten to twelve feet, *orni* will ascend the poplar trunks to a
much greater height; and his notes, as those of *hæmatodes*,
appear capable of being heard at a greater distance than those of

plebeja, though less sharp and distinct. The three species, both as regards their general anatomy, as in respect to the structure of their organs and habits are thus widely differentiated. They are all tolerably plentiful in Southern Europe in the height of summer, *Hæmatodes* appearing in June and the other two in July, but only *Plebeja* extends as far north as Fontainebleau, where it is rare. The smaller Continental cicadæ also sing ; for the little *C. argentata*, Oliv., of Sardinia is described as mounting the lower oak-bushes and ringing out a clear sharp and metallic "Tick! tick! tick!" repeated fifteen times, and this often with great rapidity. *Orni* is made a plaything by children, who tickle it with a straw, when it begins to sing in a laughable fashion.

While the music of the Cicadæ presents traits in common with the plaint of the Hymenoptera and whining of Diptera, the specialisation of the productive organs produces phenomena and determines the biology in uniformity with stridulation. Thus the incentives of love and rivalry are evinced by the males consecutively responding to the fitful notes of a congener, as by the sexes, even diverse as to species, being so induced to congregate on one spot or on the same bough, a feature peculiarly striking in the tropical zone, where it has often intruded on the notice of travellers. "On my march with the army of the Indus," says a writer in the *Journal of the Asiatic Society of Bengal*,[*] "from Kandahar towards Cabool, I observed a remarkable congregation of Cicadæ, composed of more than one species. The branches of the tamarisk were covered with them, as hardly to be able to distinguish a particle of green, and their noise all day was unceasing; the jewassee bushes being at the same time covered with the empty scales of the pupæ. Our tents and tent-ropes as soon as pitched were covered by these insects, in fact, everything looked yellow. The only enemies they appeared to have were some large dragon-flies, which pounced upon them and carried off what appeared to be double their own weight." This circumstance was witnessed fifty miles from Kandahar.

South Europe produces only about sixteen species of Cicadæ, but in countries where they are more numerous and of larger size, as Australia, their noise is deafening. Dr. G. Bennett[†] has

* *Asiac. Soc. of Bengal Jour.*, Vol. IX., p. 441 (1840).
† G. Bennett, "Wanderings in New South Wales."

adapted a notation to the notes of the New South Wales species.
He says, " The most common is the incessant drumming, but it
is not confined to this. The ' Ziz, ziz, ziz ' is often interrupted
by ' Ohoi, ohoi, ohoi,' varied to ' Whoeky, whoeky, whoeky,'
and the noise ceases. Sometimes a prolonged note of ' Alrite,
alrite, alrite' is heard, varied to ' Ohoé, ohoé, ohoé,' the last note
being prolonged, followed by ' Whoeky, whoeky, whoeky ' in
very shrill tones; then ' Ziz, ziz, ziz ' continues for some time,
followed by a sound of ' Yocky, yocky, yocky,' after which the
din suddenly ceases." This observer noticed they were capable
of modulating the sound and varying its intensity. *Thopha
saccata*, probably the most powerful of Australian Cicadæ,
and often observed in this country in naturalists'
windows, cries " Awock, awock, awock," and then
commences a deafening drumming and chirping,
to which the vibration of the air in the little
bell-shaped drum-covers (*c*) must contribute.
Cystosoma Sandersii, another kind, resounds in
the orange groves during the short reign of
twilight, or in the gloom previous to a thunder-
storm, like a loud guttural " r," often so loud as
to be painful to the ear. The Australian Cicadæ
begin their song at the end of October, *T.
saccata* sings from December 15th to 28th.

DRUM COVER
THOPHA SACCATA.

But the maximum size and din of these oratorio performers
of the insect race is reached in districts such as Guiana, and the
islands of the Eastern Archipelago, scorched by the equatorial
sun, where the notes, no more likened to the vigour of the
minstrel's lyre, are compared to the blast of the clarion, and the
performers themselves termed trumpeters. " After lying for
a week off Panama," says Mr. G. F. Mathew, R.N., in a
late contribution to the *Entomologist's Monthly Magazine* on
the habits of *Cicada gigas*, " we were not at all sorry on the
afternoon of the 27th February, 1874, to raise our anchor,
and, favoured by the afternoon breeze, drop down under sail
to the island of Tobago, where we arrived at six o'clock, and
took up a position within a convenient distance of the landing-
place. Soon afterwards, while standing on deck admiring the
beauties of the island, with its immense profusion of tropical

trees and shrubs, and the varied hue of their foliage, I suddenly heard, clear and shrill through the evening air, a whistle as distinct as that of a locomotive. The whistling continued more or less until sunset, when it gradually ceased. Upon inquiring of one of the natives what this was caused by, he informed me it was the cry of the Tree-locust.

"This creature, considering its size, is gifted with a wonderfully powerful and peculiar voice. Let my reader suppose he is standing in some secluded spot, in a forest with lofty trees all round him. There is not a breath of air stirring, and hardly a sound save, perhaps, the hum of a wandering bee, the whirr of a passing humming-bird, or the rustle of a lizard amongst the dead leaves, to interrupt the oppressive stillness of a tropical afternoon. Suddenly, from right above, you hear one or two hoarse, monotonous cries, something like the croak of a Tree-frog, and looking upwards, wonder what it can be; but wait a moment, this is merely a signal, for the *next minute* everywhere above and around you *these croaks are repeated* in rapid and increasing succession until they merge into a long shrill whistle, almost exactly similar to the whistle of a first-rate locomotive; this continues for nearly half a minute, and then abruptly terminates, and everything for a short time becomes as still as before, but presently similar cries will be heard in the far distance, as if in reply to those which have just died away overhead. The whistling pierces one's ears to such a degree that its vibrations can be felt long after it has ceased."[*] This species manifestly performs in the fashion of the *Tettigonia plebeja*.

The time of song of the European cicadæ is the summer solstice; they take up the song of the crickets of the vernal

---

[*] *Ent. Mon. Mag.*, Vol. XI., p. 175:—"In the forests of the Amazons the species of cicada sing from sunrise to sunset, and at intervals through the afternoon heat. At this time the vocal music of bird and mammal is hushed, leaves become lax and drooping, and flowers shed their petals.—Bates, "Natural. on River Amazons," pp. 105, 230, and 26. In the north of India the music of the cicadæ begins at sunrise, like the alarm of a clock or tic-tac of machinery. In Pennsylvania the cicadæ appear in incredible numbers in the middle of May, and are described as "bending and even breaking down the limbs of the trees by their weight, while the woods resound with the din of their discordant drums from morn to eve."

equinox, and prolong their notes until the arrival of the autumnal
grasshoppers, as has been rightly observed by Pliny, Pseudo-
Aristotle, and others.   In northern England their woodland
melody has not yet fallen on the ear of the entomologist, but
it must not therefore be inferred these musicians are wholly
absent, for among the rich and beauteous southern fauna of
Hampshire and Surrey we still reckon one outlying waif of
the cigales, baptised by Curtis *Cicada Anglica*, seemingly the
Montana of Scopoli, if not *Hæmatodes in propriâ personâ*.   The
male, usually beaten in June from blossoming hawthorn in the
New Forest, is provided with instruments of music, and the
female, more terrestrial, is often observed wandering with a
whirring sound among bracken wastes, where she is thought
to deposit her ovæ.

The tymbals of the cicadæ were implicitly received by the
writers of Greece and Rome as the true organs of sound, and the
males, from the shape of these instruments, were called achetæ,
the shrill-sounding ἠχεῖον being a small kind of kettledrum,
which these membranes resemble.[*]   These lenticular, shell-like
membranes were also received as the producers of the sound by
mediæval Italian anatomists, and as such have passed unchal-
lenged by more modern naturalists.   The next advance was, recording the action of these tymbals during the production of the song,
a matter somewhat anticipated indeed by Julius Casserius of
Ferrara,[†] who, as far back as the year 1600, described them as
sounding by being drawn in and out by a muscle, and rustling as
goldbeater's skin.   But, at all events, the subject was taken up
and further elucidated by Réaumur,[‡] who detected the tendon *l*,
attached at the inner side of these kettledrums or tymbals, at
a short distance from their lower edge, proceeding from the
muscle *m*, which, running obliquely downwards, is inserted into
a horny internal Y piece.   This author also promulgated the

---

[*] C. G. Carus, "Analecten zur Naturwissenschaft und Heilkunde," Dresden,
1821, s. 151.

[†] Carus gives a list of Italian authors who have remarked on the singing of
the cicadæ between A.D. 1600—1700.   See also Hagen's "Bibliotheca Entomologica," B. II., p. 477.   Reference also to Latin and Greek authors.

[‡] Réaumur, "Mem. pour Ser.," T.V., Mem. 5.   All succeeding writers
on this subject have followed this illustrious entomologist in his views respecting
these organs.

received action of the tymbals theoretically; namely, that when the insect sings the tymbals are alternately deflexed and relaxed, and each time regain their shape with an elastic sonorous spring.

MM. Solier and Goureau in France, and Dr. Bennett in Australia, have established the tymbals to be the source of the sound by direct experiment, and they each state that when these are torn the sound is diminished; but if they be entirely removed the song ceases, although a moment before of deafening intensity. Carus reproduced the sound by moving the tendons with a pair of forceps.

But let us inquire experimentally what are the stimuli which prompt the Cicadæ to thus congregate and clamour. Can it be, indeed, that insects without the threatening jaws of the cricket and grasshopper kind, but having in lieu a slender rostrum to suck the sap of trees, or as the Athenians thought the dewdrop, considered as a weapon unaggressive, are incited by kindred emotions of love and jealousy to emit a music that is to be regarded as the impulse to association? The solution of this question lies in the workings of the music, and on this subject Dr. Hartman, in speaking of the well-known *Cicada septemdecim* of the United States, whose name indicates an abundance every seventeenth year, says, writing June 7th, 1851, "The drums are now heard in all directions. This, I believe, is the marital summons from the males. Standing in thick chestnut sprouts about as high as my head, where hundreds were around me, I observed the females coming around the drumming males." He again observes, in the month of August, 1868, "This season a dwarf pear-tree in my garden produced about fifty larvæ of another species (*C. pruinosa*); and I several times noticed the females to alight near a male who was uttering his clanging notes." The females are, then, attracted by the music of the males; there is reason also to infer the song of the latter is stimulated by the principle of rivalry. "Fritz Müller," says Darwin,[*] "writes to me from South Brazil, that he has listened to a musical contest between two or more males of a cicada having a particularly loud voice, and seated a considerable distance from each other. As soon as the first had finished his

---

[*] "Descent of Man," Vol. I., chap. x., p. 351.

song a second immediately began, and after he had concluded
another began, and so on."

Our European species similarly behave. Near Turin I noticed
Plebeja and *Hæmatodes* to congregate in the vicinity of a water-
wheel, on the same knoll, at a spot along an avenue, as likewise
the partiality of the latter for the Bull-frogs of the Po, to whose
" Cro-ak ! cro-ak ! crek ! crek ! " it invariably responded.   As
regards artificial sounds, carriage-bells evoke snatches of recital
from *Plebeja*, *Hæmatodes*, and *Orni*, even after the expiration of
their diurnal period of activity; and Solier, indeed, quotes an
apothecary at Aix, in Provence, to the effect that a vocalising
male of *Orni* may be induced to descend from its arcanum by
whistling so as to command its notes.   These three ordinary
European Cicadæ likewise change position, and show tendency to
migration, the distributive sequence of rivalry.   As regards the
incentive of love, amorous modulation can, I believe, be discriminated in the recital of *Orni*, who will at times measure out his
notes without blending them—" Tip-a-tip ! tip ! tip!"—and one
day I came on two on boughs adjacent, one performing, " Derde !
derde ! " and the other, " Tip ! tip ! tip ! "   Then, at the commencement of July, when the three species are sounding on the
acacias and vines, I have observed the females of *Orni* and *Plebeja*
to fly to the vocal trellis.   Fear, likewise, is expressed in the
music of the Cicadidæ.   The males of *Hæmatodes* sreak loud and
bird-like when shaken on the bough, when chased through the
air, or when beneath the bill and talons of an insectivorous bird,
as when held in the hand ; at this juncture, the tymbals, in lieu of
their wonted tremor, forcibly oscillate between convex and concave, as Goureau has already noticed ; and the paroxysm is
marked by sharp solitary cries, " Wee ! wee ! "   The male
of *Orni*, startled from a tree-trunk, utters similar querulous
notes as he flies ; those of *Plebeja*, in captivity, become a
canine snarl.

Cicadæ, as intimated, dwell on shrubs and trees.   The female
has an auger-like ovipositor, and arranges her ova in punctures
made in the branches of trees.   The isomorphous larvæ are
white, with six feet ; and on quitting their sylvan nurseries
make their way underground by means of their strong and
dentated thighs, where they gnaw the roots of plants.   They

here increase rapidly, and the nympha state is soon attained, when their appearance and habits are not materially changed. After living about a year in this state, they leave the ground during the prevalence of warm weather, and, climbing up the stems of trees in shoals, cast their nympha skin, and then are first heard their incipient chirpings in the woodland glade.

CHAPTER VI.

THE ORGAN OF HEARING IN INSECTA.

THAT insects may communicate the emotions which stir them by instrumental music, and produce the phenomena we witness, it is necessary they should have an organ of hearing. And that an organ of hearing may be found in Insecta, we infer when we notice those provided with musical files perform in turn, or on reconnoitre, and when we find them respond to their notes, or to those of a slightly different pitch, artificially imitated.* A little observation enables us to detect this feature in the music of cicadæ, grasshoppers, and leaf-crickets, and it exists doubtless in that of the crickets or cockroaches. We may likewise observe it in the stridor of some longicorn beetles *sub diro*, and in some lamellicorns when boxed together. That the required organ should present structural variety I think we may conclude from the diverse pitch, colour, and rhythm of the notes, or from their range of expression. We shall also, perhaps, find reason to look for an auditory organ in those flies and bees whose whining, spiracular music presents kindred phenomena, or in certain butterflies, moths, Solitary Ants, and Neuroptera (?), with a capacity for music. Similar deductions may be also drawn from the wing-beating employed by some moths in pairing.

Many seeking their inference in the Vertebrata, however, have been predisposed to look for organs of sense with constant position in the cephalic ring; and thus the antennæ have come to be attributed with the faculty of hearing. This idea was once very prevalent. Professor J. Rennie has proposed to dub them ears, chiefly because insects adduct them at certain

* See previous chapters.

37 Autor

No Copriginitor

+ (inview) English

+3 Satirididade
4.3 gymmica

45 Herculean art

41.

Vivitiële
Myformation

6/3
3/3
6.3
Lot 3/0

101
11.

269
104
2er
278
300
3/0
313

vibrations or sounds; and I have already drawn attention to certain papers in the Transactions of the Linnean Society, on the structural suitability of some membraneous vesicles on their surface to the acoustic faculty. But be this as it may, it is certain any structure situated in the head or its appendages in Insecta, considered as an organ of hearing, has hitherto proved excessively minute; while, on the other hand, it is now becoming an axiom that organs of sense in Insecta—and I believe, too, in Mollusca—have neither stable external position nor internal connection, save that some communicate with the anterior ganglia of the nervous chords, these nerve-knots themselves here assuming a spontaneousness of action that in the Vertebrata flows alone from the brain.

ORGAN OF HEARING IN THE CICADIDÆ (PLATE III., FIG. 5, A, B; PLATE VI., FIG. 5.)

To commence with the cicadæ, these challenge and reply seated at a considerable distance, which, considering the low pitch, the spherical propagation and refraction of their notes, would accord an auditory adit of some dimensions; the females also "alight near"[*] the musical males. Now, the corresponding cavities at the base of the abdomen placed ventrally, and closed inwardly by a posterior, tense, iridescent, and anterior soft membrane (Plate III., Fig. 5a, *m*), answer the requirements of an external ear structurally, as by intermittent exposure in the males to atmospheric impressions, in the action accompanying their music. These cavities, taking a considerably greater development in the male, are similar in either sex, and seem to correspond with the other auditory structures I shall notice. Although considered by Réaumur [†] designed to augment the sound of the tymbals, Mons. Solier, who put the conjectures of that author to a practical test with living examples of provincial cicadæ, on the removal of their constituent parts, hesitatingly assures us "the sound became a little feebler and modified;" while Colonel Goureau,[‡] having not only removed the

---

[*] Darwin, " Desc. of Man," Vol. I., chap. x.
[†] Réaumur, " Mem. pour Ser.," T. V., Mem. iv.
[‡] " Annl. Ent. Soc. de France," T. VII., p. 403, 1838.

mirrors, but also the anterior soft membrane separating the abdominal cavity from the thorax, tells us "he did not remark any sensible diminution in the intensity of the sound."

I will therefore attempt a short microscopic description of the structure and nervous connections of these membranes, in hopes that others may more fully follow up the subject. The mirrors (*m*, Fig. 5A, Plate III.) of the cicadæ *(membranæ tympanicæ)* generally consist in a membrane surrounded with an outwardly hollow chitineous frame. Among the males (Plate VII., Fig. 5) they approximate in shape a harp-shaped triangle, situated at an angle with the median plane, so that the apex is posterior and directed inward, and the inner straight side, which is indurated and blackened, borders inferiorly the tymbal motor muscles, having an inclination of about 40° to the normal vertical. The surface of the enclosed membrane makes some 30° with the horizontal median line, thus being inclined downwards; it is also slightly directed outwards. In the females the chitineous frame *(trommelring)* is a narrower curvilinear triangle transversely situated. The area occupied by the membrane is about one-sixth of a square inch in the males of the largest European cicada, but in its females it barely measures one-half.

As we shall find in the tympana of the grasshoppers and moths, a portion of the membrane is attenuated; there this part is posterior, but here in the males it is somewhat central, rhomboidal or diamond-shaped, with the principal axis approaching the vertical; in the females, on the contrary, it is cordate, and situated more inwardly or nearer the apex of the frame. As in moths also, its superficies is rendered conspicuous from prismatic hues. A yellow disc is surrounded by a band of lake concentric with exterior lines of purple, blue, green, yellow, and lake, which are most perfect in the male. The remainder of the membrane is of a neutral tint that appears in some measure due to a fine mesh of filaments, and to these or to a waviness in the membrane I would ascribe also the bright interferential spot. A concentric striation has also been observed in the corresponding membranes of grasshoppers, and it is probable that this adaptation for light analysis shadows forth an allied capability regarding incident sound waves. Structurally, the mirror is stated to be composed of a layer of elongate, hexagonal, and lenticular cells, analogous to those of the musical tymbal, but more delicate. This, however, is no peculiarity.

Towards the outer angle of the mirror is a brown and chitineous styliform discoloration (*m*), translucent, thickened at its edges by a serrated line of tubercles, attenuated, slightly angulated, with its point tangential to the iridescent central spot, and its opposite extremity directed towards another callosity (*m'*) similarly tuberculated, situated on the edge of the membrane of the mirror, triangular in shape, and near its angle. The two chitineous pieces nevertheless are not in contact, but by transmitted light appear connected by a slight (*nervous*) thread. They are neither constant in the species in number or shape; for in the female Cicada of the Ash there are two styliform pieces inclined at an angle; one traverses the iridescent spot, the other separates it from the opaque portion of the membrane. Generally these chitineous pieces harmonise with those in the corresponding membranes of the grasshopper and moths (*Trommelfellkörpchen*).

The nervous system of the drumming Cicadidæ, consisting of double chords frequently coalescing in the thorax, presents different aspects in the various species. For example, in the female of the Classic *Plebeja*, the ganglia formed by the uniting of the chords in the prothorax and mesothorax are contiguous; and from the *latter* (?) the (*acoustic*) nerve connected with the mirror is symmetrically given off (Plate VII., Fig. 1, *a*). In the male sex it is most obvious; here it may be easily traced round the motor muscle of the tymbal to the apical angle of the frame, within which it forms a ganglion that enters a groove. The other end of this groove lies towards the callosity *m*, and here a yellowish cord may be traced from the outer wall of the mirror to the tracheal pipe (*t*) of the first abdominal spiracle; in the male *Plebeja* this is free, but in the Blood Cicada it lies along the edge of the mirror. This nervous cord likewise passes over or surrounds the triangular callosity *m*. In the females of these European species the arrangement is different, however, and the ganglion does not lie within the frame. Of the several parts of insect auditory organs, recognised by German authors, I reserve description of Muller's ganglion, which should be represented by the portion of the acoustic nerve lying on the callosity *m*.

ORGAN OF HEARING IN THE ACRIDIIDÆ (PLATE II., FIGS 1 AND 3;
PLATE VI., FIGS. 1 AND 2.)

That saltatorial insects should possess capacious ears, is in
keeping with what we witness in mammalia such as hares, kan-
garoos, and squirrels, which similarly proceed by leaps and bounds;
and yet often it is scarcely possible to say whether it be sight
or hearing that causes the grasshoppers to start up over the
greensward before our advancing footsteps. But as regards the
structure of certain auditory cavities in these last we have ample
detail, thanks to the researches of Müller, Siebold, and others.
We find in nearly all an evident recess, with an ovate, lunate, or
only linear opening (Plate II., Fig. 3, *a*), situated at the hinder
and lateral portion of the first dorsal arc of the hind body, or
abdomen, partially covered by the elytra. This cavity is closed
interiorly by a posteriorly iridescent thin and oval membrane
(*membrana tympanica*) (Plate II., Fig. 1A, *mm'*), which parts it
from the first abdominal air-bladder (*t*). On its disc (Fig. 1B),
certain brown punctuate discolorations are seen, which mark
the position of two raised chitineous pieces on its internal
surface, the larger angular, the other small and triangular in
plan. To the more projecting angular piece is attached a
snow-white vesicle distended with a clear fluid that sends off a
thin arm to be inserted in the smaller piece situated towards the
centre of the disc. This tender vesicle is penetrated by a nerve
(*a*), representing the acoustic proceeding from the third (fourth)
thoracic ganglion (Plate VII., Fig. 2). Lastly, in the chitineous
setting that surrounds the iridescent membrane where it dilates
anteriorly and inferiorly, is a minute round or oval opening
(Plate II., Fig. 1B, E), forming a communication between the
air-bladder and the external air, adapted, it would seem, to the
part of an eustachian tube by immediately introducing air
behind the membrane.

The organisation and function of these organs has been but
slowly recognised. Naturalists of the last century considered
them as the source of stridulation. Swammerdam, in Holland,
thus vaguely states the grasshopper to have two peculiar small
drums, like the drum of our ear, which being struck by help of
two lunated cartilages, vibrate the air in such a manner as to

produce the sound. De Geer, in Sweden, speaking of the Migratory Locust, says :—" On each side of the first segment of the abdomen immediately above the origin of the posterior thighs, there is a considerable and deep aperture of rather an oval form (see *a*, Fig. 3, Plate II.), which is partly closed by an irregular flat plate or operculum of a hard substance, but covered by a wrinkled flexible membrane. The opening left by this operculum is semi-lunar, and at the bottom of the cavity is a white membrane of considerable tension and shining like a mirror. On that side of the aperture towards the head there is a little oval hole (E, Fig. 1B), into which the point of a pin may be introduced without resistance. When the pellicle is removed, a large cavity appears. In my opinion this aperture, cavity, and above all the membrane in tension, contribute much to produce and augment the sound emitted by the grasshopper." These authors are followed by Latreille and Burmeister. The latter sought anatomically here to find a reason for the music as follows :—On the delicate (drum) skin, near the front margin, he says, lies a small brown horny piece, to which inwardly a fine muscle is inserted that runs over to a projection of the outer horny band above and in front of the margins of the skin. By means of this small muscle, when the swift movements of the hind legs set the skin with the body into vibration, it consequently sounds.

In contradiction to the preceding, the celebrated Johannes Müller, in a supplement to his often-quoted paper on "The Visual Organs of Insects," published in 1829, having first shown the drumskin membranes to be in direct connection with the nervous chord, postulates their function to be auditory. He finds this membrane "of an almost rhomboidal shape in *Gryllus*, now *Pœcilocerus hieroglyphicus*, five lines wide, smaller in the males, and nowhere perforated or broken by the smallest flaw. When the insect retains its wings in the position of repose, this part is quite covered by the elytra. At the inner surface of this membrane lies a very delicately-skinned bladder filled with water, elongate, and over two lines in length, covering the membrane with one extremity, and with the other directed downwards (Plate II., Fig. 1B; Plate VI., Fig. 2). This bladder (Müller's ganglion) must be plainly distinguished from the tracheæ, and according to my view is not to be confounded with an air-sac. The nerve-system has its greatest swelling in the third ganglion of the nervous chord (Plate VII., Fig. 1). The brain (first knot) is smaller than the large ganglia of the chord, and these are all smaller than the third, which is of a flat shape, and from whose posterior circumference arises a number of nerves to supply the muscles of the sternum, the posterior pair of legs, and ventral parts. The fifth of these nerves (a) on either side of the third knot of the nervous chord runs to the afore-described bladder, and attaches itself to its fore upper part, where it lies on the elastic membrane. May this part be the organ of hearing of the grasshopper? Nothing gainsays it, save that the sensory nerve arises from the third knot of the spinal chord." Colonel Goureau, independently, it would seem, arrived at this identical conclusion a little later, although he does not mention his reasons.

Von Siebold and subsequent investigators have directed their attention more especially to the drumskin itself, and its attachment to the accredited nerve of hearing. Von Siebold, in Germany and in Istria, carefully examined these parts in various species, and observes the third pair of spiracles, which he considers the first abdominal (E), want the usual horny lips, and thus constantly stand open, performing as he conceives the office of a eustachian tube. They are situated anteriorly and inferiorly to the drumskin, in a triangular enlargement of the horny ring enclosing it. Beneath this the ring itself is discontinuous with a short projection, as noticed by Burmeister.

Externally, the drumskin is often more or less covered by folds of the integument. There is an anterior (semicircular) projection of the thorax, and a posterior elevation of the horny ring. These in various ways arch over it, so as to reduce the exterior adit to a more or less lunular opening. Von Siebold remarks the drumskin is tolerably exposed in some species, and perfectly so in others, and that the elytra, when closed, cover the adit partially or wholly. Interiorly, the drumskin forms the boundary to a tracheal bladder, comparable to the tympanic cavity.

The Müllerian ganglion, again, is connected with the drumskin by means of intermediate horny pieces which Von Siebold considers analogous to the bones of the labyrinth. One (Plate

VI., Fig. 1, c) is central on the disc, small and triangular; the other, situated near the anterior edge, consists in two dissimilar thin limbs (b), concave inwardly, that meet it in an obtuse angle, the upper being the shorter; and in a third darker and thicker tongue-shaped process (b') from their point of intersection, convex anteriorly and concave posteriorly, which Von Siebold compares with the cochlea. On account of its three parts, it has been proposed to call this the composite piece.

The Müllerian ganglion emits two processes. The shorter is inserted at the upper extremity of the composite piece; the other, longer and thinner, proceeds in a gentle curve to the central triangular callosity, and, according to Von Siebold, it also sends a broad band to the edge of the drumskin posterior to the spiracle. Beneath the microscope, the Müllerian ganglion appears to be really a bladder filled with a watery fluid, which, if pricked with a needle, immediately collapses, nor can it be isolated from the horny pieces without injury. The nerve penetrating it swells internally into a true cylindrical and proportionately large ganglion as it nears the tongue-shaped process, where the bladder above, covered with white pigment, presents a limpid termination. In this terminal translucent portion (Plate VI., Fig. 2), Von Siebold was enabled to distinguish little stalked bodies, presumably the enlarged terminations of the nervous threads (m); nucleated cellular bodies (l) also characterise either part.

Leydig, Hensen, and Oscar Schmidt, on the Continent, have subsequently directed their attention to the minuter structure of the horny pieces submitted to high microscopic powers, and to their attachment to the Müllerian ganglion. Leydig describes the central triangular horny piece as hollow and penetrated by numerous pores that confer on it a striated appearance; and Herr Schmidt finds nerves radiating hence over the surface of the drumskin in a fine membrane, that probably serve to conduct the sonorous vibrations. The process of the Müllerian ganglion proceeding to this capsular piece, Herr Schmidt alleges in some species, after forming a small ganglion, penetrates the membrane of the drumskin previously to its junction. These authorities notice the attenuation of the drumskin posteriorly, its concentrically striated appearance as the mesh of black or reddish cellular and pear-shaped bodies which covers the anterior portion, and also encircles the horny pieces in a curve (Plate VI., Fig. 1). These are noticed by Siebold as brown punctures, and Oscar Schmidt traces their development from their rudiments in the larva.

Herr Vitis Graber, in an intricate paper, seeks morphological harmonies in this organ, noticing in it some obvious variation and degrees of development. He observes, likewise, the anterior denser portion of the drumskin is wrinkled and sometimes hirsute, hairy, or minutely scaly, a feature which vanishes posteriorly. Hence he concludes it may be regarded as a modification of the integument, where these characters are commonly strongly marked. The hairy aspect extends to the arm of the compound horny piece in species, which, as, Leydig remarks, is itself at its extremity gradually attenuated, thus passing into the substance of the membrane. The tongue-like projection of this piece in one kind is stated to be hollow, with an exterior orifice. When treated with oxalic acid, the drumskin shows a structure of nucleated polyhedral cells. Herr Vitis Graber likewise describes the myology in the vicinity of the drumskin, and specifies one muscle as the tensor tympani.

Yet, after all, a far readier proof of the function of the auditory organs may be gathered from observation than descriptive specialisation and comparison. For if any one will observe the grasshoppers in the meadows, it will be noticed how the male, on the conclusion of his music, lowers one or both femora horizontally, retaining the elytra somewhat raised, thus exposing these membranes until he receives a response, or how, when he seeks to allure the female, he places himself so that the stridor shall impinge directly on one or other of her cavities, which she voluntarily exposes by lowering a femur (Plate III., Fig. 8). The cavities in the latter sex are usually of greater dimensions than in the males, and the drumskin more attenuated. External structures similar, apparently to these, may be observed in male cockroaches, laterally, at the hind margin of the metathorax.*

* Regarding the hearing of the Orthoptera; see John. Müller, Für vegleich. Phisio. des Gesichtssin, 1826, p. 438; Nova Acta. Acad. Leop. Carol., XIV., p. 157; Siebold, Erich. Archiv. für Naturgesch. 1844, s. 52—81; Hansen, Gehör bei Locusta, Zeitschr. für wiss. Tool. 1866, s. 2; also, De Geer, Latreille, Burmeister, Goureau, Kirby and Spence, Landois, Fischer, &c.

ORGAN OF HEARING IN THE LOCUSTIDÆ AND GRYLLIDÆ (PLATE II., FIG. 2; PLATE VI., FIGS. 9 AND 10).

The auditory organs of the leaf-crickets and crickets are commonly on the fore tibiæ, but in the latter subterranean species these organs are found less complex, and often imperfect or absent, and here occur supplementary structures at the abdominal base claiming analogy with the auditory organs of the grasshoppers. Dr. Landois, who appears to have been one of the first to recognise these and describe them as they are found in the mole-cricket, reputes them to be obsolete organs of music of the kind extant in the Cicadidæ, in support of which he notices a muscle attached to their anterior edge, which he compares to the motor tympani.

They are further described by Herr Vitis Graber, who, nevertheless, does not find them to be efficient auditory structures, although he points out their many analogies with the acridiideous organs. He commences by observing the first abdominal ring of the crickets, as in other saltatorial Orthoptera, is little developed, and that the organ in question when present is commonly found between the first and second abdominal spiracles, but the position otherwise does not seem very determinate. Its origin may be traced in the partial modification of one of a row of small plates existing in the lateral membranes connecting the dorsal and ventral arcs, to which cause the encompassing integumental ring owes its peculiar forms (Plate VI., Fig. 10, τ). It is also present in various degrees of development in the species. In the mole-crickets (Plate VI., Fig. 9) the tympanum τ, seen from within, has an oval shape, truncated anteriorly; it is excessively thin, talc-white in colour, smooth, and slightly convex. Towards the inferior edge one notices a linear impression and slight ridge, where Landois' muscle (m) is attached. In other crickets the tympanum is elongate and externally strongly convex, strewn with small callosities; or it is glassy clear and chitinised at its hinder edge. We likewise find it hairy, or possessing other character of the integument.

The drumskin generally is covered by a polygonal cellular matrix, and behind it is situated a tracheal bladder. The distribution of the nervous system is as follows. The three thoracic ganglia have their usual position, and the first abdominal knot is close to the metathoracic. The ensuing ganglion is situated in the centre of the first abdominal segment, and the next in the centre of the third. The fourth and fifth ganglia are similarly placed in the fifth and eighth rings. Of these nervous centres the metathoracic furnishes branches to the first ring and the greater part of the second, sending special nerves to the tympanal muscles, the mechanism for closing the spiracles, and the dorsal muscles of the second ring. Similarly here, as in the grass-hoppers, a nerve (Plate VI., Fig. 9, *n*) proceeding to the tympanum parts into two principal veins, one of which runs to the closing muscle of the fourth spiracle, and the other directly to the drumskin; but this latter, instead of there forming a definite termination, ramifies over the tympanum. So that, on the whole, we find in certain crickets organs with much analogy to the abdominal auditory structures of the grasshoppers; but whether they have actually effective supplementary ears must be left to future investigators to decide. Otherwise, in tunnelling insects such as constitute this group, auditory membranes posteriorly directed appear quite consonant with the habits.

If we turn now to the tibial structures of the crickets, we find that these, although they were noticed by Colonel Goureau, who also surmises their function, have their first investigator in Von Siebold, who, having recognised an organ in the Acrididæ structurally adapted inwardly and outwardly to receive the aerial sound-waves and conduct them to a percipient nerve, turned his attention to the Locustidæ. Having failed here to find a counterpart in the required situation, he on further exami-nation became struck by the fundibular tracheal openings (Plate II., Fig. 2E) present in this group beneath the hinder edge of the prothorax, a circumstance which was followed by his recognition of the double clefts in the fore tibiæ, which on examination proved in direct communication with these (Fig. 2, *a, m*), the two parts affording him the required analogies of the acridiideous organs.

Behind the posterior edge of the prothorax, then, according to Von Siebold, lay on either side two tracheal openings—one small and closed with a valvular lip, the other situated just above, a large funnel which projects with a singularly broad opening directed posteriorly. This orifice (E) having no lips, stands constantly open. In *Hetrodes pupa* the hinder edge of the prothorax has inferiorly a lunate notch, whereby these openings are exposed. The fundibular air-tubes give out no branches in the thorax, but serve as an adit to two tracheal pipes which enter the fore-legs, where they part with some ramifications. At the femoral tibial joint these become somewhat constricted, but almost immediately afterwards dilate and form an elongate bladder exactly at the place where the external clefts in the fore tibiæ are found. Beneath this they ramify and disperse.

In that subdivision of the Leaf Crickets placed first by Burmeister, the tibia is dilated superiorly with an ovate opening at either side, closed by a membrane (*membrana tympanica*) (Plate II., Fig. B, *m*). The drumskin consists of two parts, namely, an elastic silvery membrane, with an anterior lunate notch, and a black and brown oval chitineous disc situated in this notch. In another section there is present on either fore tibia a double capsule, one covering either drumskin (Plate II., Fig. A, *a*); and in some species these prolongations of the integument arch themselves strongly outwards, forming two capacious cavities with oval openings anteriorly; but in others these capsules are not so remote from the drumskins, and the cavities are less capacious, presenting only two small clefts as entries.

Von Siebold states the internal construction of the tympanal organs may be determined in large Leaf Crickets when in a fresh condition, by detaching the capsules with a knife, and cleaving the tibia right and left, so as to raise its anterior walls. Dr. Vitis Graber, more recently, has recommended placing a leg in alcohol for a year, bleaching it with potash, and if needs be, applying a drop or so of acid, a method possibly economising labour in a necessarily difficult dissection. The successful performance of either operation reveals that the elongate tracheal bladder (Fig. C, Plate II.) aforementioned occupies at this place almost the entire capacity of the tibia, only allowing the muscles, nerves, and sinews intended for the inferior part of the legs to pass posteriorly. It presents four superficies—an anterior narrower surface, with a longitudinal boat-shaped excavation; a hinder broad and arched one; and

Q

inclined lateral walls, which correspond, and are gently concave.
With either lateral wall of the bladder the tympana stand in
closest contact, isolating the anterior portion of the tibia where
is the boat-shaped excavation.

With the large tracheal tubes of the fore-feet pass down two
nerve-threads of dissimilar thickness. The thicker lies posterior
to the trachea, ramifies in the upper leg, and inferiorly beneath
the large air-bladder. The thinner behaves somewhat similarly,
but differs inasmuch as it sends a branch direct to the boat-
shaped excavation of the air-bladder, that widens just above into
a flat ganglion (n), whose under-end enters the boat-shaped
excavation in the form of a band, attaches itself at the inferior
part, and terminates. Covered above with a whitish pigment,
this ganglion contains numerous oval granular bodies, apparently the nuclei of globular cells, between which are numerous
little rods, or pear-shaped bodies, similar to those in the acoustic
ganglion of the grasshoppers. Its posterior superficies tangential to the bladder is smooth, the anterior wavy, a feature
owing to a line of thirty or forty contiguous transparent vesicles,
each containing a small pyriform nucleated mass, probably
the anterior termination of nervous threads.

Von Siebold, in conclusion, observes the suitability of the
tracheal pipes for the function of a eustachian tube in respect
of the drumskin; the division of the latter into a posterior
thin membrane and anterior chitineous thickening analogous to
what we find in the grasshoppers, as also the similar structure
of the acoustic or Müllerian ganglion, which appears likewise to
contain a watery fluid. We also have here a more or less complicated external ear. Von Siebold recognises auricles similar
to those of the Leaf Crickets in the House and Field Crickets.
In these there is apparently but one effective tympanum on
each foreleg, placed at the outer side, instead of two, as we find
in the Leaf Crickets. This is certainly the case with the House
Cricket, where we find only one opening; but in a kindred sort,
G. achatinus, there also exists an orifice on the inner side, similar
to that on the outer but smaller, and in the Field Cricket a
small round opening closed with a membrane, presenting us,
we surmise, with the rudiments of a drum.

From Dr. Vitis Graber we gather that certain Locustidæ and

Gryllidæ, reputed mute, want the tibial organs. Some of these appertaining to the former group are apterous; others have no musical vein; and one has corneous elytra. As regards the Leaf Crickets, these organs are absent in some, and these very species are those that want the serrated elytral vein and tale-like spots. I may mention lastly, in conclusion, that as regards experimental proof of the employment of these cavities for hearing, I have noticed that the male Great Green Leaf Cricket, when it performs, extends its fore-legs, as has struck me, in order to adjust these cavities, and receive the music of its rivals (Plate III., Fig. 3).

ORGAN OF HEARING IN THE LEPIDOPTERA (PLATE IV., FIGS. 6 AND 9).

An ear will, at first sight, appear a superfluity in the butterfly and moth tribe, whose males are not heard over the meadows contending in song, and whose females are not noticed following on the trail with meek attention. Nor, indeed, to supply the requirements of the little obvious instrumental music present in a few species, even when taken in conjunction with the phenomena of wing-beating, would the general prevalence of an organ of hearing in Lepidoptera be demonstrable, for the auditory might be supposed supplanted by the other senses, as we have seen the selective pageants of love and rivalry chiefly manifesting themselves in other directions, such as the aggregation of the males in alluring wing-dances, and the sedentary colour display of the female, which would postulate an adaptation of the visual rather than the auditory faculty; or, again, the sexual provision of odorous fans would have relation to the sense of smell. But, on the other hand, in seeking for an auditory organ protective and generally intimative, we have greater prestige, especially among the dull night-flying groups, where we find the many-faceted eyes small and disadvantageously placed, though phosphorescent at night and sensitive.

However this may be, a counterpart of the auditory structure of the grasshoppers may be found in Lepidoptera, corresponding in parts and structure, save that we find often here an adjunct of empty convoluted cells, parted with mirror-like membranes, which, from their introducing air behind the abdominal cavity, may perhaps be regarded as representing a eustachian tube. I have noticed this organ developed in moths of the Bombycina, Noctuina, and large Geometrina; but in Sphinx Moths and Butterflies it appears as a rule absent, or barely indicated.

Among these, it is found most readily in the night-fliers, where its position is unvarying, and its parts are identical and readily recognisable, with only a variation of form according to species; so that here it seems to impress a constant anatomical, in addition to a well-marked superficial, character.

To examine this structure, we may capture any large summer species of these moths, such as the Dark Arches, the Cabbage Moth, or the common "Dagger," the former to be found commonly reposing by day under copings in gardens, and the last, equally frequently, concealed on untarred and licheny palings—or, still better, if we can procure from willows in the neighbourhood a supply of the larger autumnal Red-underwing Moths. Then, having killed our insect, cut off the wings, and denuded the posterior portion or abdomen of its feathery scales and hair, on proceeding to examine this part in front, at its attachment to the last segment of the thorax, we are at once struck by a remarkable constriction in the segments (Plate V., Fig. 8), that seems to have had its origin during the metamorphosis of the crawling caterpillar to an insect fitted for aërial locomotion. As to how this has been accomplished, it would seem there may be some latitude of opinion; but from a fortuitous dissection made of a caterpillar just previous to transformation, which allowed me to remove, as I believe, a sentient chrysalis from the larval body, I myself incline to a notion of obsolete generation in opposition to the little-understood theory of development by Ecdysis, or the mere shedding of envelopes; and the transformation, therefore, I would seek to explain in this way. Beneath the intestinal canal, or from the stomach (?) of the caterpillar, with its thirteen equal segments, there originates a chrysalis, which, nourished on the substance of its parent, who leaves off eating and mopes, and supplied with air by a set of tracheæ from its spiracles, grows and increases in size, till it is sufficiently distended to totally occupy to the skin of its progenitor who, his internals being then consequently consumed, becomes defunct.

The chrysalis now emerges at a rupture at the anterior upper part of the integument, with the third or fore-wing-bearing ring of the body most capacious, an enlargement effected by a contraction of the following four segments. From this torpid pupa, in due time, issues the active imago or perfect insect, exhibiting a further metamorphosis, for the fifth and sixth segment have been reduced to dorsal arcs, forming a pedicle by which the abdomen is united to the thorax, and the skin of the lower arc of the seventh has become consolidated to protect the lateral front edges of the ventral region.

In all the aforementioned moths, the organ attributable with hearing (*a*) is found between these contracted segments and the metathoracic that immediately precedes them, and projecting posteriorly, it may be said to occupy the transverse section of the fourth, fifth, and sixth rings. It is bounded in front by the muscles of the metathorax; behind it is encased in a saddle-shaped tube, varying in consistency from the most delicate white membrane to a hard, yellow, opaque substance which the needle-point chips with difficulty, as seen in the Red Underwing. The external ear (*a*) is at once recognised in a largish canal, or meatus, that here penetrates the body at either side, somewhat oval in section, with a posterior concavity or conch, that occasionally is seen to extend as far back as the termination of the seventh segment, or third of the abdomen, conferring to the mouth in the female Dark Arches an extreme length of $2'''$, and depth of $1'''$.

The adjuncts of the auditory canal are uniformly present, and consist in a minute protuberance in the conchoidal cavity posteriorly (Plate IV., Fig. 9, *w*); semicircular in profile in the Dark Arches (*X. Polyodon*); lanceolate, clothed above with smooth hair, in the Red Underwing (*C. Nupta*); or very elongate and spatulate, as in the Silver Y, or Gamma Moth. Placed behind at the origin, and concealed by this protuberance, is the opening of an air-tube or spiracle, which perhaps should be considered as belonging to the metathorax. The other adjunct is somewhat singular; it is a little membranous valve, with a fringe of hair (*v*), triangular in shape, attached at the front of the auditory canal, and connected by a muscular ligament to the

base of the hind wing, and in a way to participate in its movement. When the insect takes wing, this valve, like the poisers
of the fly, leaving its horizontal position of rest, beats up and
down above the ear-hole, so that the tip of its fringe just grazes
the round protuberance (w), suggesting its purport as a mag-
nifier of the sonoral atmospheric pulses, either by a sudden
condensation of the air in the ear-hole, or as a deaf man puts his
hand to his ear to collect the vibrations.

The auditory canal (a) is closed by a tympanum (m), very
similarly in all these moths, and shows many striking points of
agreement with its counterpart in the grasshoppers, but with
this difference—that its structure is far more delicate. The
tympanal membrane (Plate IV., Fig. 6) proceeds from the front
of the auditory canal near its entry, and closes the external ear
vertically and obliquely, so as to be somewhat presented backwards. Its surface is convex, and its outline elliptical. It is
divided vertically into two distinct portions, indicated by one
(*Nupta*) or more (*Polyodon*) horny, chitineous pieces (b), placed
on the disc. Of these two parts, that outwardly from this projecting knob, or knobs, is distinguished by a milky opacity in
the membrane; the most inward (m), on the contrary, is extremely tense, and of great tenuity, the slightest touch causing
it to rend and collapse like a withering flower-petal. Another
striking character is seen in a beautiful iridescence, or mother-
of-pearl refraction of light, that appears to portray a rough
surface and concealed acoustic power, for this is doubtless the
sentient portion of the *membrana tympanica*.

In a careful dissection it is just possible, in a large moth,
when all the other parts have been removed posteriorly with
the scissors and needle-point, to successfully detach a portion of
the tympanum from the muscles of the thorax, so as to reveal
the structural mechanism of the internal ear (b). Fig. 6 shows
the operation performed in the Red Underwing, and the various
parts then exposed, which form a striking resemblance with the
corresponding elements of the internal ear of the grasshoppers.
It is found that the removal of the tympanum has disclosed a
small air-vesicle it shuts off from the auditory canal. To the
inner side of the elongate club-shaped horny piece on the disc
we see attached a slender white cord, which it is easy to trace

to a little membranous sac of a cylindrical shape (*m*), distended with a fluid to which it is united superiorly. This membranous sac will be directly recognised by the experienced eye as the "membranous labyrinth" of Siebold, and "the thin-skinned bladder filled with water" of Müller (Plate II., Fig. 1B) ; but what is most singular is the position it occupies, as it is not, I think, as in grasshoppers and locusts, attached closely to the horny piece or pieces on the disc of the tympanum, but only connected therewith ; while, at the other extremity, it lies on the elevator muscle of the hind wing. From the superior extremity of this membranous labyrinth, proceeds another short process to the upper horny margin of the chamber, where it appears to ramify ; and from the lower issues the auditory nerve (*n*), which proceeds in the direction of the fourth (?) ganglion or nerve-knot, obliquely across and round the elevator muscle (Plate VII., Fig. 3, *n*).

By dissevering the abdomen at the junction of the third and fourth segments, and then laying bare the posterior saddle-shaped incasing of the corresponding transverse organs of hearing, a judicious use of the needle-point may be employed to reveal some accessories to the parts already described (Plate IV., Fig. 9). It will then eventually become clear that above the chamber containing the membranous labyrinth and auditory nerve, parted by a membrane sometimes white with a shining spot, sometimes wholly glassy and iridescent (*m*), is a second air-vesicle (*c*), contained in a more or less ovate process, which on either side forms as it were the pommels of the saddle (*b*). This second cell will be found, I think, on the one hand, to be in connection with the chamber containing the membranous labyrinth and auditory nerve by means of a little orifice ; while, on the other, it communicates with the external air by a lateral chitineous tube that opens at the side of the abdomen at the junction of the seventh and eighth dorsal arcs, immediately above the conch of the external ear (E) — a premise which leads us to suppose we have here the counterpart of the eustachian tube, designed to balance the atmospheric pressure on the delicate membrane of the tympanum. (See Fig. 6.)

Lastly, there is above or in front of the two central ovate cells, and communicating with them, a third air-cell (*d*), which contains an intricate structure of membrane and tracheæ; and here the main air-pipes, if I mistake not, enter the abdomen by a single tube, which again separates into the two main pipes of the abdomen immediately behind the auditory organ. In the Cabbage Moth there are just beneath this cell two horizontal semicircular membranes (Fig. 9, *d*) attached posteriorly, with a fringe of minute *nervous* (?) filaments on their edge. The internal convolutions of this assemblage of cells that on either side cap the auditory chamber, are as grotesque as their outward configuration; and their rude representation of two hunting-horns, placed mouth to mouth, would remind us of the cochlea, were not we already convinced of the different situation of the membraneous labyrinth. Their function is in part, I suppose, that of the eustachian tubes, and they possibly reproduce the mastoid cells.

I had for some time been familiar with the aspect of the auditory organ in the Noctuina before I began to attack the silk-spinning group of the Bombycina. In these the accessory of convoluted cells is obsolete, so that the mirror bearing on its disc a horny styliform piece may be wholly exposed by dissecting the abdomen as previously, and then cutting it longitudinally, so as to view the structure from within. The component parts, however, in a majority of our indigenous species are exceedingly minute, and little more is to be gained than that the ear here presents a transition from the complex form of Noctuina to the far simpler structures of the grasshoppers and Cicadæ. This is, too, what we might expect from a comparison of the habits of the moths composing these two great divisions. The Noctuinæ when disturbed are restless and fleet of wing; the Bombycinæ slow of perception and apathetic to touch; many, as the beautiful Tiger Moths and Ermines, remaining sunk in coma with their limbs limp and listless, even after impalation. Yet in the latter group alone, we have remarked certain individuals to possess sound-producing organs, a perquisite we previously have associated with this simpler form of ear, and from whence we might likewise infer their noises are intended for communication, being probably employed by the sexes in pairing.

In the stout-bodied moths with looping caterpillars these organs are likewise rudimentary. In the Scolloped Oak (*Cro-*

callis elinguaria) they project into the abdominal air-bladders in the form of two isolated pear-shaped bodies, partially divided by a septum, and terminated by an iridescent mirror-like membrane. It has been fancied the more slender butterfly-like species of the Geometrina have auditory power, and certain experiments are recorded in Nature that were made on the acoustic susceptibilities of the large Magpie Moth. In the Sphinx Moths indications of the external auditory cavity appear commonly wanting, yet possibly the stridulating kinds have powers of communication. It may be remarked the late Colonel Goureau, in searching for the musical organ of the Death's Head Moth, mentions a white muscle, inserted into the borders of a slight cavity covered with smooth transparent membrane, on either side of the first abdominal segment, just beneath the insertion of the fan.

Among the flies and bees, which are commonly unprofitable objects for such investigation, I fancied I once found something corresponding to this structure in certain Crane Flies; and Dr. Scudder finds a thoracic nerve in Bumbles which he thinks may serve for hearing, but there is in the latter instance no external indication.

ORGAN OF HEARING IN THE COLEOPTERA.

With regard to those longicorn beetles that stridulate and reply when perched many yards apart, in the fashion of the grasshoppers, leaf-crickets, and Cicadæ, we might surmise the existence of auditory organs of some dimensions, although further than certain peculiarities in the structure of the forelegs I find nothing suggestive of such. The species of Dung Beetle (*Geotrupes*) stridulate in concert when boxed together. The male does so when chasing the female; and on approaching an individual of this genus walking on the road, it is seen when the observer arrives at a few paces' distance suddenly to contract its legs and remain motionless, so that these likewise might be with little doubt attributed with hearing. Here too, certainly, structural indications are present, though minute, of organs presumably sound-perceptive. For in this genus we find anteriorly four ganglia (Plate VII., Fig. 4). The first is oval, situated at the posterior part of the head; the second diamond-shaped, and

centrally placed in the prothorax; the third, uniting the two
forms, lies at the anterior part of the mesothorax, and surpasses the
first and second in bulk; a fourth ganglion lies on a horny sternal
piece in the metathorax.   The position of the structures claiming to be auditory, as in Lepidoptera, Acridiidæ, and Cicadidæ, is
at the opening of a spiracle, here the metathoracic.   Forming the
internal superior boundary of the prothorax is a chitineous saddle-
shaped piece, as in moths; but the cellular structure is wanting
centrally, where it merely forms a *point d'appui* for the insertion
of the thoracic muscles.   At its extremities, where it abuts on
the metathoracic bladders that lie at the wing-roots and serve
their inflation, it is intimately connected with small chitineous
rods directed inwardly, while a white bladder (u), with a pitted
surface and seemingly enclosing a fluid, arises as a white tube
that communicates with the external air at a point where the
external angular pieces of the thoracic dermis unite their apices,
and then, passing along inferiorly to the saddle-piece, joins the
apex of the chitineous rods and distends to an ovate form, after
which, again attenuating, it ends in a nerve which runs to the
ganglion in the metathorax.   A minute membrane lies just
superiorly to the termination of the white tube, and contiguous
to the spiracle, concealed in a depression.   It is ovate in form,
and pellucid.   The only well-developed part here, nevertheless, is
the ganglion, or bladder.

In conclusion I should mention the implement I have found
most essential in those investigations of these auditory structures
I have myself undertaken, is a pocket lens of three glasses; a
microscope, from its obtuseness in regard to opaque subjects, has
proved of little service unless of the very best, and then it in no
ways stands to the former in the relation of an astronomical
telescope to a star-finder, as here the chief labour is manipula-
tion, and the higher power can only be employed to reveal the
tissues of the parts, not to unravel their composition and
relations.

Should any seek to know more of the capabilities of these
lowest of ear-structures than is to be gleaned from crude
anatomical description, resort may be made to observation and
cautious experiment.   And having notated a series of insect
passages, and the excitation under which they were produced,

we may, if our ear be fine, even venture on a rendering into sentiment, by the formula of Mersenne, an old and sage philosopher and mathematician who flourished during the earlier portion of the sixteenth century; and the result, I think, will fully justify the assumption of a common sound-perception to our own, participated in by these humble instruments. To this end the vowels *a* and *o* must be interpreted physiologically to signify what is grand and full; the vowel *i*, that which is small and penetrating; *e*, subtlety and sorrow; *o*, is expressive of strong passion; *u*, belongs to things secret and hidden; *f*, *th*, *wh*, and the like, frequent with insects, denote sharpness or vanity; *s* and *x*, bitter things; *r*, the canine letter, violent and impetuous emotions; *m*, magnificence; *n*, things dark and obscure; and so on.* Take now in illustration that pretty ballad of Percy's, "O Nancy, wilt thou go with me?" and compare it in the fields with grasshopper stridulation, then it will at once strike you that while the flaunt of the first six lines will suggest the translation of many a rival challenge around, the refrain of the last couplet is nothing less than a rendering of the common pairing note. So our ballad, be it noticed, first speaks of colours and danger, rus*set* gowns, *sil*ken *sh*een, a wi*sh* behind, peri*ls* keen, mi*sh*ap to rue, and such-like; and so we think we hear the grasshoppers, each in their own dialect, defy their mates. After this comes the tend*er* te*ar*, *regret*, *scenes* so gay, and w*ert* fair*est* of the fair; harsh retrospect, at least, it must be allowed, represented in the grating rhythms of those saddest of little lyres when death follows fast on the reproductive gatherings. And on such pleas is it we would claim for these insect ears analogous structure and perception with our own.

BIBLIOGRAPHY.

1829. J. Müller, "Zur vergleichen den Physiologie die Gesichtsinnes," p. 439.

1832. Burmeister, "Handbuch der Entomologie," b. 1, s. 512 (Berlin).

1837. Goureau, "Annal. de la Soc. Ent. de Fr.," p. 31.

* "Wonders of Acoustics," by Robert Ball, M.A.

1844. Von Siebold, "Archiv. für Naturgesch.," s. 52–81.

1853. H. Fischer, "Orthoptera Europea," Lipsiæ.

1855. Leydig, "Archiv. für Anat. Gehörorgane von Oedipoda cœrulans."

1866. Hensen, "Zeitschrift für wiss. Zoologie."

1874. Vitis Graber, "Mittheilungen des Naturwissenschaftlichen Vereines für Steiermark," s. 22–31, Graz. (Gryllodea und Locustina).

1875. Oscar Schmidt, "Archiv. für Mikroskopische Anatomie," b. ii., s. 195–215, Bonn : " Die Gehörorgane der Heuschrecken."

1876. Vitis Graber, "Denkschriften der Kaiserlichen Akademie der Wissenschaften " (Wien).

" Die Tympanalen sinnes apparate der Orthoptera." "Die Abdominal Tympanal organe der Cikaden und Gryllodeen."

CHAPTER VII.

THE FOREGOING PHENOMENA SUPPLEMENTED BY MIGRATION, WHICH INDUCES VARIATION AND NATURAL SELECTION.

> "Densior his tellus, elementaque grandia traxit;
> Et pressa est gravitate sui.   Circumfluus humor
> Ultima possedit, solidumque coërcuit orbem.
>   Sic ubi dispositam, quisquis fuit ille deorum,
> Congeriem secuit, sectamque in membra redegit,
> Principio terram, ne non æqualis ab omni
> Parte foret, magni speciem glomeravit in orbis.
> Tum freta diffudit, rapidisque tumescere ventis
> Jussit, et ambitæ circumdare littora terræ.
>   Addidit et fontes, immensaque stagna, lacusque;
> Fluminaque obliquis cinxit declivia ripis:
> Quæ, diversa locis, partim sorbentur ab ipsâ,
> In mare perveniunt partim, campoque recepta
> Liberioris aquæ, pro ripis littora pulsant.
> Jussit et extendi campos, subsidere valles,
> Fronde tegi silvas, lapidosos surgere montes.
> Utque duæ dextrâ cœlum, totidemque sinistrâ
> Parte secant Zonæ, quinta est ardentior illis;
> Sic onus inclusum numero distinxit eodem
> Cura dei: totidemque plagæ tellure premuntur.
> Quarum quæ media est, non est habitabilis æstu:
> Nix tegit alta duas: totidem inter utramque locavit."
>
> *Ovid*, "*Metamorphoses*," i. 5.

THE cosmogony of our globe has engaged the attention of the philosopher in every age; but as the stream of Science rolled onward new powers accumulated to be repeatedly enlisted in its investigation.   The astronomer of the day, who at midnight, looking out on the confines of space and time, observes a nebulous vapour gyrating in interstellar ether, first descries the portentous nucleus of a phantom world.   The spectroscopist, in the rainbow hues refracted from his battery of prisms, then seeks certain dark lines, where he reads the qualities of some of the hundred simple elements entering into the composition of our globe, in gaseous, liquid, or solid form; which the chemist, from acquaintance with their laws of mutual attraction, evolves out of a glowing mass into earth, air, and water. On the new-born orb, shaped by the laws of relative attraction and revolution, the anatomist postulates but the advent of a germ of physical life, whose Promethean and yeast-like ignition in dead inorganic matter corresponds in effect to electricity and magnetism, and which some think may be imported by a meteorite; and materials are at hand, accumulated by botanist, zoologist, and surgeon, to weave the tissues of species with almost unbroken thread leading up to man.

But this speculative research, mainly accrued from the destruction of objects surrounding us, is too often undertaken with the alien aim of providing for the increasing exigencies of existence, rather than that of establishing physiological laws and their operation, besides being totally irrespective of the globe's actual history, which remains recorded and sealed up for us in the disposition of the materials constituting its surface, that indicate each successive operation by which the present organic arrangement has been assumed. It is on these grounds geology synthetically confirms the actual formation of the earth's varied prospect by igneous and aqueous agency, and this with an evidence unquestionable; comparing each old weather-worn mountain and its radiating dykes with the glowing volcano and lava streams of a Hecla or an Ætna, and their smoothly excavated gullies and boulder clays, with the chilly Alpine glacier and its terminal moraine heaps; or, again, by paralleling lamellar aqueous rocks and their embedded fossils with the creeping ocean beds or river bottoms; recording as measures of forgotten time the present swelling and falling of the earth's still seething crust and the eroding power of all fostering heat, causing the perennial lapse of streams and winds. Similar uniformity is presented in the creation of the bright mantle of life which in the practical theory of geology, now worked out in various parts of the globe, springs into existence in progressive groups during the reign of these chronic laws, harmonising with that now distributed on the teeming globe in form, structure, and even colour and evident habits of life. Here the vegetable world, dawning as sea-weeds in slaty Silurian rocks, terminates with the familiar twinkling forest foliage in the marls of the European Tertiary;

and the animal world, first recognised as gelatinous zoophytes, concludes with indications of star-gazing man himself. Species uncouth like the ichthyosaurus and pterodactyle, intermediate between existing orders, are also proved to have once abounded.* The proximate theory, acknowledging generally that the more simply organised life forms had existence previous to the more complex, is the received notion or development view of geologists. Then in order to facilitate the comprehension of this progress of Nature, which in well-stocked cabinets of extinct as of recent life, from close sequence in kinds, appears zoologically to be but her phases, many engaged in systematic classification have applied themselves to determine whether indications of divergence occur in existing organic forms, themselves intrinsically inconstant in the incessant phenomena of nutrition, assimilation, circulation, and subsequent death. The attention of naturalists has thus been very generally directed to the variation, distribution, and reproduction of kinds ; investigations which, as regards the lower eccentric tribes, cannot be pursued without important result.

Firstly, as regards the variation of species. No one who walks in his garden or visits his pigeon-cots, dog-kennels, and rabbit-hutches—nay, strays as far as his farm and poultry yard—will, I suppose, deny the existence of an enchantment that has been divinely evoked to minister to his daily wants or to afford him innocent enjoyment : and with those who study and form extensive and special collections of natural objects, the universality of the laws here rendered prominent and embryonic by cultivation originates a reluctance to define kinds, which in all departments of Natural History take rank in groups as typical species, that, as regards the lower orders are far outnumbered by their numerous variations and aberrations. The discovery of the successful propagation of one of these varieties in a favourable locality to the replacement of its ordinary type, has been supplemented by the recognition of a struggle for existence, survival of the fittest, and of sexual and natural selection ; laws existing, that we see invoked to render incidentally acquired character permanent, and which we have at length every reason to believe produce progres-

* *Trans. Geological Soc.*

sive descent by successive modification and consequent acquirement
of new powers. Yet it must not be denied this, the theoretical
view of the question, since its advocation by Dr. Darwin,* has
met with notable opposition from the professional and scientific
world. There is the specialist, who on the one hand acknowledging a grandeur of conception and limited certain application,
yet hesitates in its adoption, from the apparent introduction of
typical forms, their persistence, or inverse disintegration, in the
geological strata, or from a tendency of varieties to recur ; or, as
regards man, from some negation to volition in connection with
his progress. There is also the cosmological dogma, which
assumes each species of organic life endowed with an invariable
structure and properties, adapting it best for the kind of life for
which it was designed, but which advances nothing to account
for its existence, and a spontaneous evolution theory, which
supposes organisms to arise without germs; laws which, up to
the present time, have afforded mankind no evidence of their
existence.† On the other hand, the more modern evolution
theory finds support in the embryological, which is now very
generally accepted among students as harmonising with the
geological evidence.

As regards reproduction, the embryological theory which traces
the descent of the organic world in its growth is well known.
How the various forms arise in the germ from the uniting and
increase by division of organic cells to a frothy structure termed
" blastoderm," that separates into two or more layers according
to the complicity of organisation of the individual, and in which
the several parts systematically appear : the spinal cord, observable from the Ascidian upwards ; the similar division of the
embryo into segments ; the bars posterior to the mouth, which
in fish become gills, but in reptiles, birds, and mammals form
very different parts ; or, lastly, the bone, which, suspending the
lower jaws of birds and reptiles, is converted in mammals into

---

* " Origin of Species," " Descent of Man," and other works.

† See " Insect Transformations," chap. i., " Lib. of Entertaining Knowledge,"
Lond., 1830; also " Fragments of Science," by Professor Tyndall, Vol. II., pp.
253—336. The mooted points are the phenomena of fermentation and putre-
faction. As some theologico-physiological difficulty appears to have arisen
with regard to the signification of the word " Bara " (*created*) in Genesis, the
writer may be forgiven if he refers to its use in Ps. civ. 30.

the hammer-bone of the ear. Independent of the season of adolescence, man in the fetal state has been said to pass through the successive forms of fish, reptile, and mammal; and other organised beings prolong such season of progression into a period subsequent to exclusion, the tadpole relinquishing its pisciform branchiæ and tail, and becoming an atmospheric-breathing reptile; and lower still in the scale we may instance our vermiform insect larva or caterpillar, who sometimes similarly leaves an aqueous life, and by casting successive coverings evolves a winged bird-like form. The parts of plants arise in like fashion, and generally organic species spring with common features and pass through the stages of those inferior in rank previous to assuming the perfect form. The animal kingdom, according to Professor Haeckel, thus indicates division into two classes: those in which there is no internal body-cavity—the Protozoa—and those in which such cavity exists—the Metazoa. From the latter appear evolved on the one hand Echinodermata and Cœlenterata, and on the other Vermes; from which latter, again, all the other groups of both invertebrates and vertebrates may have arisen.*

Lastly, we have the distribution theories lately propounded in this country by Messrs. Sclater, Wallace,† Bates, and others. These, in their wider acceptation, seek characteristic groups presented by the life aspect on the globe's surface, as confined to certain natural regions or centres of distribution. Consider in what measure a predominance of forms in each locality, as that of the Edentata in South America, or Marsupialia in Australia, may be due to the isolation of oceans, seas, deserts, and mountain chains, whose comparative chronology is known to the geologist, and which modify climatic dispersion; and also examine what claims eccentric species like the Marsupialia of South America have to be treated as remnants of past geological distributions. And, secondly, to determine the laws active in modifying existing

* Abridged from Prof. Allen Thomson's address, "The Development of the Forms of Animal Life," Brit. Assoc., 47th meeting. In respect to insects, German anatomists have found that the embryos of Mantis, Water Beetles, Bees, &c., have an extra pair of legs that are never perfected; thus indicating the descent of these insects from a form with eight legs. See Dr. Vitis Graber. "Die Insekten."

† A. R. Wallace, "Distribution of Animals," and other works and scientific papers.

R

life on these centres, and apply these in accounting for the
progressional aspect of extinct life on the same and other areas.
In this way it is proposed by the co-operation of zoologists and
geologists to localise the various genera of life on the earth's
surface; trace these to some original general dissemination, like
that of the elephants which in Tertiary time roamed the old
world from Northern Europe to Australia, or to some original
centre of distribution; and to consider how far their tendency
to variation will allow of their being allotted a single typical
ancestor, then or there originating.

In regard to our present topic, Entomology, these theories
have been worked on by students with more or less success; but
as they are all dependent on the geological evidence of an antiquity and progression of species, for which they seek to afford a
reason, it will be well to examine how we come to obtain a
history of a past succession of insect races on the terrestrial
surface, and give some idea of the extent to which this record
has been deciphered. To attain this end it will be necessary to
divest our mind of any bias to the notion that these familiar
woods and well-known hedgerows, along which we stray of a
summer's day to catch Fritillary and Hair Streak Butterflies,
are more substantial than the transient and highly-coloured
images of dreamland. There was a time when they physically
were not, and a time may arrive when they shall not be. On
their drapery are inscribed thick the imperfect characters of
ceaseless change; periodically confessed in the burning green
of soft spring, or crisp and gold ink of sapless autumnal suns.
From the bosom of the maternal soil and darkness they shed
their odours and arise at the first whispers of the northing
beams, speed eddying in their solar life circles, fade and dissolve,
and then stand slumbering and mouldering on the chilled globe,
until again called to exuberance by the glowing atmosphere.
And so likewise in regard to the light, fairy, and elfin life that
sports in their tangles. On an iron globe, composed of hard and
sterile minerals, where atoms are thought to combine by affinity,
or attract and repel each other, where annihilation is not recognisable, and nothing is virtually lost, breathing life, atmospheri-
cally pent in its varied sphere, clothes by incessant assimilation
its beautiful forms with the scant organic mould ushered in by

the decay of the rock-incrusting fungus and fucoid, that earliest
dawn of aërial and marine existence. These, advancing towards
perfection, attain their most attractive garb chiefly at the reproductive season, and then as insensibly fade into the venerable
matrix whence they sprung, yielding to the energy that previously sustained them, and leaving their embryos to carry
onward the torch of life. But neither the thick forest shades
nor the undulating sea-tangle have thus uninterruptedly grown
on the acres they now cover, forming ever insensibly increasing
beds of brown or black vegetable mould, the mingled *débris* and
germs of things that were. Even the solid ground over which
we ramble was certainly at some epoch a wide ocean, and not
land, where slimy sea-shells and populous corals, not trees and
insects, multiplied—information which is generally attainable
for us at the nearest roadside cutting or wayside quarry.

Indeed, at this day the insensible interchange of sea and land
which awoke the wonder of lyric poet and mediæval troubadour,
has passed into the domain of stricter science; so that, proceeding back to the earliest hours of terrestrial life, we may
confidently see a circulation of water, as clouds, rivers, lakes, and
seas, wearing down an early volcanic land of silica, felspar, and
mica, and thus forming muddy sediments that gradually become
cemeteries of the life that crawls in the lymph, or luxuriates on
its borders, and which, by lapse of time, are to be converted into
hard fossils entombed in leaf-like layers within the great book of
sandstone, limestone, clay, and slate. We may then turn to
consider the other co-operating agent, the slow upheaving and
depressing, insensibly oscillating, and very clearly traceable
along all lines of coast, lake, and river, in Europe, Asia, Africa,
or Polynesia; that in one spot is slowly raising the beach with
its shelly muds from the teeming water; and in another often
not far distant, gradually depressing the land with its garb of
woods, lakes, and their accumulations beneath the sea waves,
and so producing alternations of fresh and salt water sediment
or strata on the surface of the globe, which thus becomes coated
over as a candle repeatedly dipped in the tallow-vat. At the
present period this mechanical work of heat is easy of illustration.
We witness the land surface everywhere tracked by water-
courses, which wear for themselves distinct hollows, as that

occupied by the Thames and confluents. We may show that
generally the eastern shores of England and North America are
sinking, while that of the Baltic is rising, and this also at a
definite rate of a foot a century, and so forth. We may also
mentally work out through long geologic ages, the action of this
twofold minister of change and destruction, ever silently heaping
here, the fresh-water accumulations of lakes and rivers on old
raised and rising oceanic bottoms, as now over the flat pampas
and llanos of America, or over our older valleys of sea limestone
and chalk; and there, salt-water muds with shell and sea tangle
on sinking landscapes, as now along the depressing Sussex coast.
Or we may note its chronic accelerations and retardation, by the
throes of volcano and earthquake, recorded in contortions and
upturnings of these sedimentary or stratified rocks that now en-
crust like coats of an onion the entire face of the globe, and even
infer the comparative age and period of activity of each weathered
mountain chain—erst volcanoes, granitic, basaltic, or volcanic—
which, at successive periods and in various places, has burst
through some of these aqueous rocks, and risen into existence, as
in modern time the cindery Monte Gaurus or Jorullo in a night.
Gravitation subsequently co-operates with heat in levelling the
earth's surface; and as lunar attraction sweeping the tidal waves,
it becomes a permanent agent in sculpturing and bevelling the land
each time it rises above water-level into islands and continents.

We will now review the strata containing insect remains in
order. From the absence of unequivocally marine fossils and
from lithological characters, it is inferred that the Scottish Old
Red Sandstone probably originated in inland sheets of water,
which Professor Geikie distinguishes as a northern Lake Orcadie,
occupying the site of Caithness and Orkney Islands; a Lake
Caledonia, occupying the central valley and basin of the Clyde;
a Lake Cheviot; a Lake Lorne, in the north of Argyllshire;
and the Welsh Lake. The land surface diversified by these
ancient sheets of water is stated to have arisen from the shallow-
ing of the sea, which became converted here and there into
salinas, or inland seas, by a series of subterranean movements,
which have left their indelible traces upon the upturned Silurian
rocks, while the now rock metamorphosed sandy material was
deposited in their basins during a subsequent general subsidence.

At this period, by tracing the deposits of sandstones and muddy limestones, replete with the earliest types of cosmopolitan shells, geologists delineate an ocean studded with coral-fringed islands occupying the future site of Europe and North America, whose volcanoes and larva streams, seen in the conical mountain peaks, corrie, and dyke, still form recognisable features in our Highland scenery. The first vegetation of these islands faded on the

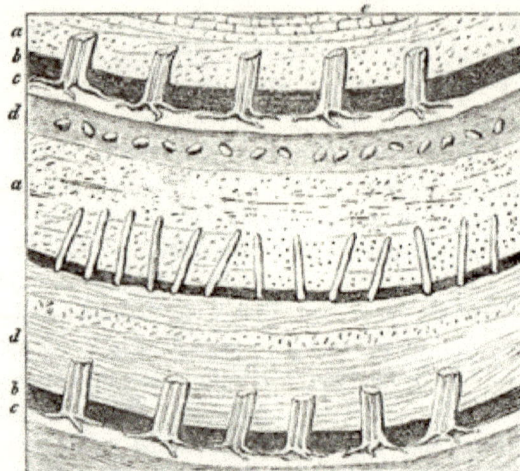

DIAGRAM OF A COAL BASIN, OR ANCIENT LAKE BOTTOM.
Fresh-water layers—a, sandstone; b, coal; c, under clay or ancient soil; d, shale and iron-stone balls. Salt-water layer—e, limestone.

stranded districts once its own, must be sought over barren and thickly-populated spots where seams of sooty coal (See Fig. b) constitute the perennial falls, matted leaves, seeds and bark of this rustling land that, originally a raised sea-bed, again and again slowly sank in some lake or river, which as the land subsided buried its rotting columnar trunks of sigillariæ, slimy with land shells and millipeds, and its rustling beds of tall and reedy calamites in drifting layers of sand (a) and clay (d). And this went on until this sediment accumulating, or raised (as has been the opinion), reaches the surface,* so as to allow another and another thicket to spread from the land, spin a light tress of

* Some have thought the deposit to have been successively *raised* and depressed. (Yet why ?) Anniversary address of Leonard Horner, Esq., Journ. Geological Soc., Vol. II., p. 171, 1846. Sir Charles Lyell, "Principles of Geology."

fern and reed, sink and become covered over as the former, and so by continued depression the land surfaces eventually alternate vertically with aqueous beds like the lettered sheets of a ledger. When, lastly, the wreck of these zephyry scenes now lowered to the sea level plunges bodily under the moaning waves and becomes covered with drift-wood and sea-shells, sands and muddy limestones ; then on cessation of the depression, a reflux of the elevating force to this area acting on the teeming sea-bed, raises the sunken surfaces as dark seams of coal parted by layers of shales and sandstone (as shown in the diagram). This, the common history of carboniferous accumulations in the northern hemisphere, is specially interesting to the entomologists, as here in excavating the nodules of clay ironstone at the ancient tree-roots for the smelting furnaces, the workmen virtually dig out the rusty infusorial scum left by this perpetually shoaling water, enclosing impressions of its floatage of ferns and fruits, king-crabs, myriopods, and insects. So, curiously enough, in seeking a limestone flux to separate the metal, they blast the adjacent marbles (e) replete with the well-known ammonite and sea-lily, and once redolent of the foam that broke on the advancing coral reefs.

Standing on the purlieus of the black countries we may with little difficulty evoke these ages of fern-crocheted glades and shadow-chequered dells, embroidered and twinkling with light fronds of *Pecopteris* and *Sphenopteris*, overshadowed by tall gloomy beds of chafing reeds fifty feet in height, bare leaf-scared stems and fir-like Lepidodendron, the giant analogues of the ditch-side horse-tail and club-moss—dells and glades where early insects crept and flew, such as tinkling leaf-crickets, cockroaches (*Orthoptera*), painted May-flies, dragon-flies, decay-loving white ants (*Neuroptera*), beetles (*Coleoptera ?*), mimetic shepherd spiders and scorpions (*Arachnida*), that lent animation to a bright sunlight. Or we may look from out these dark thickets, ferny brakes, and beds of wind-swept reeds, over a vasty sea, unfurrowed as yet by the keel of the mariner, whose summer calm was alone broken by the looming of a floating nautilus, the whispering cat's-paw, or the ripple of a bony scale-fish—a scene of peace slumbering around early volcanic centres that as ages glided by told out the course of time by chronic convulsions, vomiting pillars of smoke and scoriæ by day, and red lava reflecting against the midnight sky.

The first evidence of this carboniferous forest and its insects on northern areas is found in Devonian beds. Its plants are stated to present the essential structure of their living representatives. The ferns are of existing type with some now extinct. The Equisetaceæ were arborescent and large-leaved, with fruit cones protected by scales containing small spores similar to those of living horse-tails, and alike furnished with hygrometric elators. With these vascular cryptogams of the swamps coexisted some true conifers yet little known. Its insect tenants, for the most part recovered as the merest wing fragments, whose identification is precarious, are earliest detected by Dr. Scudder in a dark shaly bed of Devonian age near St. John's, New Brunswick, presenting us the pristine marsh or esturine slime where marine shell and trilobite crawled among decaying reeds and fern. These oldest imprints are thought to belong to the Neuroptera, a tribe mostly aquatic in its earlier wingless state, and which, represented by the May Fly, Caddis Fly, and Dragon Fly, still lights up with its glitter the umbrageous nooks of river and ditch. The best determined, however, of Palæozoic insects are certainly the terrestrial cockroaches, of which the coal measures of Arkansas in America, and those of Saarbrücken and Westphalia in Europe, have yielded species in tolerable preservation. These by habit are well suited to increase and multiply on the damp leaf-fall and succulent stems of such plants as ferns and sigillariæ. On the whole, it may be said we are presented with a flora and fauna most akin to that of the moist and warm islands of Australasia, which then threw out its lines in the Northern Hemisphere to the snows of Melville Island and the Arctic circle ; and they who, like the writer, have basked an hour beneath the whispering tree-ferns on some southern volcanic eminence washed by a bright coral foam, among whose newer complex life lie enshrined the last lingering relics of this past, have there doubtless mentally reconstructed much that a museum is powerless to reveal.

If we examine the subsequent Permian rocks in England and Germany, we find them everywhere deposited on the upturned edges of the Carboniferous, while in the ancient kingdom of

Perm, the plain situated between the Volga and Ural, they are
conformable. Thus, according to Professor Dawson,* we perceive
at the western side of Europe and eastern side of North America
great disturbances inaugurating the secondary period; and in
the interior of both, over the plains between the Volga and
Ural in one, and between the Mississippi and Rocky Mountains
in the other, entire absence of these disturbances. During the
Permian period there remained in each of our Continental areas a
somewhat extensive inland sea. That of Western America was
a northward extension of the Gulf of Mexico; that of Eastern
Europe a northward extension of the Euxine and Caspian. In
Europe the land was interrupted by considerable water areas, not
lakes, but inland sea basins, sometimes probably connected with
the open sea, sometimes isolated. In England the thick yellow
magnesian limestone, the outcrop of which crosses in a nearly
straight line Durham, Yorkshire, and Nottingham, marks the
edges of one great Permian sea extending far to the eastward.
Yet while this age of land depression and change in character
of the European flora lasted, when much of our nascent northern
continents were plunged beneath the sea, we have yet evidence
to show that insects resembling large Hemiptera, and cockroaches,
enlivened the forest stretch, where as in Saxony and Cassel they
remained unsubmerged, although such recovered casts are few,
till in the thin, clay-parted marine and fresh-water limestones
of the Lias age, deposited along a sinking shore line in
Gloucester, Warwick, and Somersetshire, or in the later litho-
graphic limestones formed in Bavaria during the Oolitic, or
those deposited at Swanage during succeeding Purbeck time,
we are transported to other tranquil marine marsh-lands in
Europe, and to river banks, fringed with their feathery equiseta
and crowned with landscapes of stunted palm, Cycads, and
Zamias, or to fern-carpeted woods of tree-ferns and palms.
Here we hear the surging summery sound of gnats, see
the larger dragon-flies of the family *Æschnina* multiply and
hawk to and fro, or mount high over boughs of arbor vitæ,
while smaller brilliant *Agrionina* sun on the watery mirror and
divide the prey with their sub-aqueous larvæ and diving spiders.

* The disturbance would seem to be owing to the general depression of these
areas.

Sphinx moths (*Lepidoptera heterocera*) now poise in the twilight, appropriate food for the bat-like pterodactyles.

During these ages of the Jurassic Sea, when coral reefs fringed the coast-line of Professor Heer's* Gulf of Alsace, there reigned not yet over Palæarctic and Neartic areas that death-like stillness of summer so felt by the tropical traveller. For from the earliest days of the Devonian flora, the *shrill* of arboreal crickets echoed in the coverts of our atolls, and now again we catch from bush and dell the dream-inviting dirl of huge and voracious Dectici at autumn, while the vernal ground echoed to the "crink-crink" of Grylli, near akin to the present field and wood cricket, or the woodland hush was broken by the startling calls and whirr of large cicadæ (*Homoptera*). This quasi-tropical harmony presented by our island eastward, emerging from the Jurassic waves, is further attested in the character of its wood-destroying beetles, the Longhorns, Weevils, and *Buprestidæ*, the great size of its Dragon-flies and Leaf-crickets, or we may note it in the Cycads and Zamias that abounded among the coniferous forest. The stems of the latter in the Isle of Portland and on the denuded face of a cliff at Lulworth Cove were found, together with these carboniferous trees, actually rooted in the "dirt bed" or soil in which they grew, still exhibiting their petrified buds and at least evidence of their plumy foliage. This palmy thicket is covered over successively by the deposits of a lake and estuary, capped by the organic sea muds accruing from the continued land depression and consequent westward reflux of the cold cretaceous ocean, which as the land sank and shore line receded, threw down its clays and green sand, the spoil of ground swell and tides, and over these white chalk and lines of black flints, the long slimy accumulation of deeper sea microcosms, the round Foraminifera shells and flaky vegetable Diatomaceæ.

In the Cretaceous period, according to Prof. Dawson, the great continental plateaus were brought down, for the first time since the early Laurentian, to the condition of abysmal depths. This depression affected Europe more severely than America, the depression of the latter being not only less but somewhat later in

* Professor Heer, "Primæval World of Switzerland."

date. In Europe, at the period of greatest submergence, the hills of Scandinavia and Britain and the Urals perhaps alone stood out of the sea. In America the Appalachians and the old Laurentian ranges remained above water; but the Rocky Mountains and the Andes were in a great part submerged. A great Cretaceous sea extended from the Appalachians westward to the Pacific, and southward to the Gulf of Mexico, opening probably to the north into the Arctic Ocean. Towards this close of the secondary or Mesozoic period some representatives of the exogens, very like our ordinary forest timber, were introduced. In America a large number of the genera of the modern trees are present, and even some of those now peculiar to America, as the tulip-trees and sweet-gums.

According to Sir Charles Lyell, at the dawn of the Tertiary period, when the honeyed flowers that deck the maypole first bepaint our land, the southern portion of Europe and Northern Africa formed a large island with deep inlets. These inland waters, as the land emerged from the waves, seem to have given place to extensive swamps, attested by fresh-water strata, which in Provence, Auvergne, Croatia, or the basin of the Rhine, entombed the insect life that animated the existing streams, woodland, or flowery waste. These, curiously enough, prove chiefly of the existing Indian and South African types. Nine species of butterflies have been recovered from such beds of gypsum, marl, and lignite; and of these a species allied to the south European *Thais*, one, perhaps, allied to the extra European *Neorina*, a "brown butterfly" and "skipper," afford, as shown by Dr. Scudder, tolerable generic characters. The dappled south Asiatic swamp-cricket, Gryllacris, from Radoboj, preserved on the same slab as a piece of floating water-grass; the brilliant locusts (*Œdipoda*), or grasshoppers (*Gomphoceri*), in one case prettily scattered round a wind-fallen Cenothus leaf from Œningen; a Phaneropteron, representing a genus now scattered over the warmer temperate zone of either hemisphere, and others, exemplify the Orthoptera. Some of these retain not only their ancient forms, but also present faded pigment markings. There has also been a wealth of more or less perfectly identified bugs, beetles, bees, and flies recovered from these interesting strata.

Generally speaking, the Tertiary plant and animal life

approximates that of the modern world in all save distribution. The first, or Eocene, flora dates from the Cretaceous or Chalk age, and, as evidenced by the London Clay drift beds at Sheppy, shows fruits of fan-palms (*Flabellaria*), with those of Australian banksias, silver-trees, wagonbooms, whose leaves in Hampshire gravel banks of similar age mingle with the fig and cinnamon. In the sandstones that lie at the base of the Alps similar plants, and also palms of American type, occur. This clearly indicates a flora akin to that of Australia and South Africa over Palæarctic areas, while the second, or Miocene, flora shows a gradual influx of North American vegetation. Thus the leaf-beds of Mull and those of Bovey in Devonshire, covered by widespread eruptions coeval with the cliffs of Antrim and Staffa, show the prevailing plants to have been *sequoias*, or red-woods, vines, figs, cinnamons, &c. At Bovey, according to Prof. Heer, the woods that covered the slopes consisted mainly of a huge pine-tree, resembling the Wellingtonia of California. The leafy trees of most frequent occurrence were a cinnamon and ilex oak, similar to those now seen in Mexico. The evergreen figs, the custard apples, and allies of the Cape jessamine, were rarer. These glossy trees were festooned with vines, beside which the prickly rotary palm twined its snake-like form. In the shade of the forest throve numerous ferns, one species of which shot up into trees of imposing grandeur; and there were masses of underwood belonging to various species of *Nyssa*, like the tupelos and sour-gums of North America. At Œningen, on the embouchure of the Rhine into the sea of Constance, a marly deposit of Miocene age eroded by the stream, that indicates a former extension of the lake, has furnished abundant remains of maples, plane-trees, cypress, elm, and sweet-gum. Like the Carboniferous flora, the Tertiary in the Miocene age attained high latitudes; thirty species of sombre pine, with rustling stretches of beeches, oaks, planes, poplars, maples, walnuts, limes, magnolias, and vines, extended northward to Greenland, and there left their remains in beds of sand, clay, and peat. Similar plants are found in Spitzbergen lignite strata in Lat. 78° 56′, testifying to a large amount of heat and light, with mild winters for the evergreens.

The migration and change that closed this genial epoch when swarms of *Tipulæ* and gnats hummed around the incense from amber-dropping pine forests, the resort of small cicadæ and wayward bee-flies, long since buried beneath the Baltic and Vistula sands, is due to the chill of the glacial period advancing from the north-east, that scored and rounded the verdurous mountains of Europe and America, coastward sinking in the sea, with glacier or iceberg, leaving as witness huge moraine heaps, entombing Arctic shells and boulders in our sunken valleys. At this period we see the present fragile Arctic flora bud forth on European areas. At Bovey Tracey, above the great series of clays and lignites containing the Miocene plants, is a thick covering of clay, gravel, and stones, evidently of much later date. This also contains some plant remains; but instead of figs, cinnamons, and evergreen oaks, they are said to belong to the dwarf birch of Scandinavian and Highland hills, and to three willows, one of them being the little Arctic and Alpine creeping willow. At the close of the glacial epoch the continents of the northern hemisphere again rose from the waves, as may be seen in the old raised beaches, strewn with rounded pebbles and Arctic shells at different levels along the lines of coast and estuaries. At this conjuncture the influx of a milder climate drove the plants and insects of the two continents northward to the Arctic region; and forced them to climb in successive waves alike the summits of the bare-backed Grampians and Scawfell, or the newly-upheaved Alps, Pyrenees, and Apennines, where saxifrages and gentians yet nestle with insects terrestrial in more northern latitudes, such as *Pachinobu alpina* in Perthshire, *Erebia Cassiope* in Cumberland, the ringed Apollo butterflies and *Chionobas Aello* in Switzerland, or the golden *Carabus auratus* in the Pyrenees.

Fresh from the workshop of Providence, England and Ireland, like all the islands bordering the great continents, now stand on a bevelled plateau sunk from twenty to ninety fathoms, virtually a submerged selvage of the European area.* Around this the plummet sinks 2,000 fathoms, and brings to light a line of submerged cliffs bordered with beach shingles and shells. During the Miocene, Pliocene, and again, in the opinion of many geologists, at the close of the glacial period, this entire

---

* Henry Walker. *Leisure Hour*, Part 271, p. 421, July, 1874.

plateau bodily arose and stood out from the wash of the waves, making England and Ireland continuous with the Continent, and presenting the English Channel and German Ocean as inland plains. In the latter epoch this area was probably made drear with stretches of reed and bloomy heath, with forests of oak, chestnut, alder, and yew; and at this time, according to Professor Forbes, the Germanic flora travelled to Great Britain.

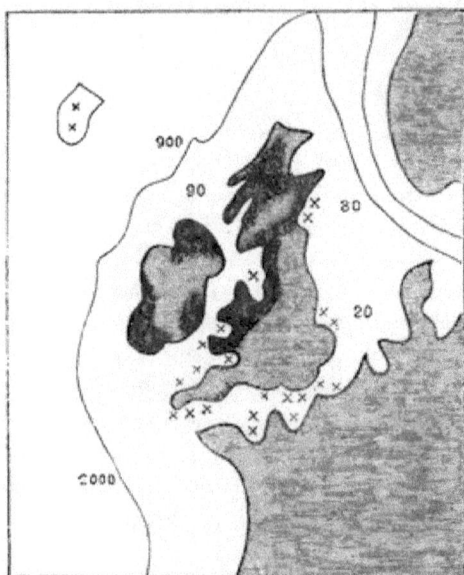

MAP OF GREAT BRITAIN IN THE MAMMOTH PERIOD.

x x, submerged forests of the old land area, rising up at the coast-line of to-day.
Adapted from an engraving in the *Leisure Hour*, July, 1874.

Submarine channels proceeding from existing rivers and trending to the then coast line have been traced by Mr. Godwin Austin, and close to the ancient embouchure of one traversing the English Channel, a fresh-water mollusk (*Unis pictorum*) has been obtained. Evidence of this plateau elevation may be deduced from existing surface depression evinced by encroachments of the sea, accompanied with land submergences and the sinking and denudation of cliffs. Mrs. Somerville relates the connection of the Hebrides with Scotland; and in a MS. from the Monastery of Mount St. Michel we learn that Jersey in the sixth century was united to the mainland by the wooded district of Quokeland, now lying submerged with its old Druidical remains. Sunk forests, in like manner, fringe the shores of Sussex, Hampshire, Dorset, Devon, Cornwall, whose trees, identified as oak, elm, chestnut, and hazel, pertain not to a seaboard, and some of these, indeed, appear mid-channel, where, covering the Goodwin Sands, again, is a deposit of clay. Cromer Forest similarly plunges beneath the German Ocean, and after a storm stumps of oak, alder, yew, and Scotch fir, are seen standing upright in the water. Old terraced beaches lie out at various depths in advance of our shores; and in the time of Tacitus the Zuyder Zee was a series of freshwater lakes. Farther south, at Sangatte, near Blanc Nez, I have myself noticed an old reed swamp full of the green wing-cases of *Donacia* to run under the water. On the contrary, accompanying this general land depression, we have local indications of partial gain of sand-flats and mud land, with the disappearance of ancient river courses, probably due to silting and land drainage, as at Sandwich, Romney, Dungeness, or Calais.*

It would, then, appear that the first indigenous flora over the northern hemisphere had insular character, with affinity to that existing in Polynesia. During secondary time, according to Professor Owen, it approached more nearly that of Australia; and towards the close of this age, according to Professor Dawson, there arose in North America a new and exogenous flora, which, after spreading to Europe, became naturalised across the Atlantic. To this flora succeeded on the European area one of Australian and South African type, which on the advent of the Glacial epoch was replaced in Europe by the present Arctic flora, the Arctic flora in turn yielding in process of time to the present vegetation of the area. We thus learn that Europe in particular has been successively covered with a flora akin to the Polynesian and Australian, with one more or less identical with the Australian and South African, the present North American, and the Arctic.

* For an account of fossil insect hunting in the West of England see the Rev. P. B. Brodie's " History of Fossil Insects." A comprehensive synopsis of the " Insect Fauna in the Geological Periods " has lately been published by Mr. H. Goss, F.G.S. It must, however, be remarked that zoological species older than the Tertiary cannot be established from the remains, and that the nomenclature employed by palæontologists merely indicates affinity in the specimen.

The older insects of the several epochs, as far as can be worked out, harmonise in character with the vegetation, while those inhabiting Europe in Tertiary time are more or less established as Indian, Ethiopian, or American in type. Agreeably too with this, a trace of the last glacial flora and insect fauna is found isolated on the tops of European mountains. The present insect fauna of the European and North Asiatic region resembles the North American, with that of California and Chili, united by the cold elevated ridge of the Rocky Mountains and Andes—a fact harmonising with the common Europeo-American vegetation at the close of the Tertiary. Although the Australian and South African insects which characterise the early portion of the third period are not generally recognised as existing on the present Palæarctic area, still we find some unmigratory species, as the Silver-lines Moth, now localised on the diverse vegetation of Europe and Australia, and these therefore should be the more ancient.

If we next cast a retrospect on the laws governing the migration of plant and insect with the present operation of nature, in modifying and adapting species thus submitted to changed circumstances, and proceed to summon the countless array of ages these laws have had force with the Articulata, since the first vermes became by immutable mandate a chitineous myriapod, or the fin-like extensions of its tracheæ took alar forms—if we, then, consider the effect of these laws at intervals of vast ages, attested in the varied aspect of fossil insects from Palæarctic areas, and weigh their relations in character and magnitude to species existing on these and other areas, here, as in allied departments of Palæontology, we are led to postulate cooling, or more probably, as Mr. Page has propounded, a recurrent heating and refrigeration of the northern hemisphere, most naturally accompanied by general land elevation and depression compelling the migration of terrestrial species. The colder circles of this life are, according to Mr. Page, evinced by the Azoic Grits of Cambria, the Old Red conglomerates, Permian breccias, and the boulder clays of the Pleistocene epoch—the intermediate warmer mark those of the Carboniferous, Oolite, and earlier Tertiaries, whose floras once stretched and have left their remains beneath the frozen Arctic. In short, there is seen a

graduated recurrence of warm and glacial periods of which we
now enjoy the transition from a more genial. The precession of
the equinox, or apparent change in position the constellations of
the zodiac have assumed in historic time, interpreted as continual
variation in the obliquity of the earth to the sun (as though its
axis of diurnal revolution insensibly moved from A to A´), has
been often advanced as ruling these geological phenomena. In
this way the sun would alternately concentrate its heating rays
(a) on the atmosphere nearer the Arctic circles during the summer
solstices (A, c), and after a time recede to the region of the

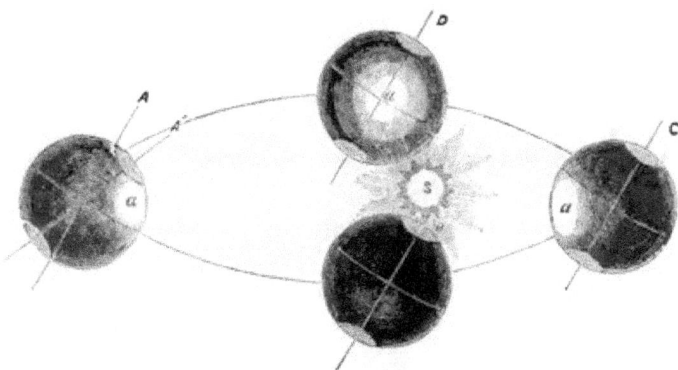

equator, inducing glacial phenomena at the poles now exposed to
the continuous action of cold interstellar space, with a collapsing in the earth's superficies, giving birth to the ensuing
wrinkling marked by earthquakes, volcanic action, and land
depression, or *vice versâ*.*

And that some such law has guided the coeval deposition of
strata, I think we may infer from the methodical sculpturing the
great continents have undergone from successive formative elevation
and depression, evinced by a fringe of submerged plateaux off
their coast, abruptly precipitous to 10,000 or 12,000 feet; and
also from the methodical way seas have everywhere invaded, and

* Sir Charles Lyell's view of the thermal distribution of sea and land
appears subordinate to this theory: but the changes in the terrestrial orbit
some have advocated do not seem of so much moment if we ask ourselves their
import on the present condition of the more remote planets. These cannot be
frozen?

deposited sand, limestone, and chalk over the land, as we have shown to be the case in respect of our own island, where around the feet of the mountains of Scotia and Cambria from the earliest times the invading waves have been stayed.

We will now turn and inquire what evidence of the geological past is furnished by a study of Insecta, and how far the modern cosmological theories of descent are vindicated. The present distribution of insects on the terrestrial superficies is conveniently treated in respect to six great zoological regions, namely, the Palæarctic Region, comprising the whole of Asia-Europe excepting the south-east of Asia, Northern Arabia, and North Africa as far as the Sahara; the Ethiopian Region, including Africa south of the Sahara, and the southern portion of Arabia; the Indian Region, including South-Eastern Asia; the Australian Region, comprising Celebes, New Guinea, Australia, and New Zealand, with the Polynesian Archipelago; the Neotropical Region, comprising South and Central America, the West Indies, and a great part of Mexico; lastly, the Nearctic Region, or North America. Since certain species may be selected as characterising either division, and others are more or less widely spread or cosmopolitan, superficial harmony is presented to the classifier, who allots his various kinds different localities thus indicating their distribution, and proceeds to associate them with the soils and similarly dispersed natural productions. But deeper insight tends to dispel this crude idea of constancy; for we notice a species invariable in one district in another assume so changed an aspect that until its transformation is understood, it is made to rank distinct. Others, again, are so prone to vary, that on one spot two specimens are seldom alike—a circumstance which has signally foiled the most expert, who experience an acknowledged difficulty in defining types from their varieties.

Insects, indeed, vary in all their stages, and in the final or perfect condition they do so uniformly or by differentiation of the sexes, producing some varieties, seasonal or of geographical limitation, that are selected or become permanent, and others resulting from change in larval diet; nor is hybridism unknown, and species vary generally on the limits of their horizontal and vertical distribution. This variation, which preceding geologists consider marks extinction or, as has been believed, mutation of

S

a form of life in the earth's strata* among insects, is ascribable
to divers causes; but these in seeming harmony with the law
that a species should vary on the confines of possible existence,
and that variation is dependent on change of circumstances. As
regards evidence of this, these laws of insect modification, as
Mr. Bates observes, are nowhere so legible as in the framework,
shape, and colour of insects' wings, especially in those of the
butterflies and moths, where these dermal tracheal extensions are
clothed with minute feathery scales, coloured previous to expansion in consanguineous patterns that, like those of a kaleidoscope,
differentiate as regards form and staining secretions, with the
slightest change in condition to which a species is exposed.

Insects firstly obtain nourishment and secrete their colours
from their food, and change of diet often produces variation.
The caterpillar of the Large Tortoiseshell Butterfly, reared on
willow instead of elm, excludes a pale dwarf smaller than its
nettle-feeding congener—the Small Tortoiseshell. Varieties of
the Small White Butterfly, with the wings of a sulphur yellow,
have occurred in England, and are frequent in Canada; these
generally prove to be males, and are obtained at Montreal by
feeding the ordinary cabbage caterpillars on mignonette. The
Drinker Moth is likewise said to vary with its food-plant. It
has been also stated that generally larvæ reared upon succulent
and overgrown herbage have *imagines* of large size and paler
colour, while dry and semi-withered food produces dark moths of
small size. The common Silkworm Moths, fed on lettuce and
mulberry respectively, are adduced in exemplification.†

Lepidoptera exhibit seasonable varieties that from experiments
instituted appear to us the result of heat and cold accelerating or
retarding the pupal stage; and as we proceed northward or

---

* Darwin, "Origin of Species."

† E. K. Robinson, *Entomologist*, 1877, p, 131. The food-plant in very
many cases doubtless is the immediate cause of variation. The Chelonidæ produce local variety of a white, yellow, or rosy colour, and in a Common Tiger
Moth the latter hue may be forced by feeding the caterpillar on lettuce and
onions. At the moment of correcting the proof sheets I have received from
Bareilly, North India, a specimen of the Buff Ermine Moth, taken during
February or March, that differs mainly from our summer examples in this very
rosy colour. Dr. Staudinger, in his Catalogue, p. 60, also notices a similar
variety of the White Ermine. These might surely be bred by some variety
fancier.

southward (?) from the centre of distribution, we witness the species producing fewer annual broods, and these varieties often supplanting the type. Thus the butterfly *Araschnia Prorsa*, common on the Continent in marshy woodland, produces three annual broods. The individuals of the first, that come out in April, are of a variety *Levana* (Frontispiece, Fig. 1.), which is above of a fulvous colour with black spots, while the summer dress is black, with a red line (Fig. 2). This wonderful disparity may be nevertheless resolved as a phase of æstival melanism, where the black spots of the early brood have run together into lines, and an excess of summer ink blotted over the wings. We see all the markings destined to enliven the warmer season as in a Chinese puzzle sketched out on the orange wings of spring, and the change in character of secretion, due to heat it may be presumed, and less duration in the chrysalis state, has served alone to expand the pattern and mark it in with black and white. Beneath, from similar cause, we find yellow has become white or brown, as the case may be. Dr. Staudinger, to further establish the matter instituted an experiment, and by placing the chrysalides of the April disclosure in an ice-box, found that an intermediate form, *porina*, rare in a state of nature, resulted.

Other examples of season metamorphism have been discovered. The spring brood of the cabbage butterflies *Pieris rapæ* and *Napi* in the Northern Hemisphere is of a purer white than the summer varieties, with the blackish markings more or less obsolete ; and pupæ of the summer generation of *Napi* kept in an ice-house produce an October brood with the winter dress. The North American Swallow Tail (*Papilio Ajax*) has three varieties. The first two appear successively in spring from over-wintering pupæ, while the last appears in summer in three generations. In this case Mr. Edwards found by application of cold, the May caterpillars of the earliest form reproduced the same variety in August, while those which emerged later on were of the summer variety. Another butterfly (*Pgriodes Tharos*) is likewise stated to be polymorphic, and in the Catskills digoneutic, or with two annual generations, the first of which is always of a winter type : but at Coalburgh this butterfly produces two more annual generations. A fourth generation produced in exceptional Canadian seasons exhibits the two forms. On the coast of Labrador, again, this species is rare, and accredited monogneutic. *Lycæna amyntula* and *polysperchon* are also summer and winter forms of the same butterfly.*

In Europe the Orange Tip (*Anthocharis belia*), by prolongation of the pupal state, produces a variety (*Ausonia*), which has the under-wings white with yellowish green blotches, instead of being green with silvery spots. The Camberwell Beauty, which in spring in Southern Europe has white wing-borders instead of buff, in Holland has them generally of a pale yellow; and in Sweden, Norway, and Lapland, white borders throughout the year. Summer varieties are also frequently dwarfed. For example, a second brood of the Queen of Spain in Italy is less in size and more intense in colouring than the vernal. On the southern limit of the Palæarctic area, we find the Orange Tip (*Anthocharis Belemia*, Esp.), at Morocco, has a spring variety (*Glauce*, Hb.), in which the silvery streaks beneath are replaced by white; the varieties *Helice*, Hb., of the Clouded Yellow and *Cleopatra*, L., of the Brimstone, at this season are not uncommon; the Meadow Brown is represented by *Hispulla*, Hb., which has the fulvous on the hind-wing increased, and the Speckled Wood by the variety *Meone*, Esp.; but perhaps some of these forms are local. *Helice* is also common in Sicily in spring, although in these islands, curiously enough, it is not found until autumn. Lastly, the autumnal brood of *Grapta C. album* is dark.† Seasonal varieties should likewise be looked for in moths. A second brood of *Hadena xylinoides* in Canada was noticed by Mr. Norman to be dwarfed; and our common hedgerow thorn (*Selenia illustraria*) in this country produces a seemingly constant summer pigmy (*delunaria*). This, as before, appears due to heat or cold accelerating or retarding the metamorphosis, and depends on the pupal duration, since Mr. B. G. Cole, as the result of two experiments, found that another geometer (*Ephyra punctaria*) bred from eggs laid by the same female, produced a *spotted* variety in July, while the remainder appeared the ensuing May in all respects resembling the parent. In spring the small Wave Moth (*Acidalia emutaria*) has a pink tinge.

---

* Vitis Graber, "Die Insekten," "Canadian Entomologist," &c.

† *Entomol. Mon. Mag.*; "Trans. Lond. and Italian Ent. Socs.," Var. Papers; Newman's *Entomologist*.

At the present time we notice these seasonal varieties not alone alternating in ordinary years, but witness their production by fluctuation in annual temperature. Thus while many butterflies and moths in the British Isles produce one or two annual broods, on certain years those ordinarily single brooded become double brooded; or those which are double brooded produce three annual generations. Similarly we find increase in production, accompanied with the generation of variety on proceeding southward. Thus an unusual third brood of Clouded Yellows, developed in England in 1878, was noticed to be dwarfed, with the black margin of the wings narrowed; those of the usual spring brood in Southern France, on the contrary, are large, with broad black borders. A second extraordinary disclosure of the Burnet Moth (*Anthrocera filipendulæ*) in September presented smaller crimson spots, with a tendency to coalesce. The South African butterflies *Callosune Evarne* and *Keiskamma*, it is stated by Mr. Mansel Weale, are varieties produced in successive years.

Polyommatus Agestis, a well-known little brown butterfly with a marginal row of rich orange spots, common in the south of England during May and August, when producing but a single annual brood, appears in July as a variety (*Artaxerxes*) that presents the black spots on the wings replaced by white ones, and which was for long on that account regarded by our entomologists as a distinct species, the Pride of Scotland. Yet the gradual selection of this "sport," that suns itself on Arthur's Seat and the eastern lowlands, may be clearly traced as we approach the north-eastern counties, where there occurs an intermediate form (*Salmacis*); and we more than fancy we witness its elimination during exceptional years in the south, as we learn from the "Manual of British Butterflies and Moths" that a *singular variety* was captured near Brighton during the July of 1857 by Mr. Cooke, which showed the characteristic white spot on the fore-wing. That it is not an aberration due to food, as Professor Zeller once surmised, has been triumphantly proved by Mr. Buckler, who succeeded in rearing caterpillars from Hartlepool on the rock rose, that disclosed butterflies pictured as the three varieties above, while conjointly he has discovered the northern larvæ merely differ from the southern in

being less bright in colour.* On the Continent Dr. Vitis Graber
speaks of three forms of this butterfly which are seasonal in their
appearance. He says there is found in Germany a summer and
winter variety; that the former in warmer Italy becomes the
winter form, and that this in turn there produces a new and more
southern summer form.

In the glacial epoch, when the summers were, there can be
little doubt, shorter, we on reflection perceive many insects would
appear alone in their spring forms in the Northern Hemisphere,
and these, supposing distribution constant, on the advent of
more genial conditions would generate their present summer
varieties. Nor is this subject theoretically sterile. Let us
suppose the case of *Araschnia Prorsa*. If we regard the form,
colour, venation, and antennæ, the usual criterion of butterfly
classification, we seem to witness in its spring variety, not alone
a form of the glacial epoch, but a further passage back from the
Red Admiral family to the stronger angular-winged Fritillaries,
that yet approximate them in habit and form of the larva.
Another minor link between the Red Admiral and Painted Lady
was long ago figured by the Rev. Mr. Bree, as found on the
flanks of the Himalayas; and could we so proceed to trace in all
the minor transitions and affinities, we might at length positively
affirm the *Vanessidi* are in a state of differentiation, and a recent
group; and then, as evidence of this, we might allude to the existing distributive migrations of Painted Ladies and Camberwell
Beauties as proving an active dissemination of the species.
Actually we find evidence of general laws appertaining to the
moulding and selection of varieties lingering in operation along
the limits of migrative distribution. Here species of all orders
betray the impress of melanochroism and leucochroism.† Dark,
suffused, or colourless varieties of moths, with confluent markings,
meet our view in the cloudy mountainous tracts of Scotland,
Ireland, the Isle of Man, and north-western counties, where

---

* See "British Butterflies," "Naturalist's Library," Stainton's "British
Butterflies," Coleman's "Brit. But.," *Entomologist's Monthly Magazine*,
Vol. IV., pp. 73—77; Vol. XV., pp. 241—244.

† E. Birchall, F. Buchanan White, M.D., W. A. Forbes, E. K. Robinson, Dr.
Jordan, and others; *Ent. Mon. Mag.* and *Entomologist* for 1876 and 1877
respectively.

insects are scarcer and produce fewer annual broods. Nor is this but a phase of a widespread agency, which a consignment of butterflies and papilionaceous moths from the temperate airs of southern France that vanish in the south of England or Scotland, or of a series of the same taken at different mountain altitudes, or of a series of the same that push their migrations outwards to Northern Africa and India, alike will prove to establish and confirm as regards the old world area.

Firstly, we shall find the forms themselves vary. In Southern Europe our clipped-wing little Clouded Yellows of railway banks and woodland Black-veined White Butterflies, float on ampler and more rounded vans, whose redundancy seems even to impair their powers of locomotion. Then light and heat, as observation and experiment alike confirm, rule the colour scale. The pale white of day-flying moths, as we see it in the Common Heath Moth of the Highland heather and in the male Bordered White (*F. piniaria*) of Perthshire fir woods, passes into a rich yellow among the spruce and furze of the Hampshire New Forest. The dirty white of an Argyleshire Speckled Wood Butterfly becomes in England yellow, and in South Europe fulvous; and here enlargement of the speckles causes the dark ground in one variety to resolve itself into lines, giving the butterfly the look of its congener of the hedgerows. Then it has become an established axiom that those colour-bands with charming ocellated spots, that so enhance the butterfly kind, should everywhere vary, and in localities vanish; and many drab and brown wings fluttering among grass and shade from time to time have exhibited the hillsman sports that have caused a cry of new species, or prompted experts to enter on description where others see but variety. The Large Heath Butterfly may be reckoned among these. This kind in the north of England, at an elevation of two thousand feet, according to Mr. T. Marshall, and in some parts of Ireland, according to Mr. Birchall, has the eyes painted on its sandy wings greatly decreased in number; and on the Perthshire mountains, conjointly with our English type, an aberration is sometimes seen even less ocellated; and this anomaly we find has established itself in Lapland as the local form *Isis* of the species, the most boreal variation. The Mountain Ringlet (*Cassiope*), another prize of our southern collectors, varies much in the development of its red spotty cincture among the slaty Westmoreland crags; and Dr. Buchanan White finds the takes on Perthshire areas are much larger, with white pupils to the black-wing eyes in the female, more like, indeed, the accredited type coming to us from Central Europe, for this mountain waif is not reputed to extend its lines of dispersion to the chilly auroral light.

The May Orange Tip Butterfly in Southern Europe is sometimes dwarfed, and has the white on the inferior surface of the fore-wings replaced by yellow, while inversely the White Ermine in the north becomes yellow. One singular example, shown in the diagram, taken in the house of Sir Patrick Walker, at Drumsheugh, Edinburgh, at the end of August, 1820, and figured by Curtis in his elegant volumes under that gentleman's appellation, approached in colour the female Buff Ermine, and presented the appearance of the fore-wing spots run together

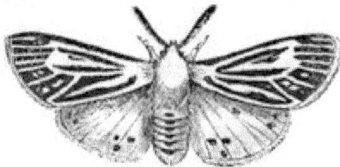

SPILOSOMA WALKERI (CUR).

A melanic aberration of the White Ermine, showing how the wing-spots become lines by uniting together.

into longitudinal dashes. On the other hand, those stray specimens of *Enperia fulrago* taken by our London collectors are of a buff colour, though farther north, where the species is commoner, they assume an ochreous tint. Brown, again, is generally intensified into black in the Highland moth fauna, and especially so in the common Dark Arches Moth and its garden congener *X. rurea;* so is it likewise in many other swift evening-flyers I have frequently taken meandering among the clammy ash boughs during the prolonged summer twilights. The warm sunlight fawn on the Fritillary's wing in Northern Europe, by enlargement of the black dots, runs into new patterns that are enhanced by paling or deepening in the ground. This will be manifest on comparing species captured along those northern limits of distribution that pass through the United Kingdom and Scandinavia, with types from more southern forest clearings. One specimen of the Queen of Spain, taken in Norway, in particular, proved to be quite black above, with the pearly spots beneath also confluent, forming streaks. In *Nemeobius Lucina*, however, I think we see the converse; for some specimens of this woodland butterfly I took in the

Hampshire New Forest in the spring of 1871 were small, pale-coloured, with rounded wings; whereas those I netted on the Superga Hills, near Turin, in 1878, were all large and dark, with produced wing-tips. There likewise the Small Copper Butterfly has its wing often suffused with dusky, indicating, as I imagine, a southern distributive limit. Indeed, the plains of Italy prove quite a trans-formation scene to the British collector, and it is here on the limit of the olive zone we find that the ova of our blackthorn-feeding Brim-stone Butterfly, according to Boisduval, produce the rich orange blotched variety *Cleopatra*, that may be met with at Avignon flying in company with our ordinary February harbinger of the primrose. In America, on the other hand, it has been long known the common Swallow Tail (*Papilio Turnus*) is variable, with a dark female *Glaucus* rarely found north of about forty degrees of latitude. Dr. Jordan further affirms that certain North American Lepidoptera are darker than the European types, citing the Camberwell Beauty, Painted Lady Butterfly, and *Melanippe hastata* in illustration; and this is the precise quality in which many other so-called representative species on the two areas differ. For example, the Nearctic *Deilephila Chamænerii* and our own *galii*, or *Phlogophora Iris* and our Angle Shades, among moths; *Vanessa Milbertii* and the Small Tortoiseshell (*J. album*), and the Large Tortoiseshell (*Thanaos brizæ*), and our Dingy Skipper, among butterflies.

We may also trace in local varieties of butterflies and moths a process of approximation or differentiation of the sexes. Thus the deep black Swallow Tail, with blue wing flecks (*Papilio memmon*), found in China and the islands of the Pacific, produces in Java a light-coloured variety, with spoon-shaped tails to its hind wings, they not existing in the ordinary form of this butterfly; and as regards colour, on the other hand, the common English Ghost Moth, in the crepuscular light of Zetland, assumes the orange markings of the female on its immaculate wings of satiny white, a very remarkable variety, which has been likewise taken by Herr Snellen van Hoeven near Rotterdam. From a note by the late Mr. Hewitson we learn another Swallow Tail (*Papilio Merope*) in Madagascar has a female resembling its male, but that elsewhere the sex is polymorphic, or takes numerous forms.

As we ascend the hills and mountains of Europe butterflies and moths become small and dark; and their sexes often lose their colour differentiation, or become discriminated anew by melanism. Mr. Goss, at a meeting of the Entomological Society, exhibited a series of specimens of our somewhat local blue butterfly *Polyommatus Arion*, taken during June, 1877, among the Cotswold Hills. One-third of these were far below the average dimensions, whereas those specimens he had previously obtained from their Devonshire and Northamptonshire habitats, Bolt Head and Barnwell Wold, I think we may presume, were, as a rule, of the normal size. I have myself taken the male of the Orange Tip Butterfly on Robin's Hill, near Gloucester, only 1″ in alar expanse, and the Fritillary *Melitaea Athalia* on Alpine heights dwindles to less. In a few cases it is the female of Alpine butterflies alone that is melanic.* A large number of these Alpine butterflies are likewise normally dark. Nearctic species present the same features on the heights. *Limenitis Misippus* on the Catskill Mountains of North America loses dark markings, and *L. Arthemis* has an Alpine form, for example.

Arctic Lepidoptera are equally variable with Alpine, and this it is considered in either case may be owing to an enforced lengthening of the period of their transformations beneath the snow, and to their isolation or segregation. At Grinnell Land, between the parallels of 78° and 83° north latitude, one month each year is the longest period insects can appear in the perfect state, and six weeks is the period in which phytophagous larvæ can feed; so that the pupal state must here be generally of long duration. In Newfoundland we find the variety *frigida* of the common American *Pieris oleracea* differs from its Canadian type in being more marked with black along the veins, both at the base and tip of the wings. *Papilio Turnus* is dwarfed, paler, and with narrower black borders. The caudal lobe or tail of the hind wing in *Papilio brevicauda*, Saund., a local form of *Asterias*, is reduced by one-half.

Woody coverts and proximity to the sea, as also the smoke of towns and manufacturing districts, are associated with variety and

---

* *Pieris Napi*, var. *Byoniæ; Polyommatus virgaureæ*, var. *Zermattensis; Argynnis Paphia*, var. *Valezina*. This may be due to the greater duration of this sex in the pupal state.

melanism in the butterfly and moth fauna. The system of variation in such localities is the same, showing the effect is directly dependent on the glandular organs of the insect, and that the law is constant, while external conditions of the environment are multifarious. Thus the shades of the New Forest afford us a constant variety (*Valezina*) of the Fritillary *Paphia*, which instead of being fulvous, is brown and spotted instead of streaked along the nervures. Another sub-variety of this lepidopteron, taken by Mr. Bates, showed a white spot on either wing. It is thought the years the species is scarce this Black *Paphia* is most plentiful. Somewhat less commonly, the sooty variety of the floating White Admiral Butterfly is seen among the briery tangles of our woods, showing coal-black wing with the distinctive white ribbon more or less obliterated from its disc. And on the Alps, altitude is found to have the same effect as shade in casting these sports from its dies, which there roam at large, as here beneath the lowland brush. In the New Forest likewise, not unfrequently the black-dusted geometers of the bushes, come into our nets and collecting boxes with the irroration on their wings increased so as to confer a dingy, if not a unicolorous appearance; and the little *Prays Curtisellus* sometimes has the white totally effaced from its fore-wing, as I have found it near Lymington. *Anchylopera subarcuana*, Wilk., also produces a grey variety. In Sherwood Forest a singular specimen of our Small Copper has been taken with the rufous colour replaced by white.

As regards proximity to the sea, butterflies are there paler or darker, and frequently want wing-spots. A specimen of the Tiger Moth (*Chelonia villica*) captured near Brighton, had the " cream spots " on its fore-wings more or less obliterated, most completely so on the right ; and *Erebia Blandina*, from Morecombe Bay, had the brown bands on the fore-wings replaced by yellow. As the female of the last species in the Lake District is dimorphic, with a bluish-ash band on the posterior wings, these markings would appear to vary between ash, ochreous, and brown. The Lepidoptera at Hastings and on the coast of Wales have been in like manner noticed as being often deviations from the types, as have some from the Fen Districts.

Butterflies and moths are next generally melanic near large towns. A Small Copper taken near London had the superior black spots run into a band; a specimen of the Yellow Shell (*Camptogramma Bilineata*) in my possession has the lines run together so as to form two black bands, and some of *Cerostoma vittella* have the dark markings extended from the inner margin half over the wing; but many more striking varieties occur from time to time.* *Noctuina* are notably blackish near towns; for example, *Noctua glarcosa* taken near Barnsley had the fore-wings chocolate-brown in place of pinkish grey. The isolation of islands is no less congenial to the establishment of variety, and these reproduce their like. In England and Scotland the aberrations of three butterflies, two of which are scarce, the Large Copper, the Purple-Edged Copper, and Blue *Polyommatus Artaxerxes* are noticeable; the second differing from the Continental type *Eurydice* in having purple instead of black wing-margins. Our yellow variety (*Stramineola*) of the Footman (*Lithosia griseola*) is also not known to occur in Europe. Indigenous species are likewise distinguished by melanism. In Ireland the Marsh Fritillary, for instance, is present as a variety (*Hibernica*, Birchall),which has the fulvous wing-colour replaced by white; and this kind is larger than the English type, and much larger than the Scottish. On small islands butterflies have been considered to have enlarged wings. This has been affirmed of those of Madeira by Mr. Woolaston, and I find a note by Mr. Butler to the effect that specimens of our Small White from Japan were found to exceed the alar expanse of English examples by one-third, and that Pale Clouded Yellows from the same extreme outpost of the Old World were likewise larger. But it should then be noticed these islands lay far south in latitude, a consideration that might cause us to hesitate in accepting the premises on Dr. Darwin's explanatory theory that the wings are acquired from battling with the winds.

Albinism, the converse of melanism, where a pale ground colour is increased at the expense of the darker tint, is attributed to the species happening to frequent localities with light-coloured soil. It is active on the English Chalk Downs, where it produces

* A singular melanic variety of the Painted Lady, bred on September 3rd, 1879, from a larva taken near the locks on the River Lea, at Clapton Park, had the wings bordered with white spots, as regards the hind wing evidently caused by replacement of the black.—*Entomologist* for April, 1880.

varieties in unstable genera of moths such as the Annulets and others. The Common Swift Moth I have taken at Guildford, in Surrey, with its pale brown fore-wings rendered almost entirely white from an enlargement of the streaks. Other moths, as some of the Tortricina, would appear inherently variable, and species in the same hedgerow exhibit endless wing patterns. This may be ascribable to atavism, or reverting to a former type, or may be doubtless due to the variety of food the caterpillars there consume.

Mr. C. S. Gregson, of Liverpool, has published a small pamphlet on the effects manufacturing towns of this country induce on the circumambient lepidopterous fauna. "The mere variation in shade of colour," he considers, is there always unreal "when it tends lightwards, or to buff or ochrey yellow, or yellowish, or to ashy drabs, from cold dark browns. It does not follow, however, that the wonderful changes in colour have always

Albino aberration of the Common Swift (*Hepialus lupulinus*), from the Guildford Hill (chalk), taken with some other similar varieties in 1875. Wings rendered white from the increase of the paler spots.

been intentionally caused. Often it appears as the insect emerges from the pupa, and here the breeder points proudly to the fact that his friend the *dyer* or bleacher bred it for him whilst he was away collecting. Such was a '*Porcellus*' case; and the word 'dyer' set me thinking that the fumes from his clothes had made the change, and it took me ten minutes to make one like it on my return home again. I am told that this light buff and this dark brown variety were taken in côp. on a tree-bole, and as certain streams of gas are escaping from chemical works not far away, shortly afterwards I have artificial varieties of a '*Betularia*' being manufactured in my little laboratory of all shades of buffs. When we consider what a great difference in colour a slight chemical action will make in many species of Lepidoptera, and the great amount of free acid (chloric) gas is ever escaping from our immense chemical works, the wonder is we have not more aberrant coloured insects amongst us, especially so when we know that a little chlorine in or near chrysalis boxes will remove any of the more fugitive colours as the insects appear."

Mr. Gregson proceeds to notice certain chemical changes that can be induced in insect colour; but the more interesting topic is a possible application of analysis to account for the selection of the geological formations. "The Large Heath Butterfly is a very light insect (var. *Typhon*) in Cumberland and Scotland on high hills; whilst on low moss land, on which the water is charged with iodine, in Cumberland, Westmoreland, and Lancashire, it is a rich fulvous brown insect, larger and stronger built; and when these are acted upon by hydrochloric acid gas they assume the exact colour of the hill specimens. The Dark Annulet Moth (*Gnophos obscuraria*) on chalk lands is a light-coloured grey or drab insect. In Carboniferous limestone districts it is a lead-coloured insect, whilst on the New Red sandstone, 'Keuper' formation, it varies from a rich ochreous colour where oxide of iron is present in the soil, to a dark, almost black, insect on the white sandstone parts of the New Red formation, thus clearly pointing to geologically-caused changes of colour. Any of these latter forms, acted upon by chlorine, appear as light-coloured grays. The same remarks apply to *Dranthacia carpophaga*. On chalk it is light buff; on 'New Red' here, darker; but all buff on 'Cambrian' at Llangollen; and at Penmaenbach darker still, buff or ochreous brown; and on quartzoise early rock rich dark cold grey-brown, as in the Isle of Man, and at Howth, in Ireland, ochrey shades being rarely observable upon them; but, acted upon by hydrochloric acid gas, they all turn to beautiful bright light fawn buffs, veritable *Carpophaga* of the chalk." The writer, however, suggests that some varieties we might be inclined to attribute to certain formations may be the result of a food proper to the soil. Thus in the case of the Welsh Wave Moth (*Acidalia contiguaria*), bred continuously on heather from the moss lands, all specimens become varieties, fumose specimens, whilst when fed upon low succulent plants they were large light-coloured specimens, rarely darkish, but never so dark as when fed upon heath from the moss.*

Some general rules of variation in European butterflies and

---

* The varieties are figured in Newman's *Entomologist*, Vol. XII., p. 65–67. In this paper, by Messrs. Fryer and Capper, it is stated that the dark variety on Bettws-y-Coed is larger, and occurs at a greater elevation than the light bone-coloured one at Llanfairfechen, and that it likewise is found, as I conclude, on the Cevennes Mountains. Guenée, "Hist. Nat.," Vol. IX., p. 464.

moths that may, when worked out by collectors, be possibly found to be due to one or other of causes assigned, deserve notice. For example, we find the female of nearly every species of the glowing yellow and orange genus of Clouded Yellows is liable to an almost white variety, and the orange species are frequently shot with a purple or violet flicker. The fulvous Fritillaries are often melanic or greenish, and those inlaid with metallic have the silvery blots replaced by yellow stains; our wood-feeding Pearl Bordered Fritillaries thus normally differ from one another. The violet-shot species of Purple Emperor have their white bands replaced by yellow, and in this and all genera of European butterflies spots and bands are proverbially obsolete or supplementary. Among the moth kind, the rich rosy red of the spotted Burnets of the long grass, and that of a Small Elephant Hawk from Perthshire, has been found changed to yellow. The Swift Moths are all variable. The dark-brown of the Northern Swift is liable to be replaced by orange colour, which in the variety *Carnus* obscures the wing, or it is suffused with ochreous, approaching the species in appearance to the common English Swift, which varies in like fashion, or to its fern-feeding ally *sylvanus*. The Loopers, or Geometrina, vary by enlargement of the transverse darker or lighter wing-lines, and by their continuation or absence on the hind wings. In this fashion the Latticed Heath Moth of clover-fields changes from a whitish ochreous, with a few transverse lines, to fuscous, with sparse light spots, and the Common Heath varies from yellow to fuscous. The Carpet *Anticlea badiata* becomes brown, and the species of *Cidaria* pass from grey to ochreous and brown. Aberration in the Blunt Wings or Tortricina, due to suffusion of the primary wing-colour or to enlargement of darker markings, would seem to resolve itself into melanism and albinism. With these little moths a yellow wing varies to grey or brown, and a grey or brown wing becomes black-brown. The genus *Peronea* has quite magical varieties among sallow bushes and roseries. The little Tineidae, or clothes' moth group, presents species that vary from bronzy to steely, and others that are dwarfed, with the metallic spots more or less obliterated. The latter has been noticed of our *Gelechia Hermannella* found in the United States of America.

We may, in conclusion, witness exterior circumstances inducing variation in almost every alar feature of the butterflies and moths. The wing is enlarged and dwarfed, and in this process it becomes prolonged, rounded, or scaleless; the caudal appendages are found to be occasionally eliminated in rich exotics, and they are frequently abbreviated. Then, as regards colour, the upper surface may be considered as diversified with shades of dark and light. A dark confused and composite ground, under favourable conditions, is resolved by an increase of light spots into clear generic dark lines and blotches inaugurating the species, which then fade away, and the wing is finally reduced to a paler hue; or on a light ground arise dark spots which coalesce to form lines, enlarge to bands, and finally overcast the wing with shade. The paler tint in many species may be traced to a fundamental white. Perhaps the earliest butterflies were white ones? And this colour we may often trace through its glandular stains of yellow, orange, or the softest rosy, into the more compound tints of slate colour and brown, or we may resolve these stains again in other specimens in an inverse fashion. The darker tint lastly passes on by a rougher transition into clayey browns and funereal black. For instance, take our English Swift Moths. We can plainly trace orange differentiated in a series of males into white, or white forming into orange bands. These run together into spots, and eventually give place to a uniform tawny, which in the fleet-winged Northern Swift is afterwards treated in identical fashion with brown, so as to evoke intermediate patterns in three colours. Again, examples of the very common garden Lackey Moth of either sex may be selected from a cabinet row of an ochreous colour with two purple lines crossing the fore-wing; and in others of the series we witness these lines filled in with purplish, and forming the edging of a band. To this central ribbon a purplish blush extends, leaving a distinct marginal line of the primitive colour, which by gentle transition of the purplish blush to a deep sandy-red stands out sharply defined in many specimens. A similar chromo-lithographic process may be observed recorded on the wings of the common Drinker and many other moths.

In butterflies that produce dimorphic and polymorphic females we may proceed in like manner to unravel the vestiges of divergence. But what is perhaps more surprising, it has been found, despite the close approximation in shape and colouring of the scale wings, the better marked forms and scientific types in a state of nature recognise and freely cross with their aberrations; and in the case of melanism, Mr. Llewelyn has found that the result of breeding in is to increase the number of blackamoors. Thus from a dark Welsh female of the geometer *Tephrosia crepuscularia*, the offspring in the first generation were one-half dark and one-half pale. In the second batch, the produce of dark parentage, they were dark in the ratio of two to one, and the third generation from these negroes were alike dark.*

It is interesting to inquire in what way the colouring matter is imparted to the wings by the glands as they lie beneath their chitineous pupal covers previous to expansion; for then it is the pattern is impressed—extension of parts subsequent to exclosure merely serving to elongate the markings and enlarge the design. This it would seem can happen only in two ways, in either as there is a prepossession: the secretory matter must flow from the thorax through the tissues of the flaccid wings, or its stains are impressed by their membranous coverings in the manner of a printing-press. That the latter hypothesis is not wholly groundless, it may be mentioned that a specimen of the nocturnal moth *Leucania conigera*, captured near the Welsh Harp Tavern, to the north of London, in 1877, was found to have the markings upon the upper wings reproduced on the upper surface of the

---

* Figures of our insular varieties of moths and butterflies will be found in the *An. and Mag. Nat. History*, Newman's "Brit. Butterflies and Moths," and the *Entomologist*. Many coloured figures of Continental varieties are scattered through German works, but the most valuable reference to the variation of Lepidoptera on Palæarctic areas is doubtless Staudinger and Woeke's "Catalog de Lepidopteren des Europæischen Faunengebiets." Dr. Staudinger remarks on insect kinds as species, chance variety or aberration, local varieties, or varieties of race, Darwinian varieties, and seasonal varieties; variation due to soils, altitude, and other causes, is also noticed. This work will more fully illustrate those few observations I have here recorded from British sources. Thus the Scarce Swallow Tail has a pale southern variety; the Clouded Yellow Palæno, a pale northern one. The female of the latter on the Alps is sometimes *yellow* like the male, and here the brown female of *L. Corydon* becomes sometimes *blue* like its male. Another Blue Butterfly is pale on the limestone mountains of Spain. Certain Hair Streaks have longer tails in Asia Minor, and species vary everywhere in wing expansion and colour according to certain laws. As regards variation of Lepidoptera on Neartic areas see "The Butterflies of North America," by Wm. H. Edwards, &c., &c.

T

left lower; and a specimen of the common Meadow Brown
Butterfly, taken near Oxford in 1878, showed the same transformation effected in the under surfaces of the same wings, accompanied with the development of an additional vein and consequent
enlargement, that seemed to afford a reason how the hind wing
came to be folded so as to receive not alone the impress of the
inferior side, but the very eye-spot of the upper.   The enlargement of spots and lines is evidently due to an excess of the
darker or lighter secretion as the case may be; but how far the
dermal darkening, usually an indication of induration, may be
the result of exposure, it is in each case interesting to inquire.
Besides the wing patterns, the markings on the bodies of
Lepidoptera are likewise liable to chromatic variation. A specimen of the Clear Wing *Trochilium culiciforme* taken in Tilgate
Forest, for instance, had the usual red ring on the abdomen of
a white colour; and from this cause likewise the gregarious
hammock-weaving Porthesia caterpillars, whose transformations
are synchronous, produce moths which closely approximate.
In colour, maybe, they slightly differ, as a milky from a
creamy white; and the front wings of the male Gold Tail sometimes take four wing-flecks—two at the anal angle, one at the
apex, and rarely another just over the trifurcation of the central
vein—uncertain and hereditary features that will also at times
spring to light in its congener.   But a main distinction has been
said to rest in the colour of the anal tuft and stain above on the
terminal portion of the hind body, that has originated the two
names which have the sanction of usage.   In the Brown Tail
this stain is more or less marked and is of an umber colour, and
these hairs vary from a vandyke brown to a light and glossy
auburn.   In the Gold Tail, where the stain is mostly seen in the
males, the tuft should be yellow, but it is not the less sometimes
of a deep brown colour fringed merely with a flaxen rim, leaving
it somewhat difficult to say which is the Gold Tail and which is
the Brown Tail species.

A darkening in the colouration of our English ground beetles
may be observed as we go northward, connected, it has been supposed, with a constitution better fitted to encounter unfavourable
conditions of life.   Thus the common *Pterostichus nigrita* has
been found dwarfed with narrower and duller elytra; the little

Loricera pilicornis, naturally brassy, becomes somewhat ferruginous. We have likewise the dark Highland forms of *Carabus cantenulatus* and the mountain *Nubigena*. Then, as regards Europe, *Carabus clathratus* in the south is large and able to fly; in Sweden and Siberia it is small and can only creep on the ground. The Long-horned Beetle, *Saperda scalaris*, frequenting poplars, in Lapland is covered with an ashy white instead of yellow pile. Coleoptera from the Alps, Pyrenees, and Apennines are also often dwarfed, and a variation of altitude obscures the colours of Alpine beetles, whether terrestrial, floral, stercoraceous, or aqueous, so that species in the inferior zones ornate with colours of metallic reflection become on the elevation uniformly black. Those green and coppery in the high Alps are pure black, while a smaller number steel-blue and deep blue, with others brown, olive, and golden green, pass into pure or bluish black. Even the yellow leaf-feeding *Chrysomela alpina* becomes black. *Carabus Rossi*, Dej., for like reasons, is dwarfed in the Apennines, and there assumes a chestnut-brown hue; and our common South of England scavenger *Carabus violaceus*, Lin., dwarfed and blackish in the Tyrol, in Central Italy is large, with violet elytra margined with golden. This melanism and dwarfing is, as in other cases, ascribed to a prolongation of the metamorphosis, passed often beneath the mountain snow. Proximity to the sea-coast likewise dwarfs beetles and affects their coloration. For example, I have observed the Coleoptera of Calais on the French coast are often smaller than those taken at Lille; so *Pimela Fairmarei* at Morocco attains but half the dimensions it takes farther inland.

As regards insular varieties, the Isle of Man, according to Mr. Rye, furnishes a dwarf form of the common Oil Beetle, and exemplifies an intensifying in the colour and punctuation of certain of the smaller terrestrial beetles. The Coleoptera of the Shetland Islands are alleged to be dark and dwarfed. On small oceanic islands that give us the Dodo and Solitaire, the insect fauna is likewise often apterous, as has been observed by the Rev. Mr. Eaton in Kerguelen's Island, and by Mr. Wollaston in Madeira. On the latter outlier of the European plant and animal life, two in every five of the beetle species proved so far deficient in wings that they could not fly—a peculiarity Dr. Darwin attri-

T 2

butes in his happiest mood to the circumstance that, during many successive generations, each insular individual which flew least, either from its wings having been ever so little less perfectly developed, or from indolent habit, has escaped being blown to sea, and that the beetle population of this and the adjacent rocky pinnacles have come to inherit a character that has been gradually transmitted from sire to son.

Although the species of Coleoptera which are mostly erected on outward appearance when in the perfect state, and seldom checked by rearing from the larva where they present less character, do not afford the same excellence of test in regard to colour variety as the Lepidoptera, yet we have seen locality produces a variation in this attribute, and on southern confines of Europe especially a decided variability of tint in some of the brighter Palæarctic kinds may be remarked. The Rose Beetles (*Cetonia*), or Ground Beetles of the genus *Carabus*, may be selected to illustrate this. *Cetonia metallica*, Fab., Mons. Mulsant remarks in his work on European Coleoptera, in the cold and temperate parts of France has the head coppery and the wing-cases marked with three fasciæ; while in the typical kind the head is violet and the fasciæ are less entire, often mere spots. Again, in the variety *Olivacea* proper to the warmer portions of Provence, the head becomes a more pronounced coppery-violet, and the fasciæ on the elytra have generally disappeared. The body also has assumed that transparent or varnished chafer lustre, which in a greater degree distinguishes those individuals from the fervid climates of Corsica and Italy that constitute a variety *Florentina*. Another sport of Fieber's of a Kermes red *Kermesina* appears not to occur in France. The *Cetonia Ænea* has a variety, *Albiguttata*, proper to the Alpine regions, which is of a semi-golden green; but in the warmer districts around Lyons, the upper portion of the body is of a shining bronzed green, and no longer presents a varnished lustre. The southern varieties of our own Rose Beetle, it may be stated on the same authority, are of a violet hue, and some of its sports are hairy. According to Mr. Rye, male beetles when dwarfed lose distinctive horns, large jaws, or "formation" of the hind femora; and on the other hand a small specimen of *Osphya bipunctata* has occurred combining the male colouring with the female form; so that

Coleoptera, as Lepidoptera, produce variety by approximation or differentiation of the sexes. Like Lepidoptera also, the brightly-coloured sorts vary by an increase of the dark or bright pigment and confluence of elytral spots, as has been observed of *Dromius quadrimaculatus* and *Panagæus quadripustulatus*. Lastly, in *Carabus convexus*, Fab., the lines or striæ on the elytra of some varieties become irregular, and in others they are replaced by punctures, as is also the case in the brown variety of another Carabus, *C. cantenulatus*. The punctures themselves are likewise liable to be obsolete or supplementary.

With regard to the Bugs, or Hemiptera, Dr. Puton found that the *Plinthisus minutissimus*, one of the *Lygæidæ*, in the south of France, where it lives in society with ants, had rudimentary elytra and wings; but in Algiers its elytra were fully developed. This local alar expansion, allowing the species to fly and migrate, is characteristic of the genus; and Mr. Douglas remarks it is accompanied with a development of the thoracic muscles that changes the corporeal forms from oblong to oval.* *Coreus spiniger* is said to be less in size at Naples than in Calabria, and another bug, *Neides tipularius*, is stated to be a form of *N. parallelus* with abbreviated wings. We may find seasonal variety likewise in a common gamboller over pond-water, *Hydrometra lacustris*, Fab.; those individuals, which jerk about and disport themselves in the spring, being reputed apterous—a circumstance M. Audouin surmises is due to the rigours of winter having checked their development.

Orthoptera, in their hordes of ravine and rapacity, are so chameleon-like in their changes, even on the same confined spots, that we find their recognition and definition extremely difficult, and still more difficult is it to classify and deduce the operating laws that rule their confused forms and colours. Still, as far as would appear, they betray a general distribution and diversification similar to what may be observed among Lepidoptera. On the Palæarctic area, which has been most worked, we notice species that spread northward from the genial shores of Southern Europe decrease in size, in alar expansion, and in richness of colour; and one large pinkish-winged grasshopper, *Caloptenus italicus*,

---

* J. W. Douglas, *Ent. Mon. Mag.*, Vol. XVI., pp. 217–219, quotes Dr. Puton to the effect that pilosity is seen in northern and dwarf forms.

Lin., conjointly drops off the richer unicolorous appearance its
protecting surfaces show among the stones of Spain and Calabria,
and in Central Europe becomes sharply spotted with mossy
brown.   Others in advancing northwards are seen to again recoil
from sight, having their markings obscured and obliterated with
sombre and earthy tints before the species finally takes its leave
in the starved and ungenial clime.   Either phenomenon may
doubtless be referred to what we understand as the phases of
melanism, as I find on comparing specimens of the active little
Red-legged Grasshopper from the heather of Western Scotland
and the cornfields of Italy, that in both cases the elytra are often
black and the hind legs reddish and unspotted; but that the red
in the southern negro is far the richer of the two.   Proximity
to the seaside is another inducement to melanism, and individuals
here and elsewhere are often found wholly or in part of an
unnatural roseate colour.   Respecting a dark-saddled Leaf-cricket
deprived of leaping power, found by Rambur lurking beneath
the stones that fringe the picturesque snows of the Spanish Sierra
Nevada, Fischer, in his " Orthoptera Europea," remarks :—" The
obscure colour in this kind is unusual, and corresponds to the
observations of Professor Heer, who has found that beetles with
metallic hues become opaque and obscure on the tops of the
Helvetian Alps."

   There are some other chromatic changes advocated by this
systematic writer that might appear to us at first sight hardly
creditable, namely, that a locust should have hind-wing pigments
like climatic litmus paper, so as to appear in some examples of a
bright red, while other specimens should come to hand with
bright blue wings.   Yet he admits this tacitly in describing the
Variable Oedipoda of Pallas, which seems to assume the latter
cold hue on passing eastward to the wide plains of Russia and
Siberia; and he more than asserts it in regard to the common
crimson-winged O. fasciata of Siebold.   Of this butterfly-like
creature the starchy-coloured variety is deemed the more nor-
thernly form, but in parts of Central Europe acidulated wings
of blood cinnabar are seen commingling on the hills with those
of blue and bluish-green, which, as we reach the warm southern
peninsulas, cast off this disguise and turn to love's proper hue—
a rich rosy crimson.   And it may be further remarked that of

the seven red, blue, and yellow so-called European sorts at present accredited to this genera, some do appear very ordinary varieties due to climate or soil; for certainly wing colours, to say nothing of stature and form, vary most deceptively in these locusts. Sometimes in our cabinets we notice the flowery wings more or less destitute of colours and diaphanous, especially as regards their front and outer margins; and then in others of the series there come a warm flush and row of dark spots blotting over the disc, which change to lines and bands, or, spreading, widen out, until they frame the gay wings in an umber or inky setting. Herr Fischer has fully illustrated this transformation, or rather the inverse, as it occurs on the Palæarctic area in that dwarfed rosy-winged waif of African extraction, the *Œdipoda insubrica* of Scopuli, whose bright tropic hues may be rarely seen glancing as far north as the grand old Alpine barrier; and in his two blue species of Northern and Southern Europe we fancy we again trace the identical lineaments of change.

The frequent grasshopper green, like all insect verdure, presents changeable chlorophyll-like properties; but here chronic disinvestment of the vernal hue in the individual for the garb of sere autumn is much more pronounced. As on the soon bleaching wing of green geometric moths, these changes are photographic, and mostly ensue subsequent to skin casting, spreading from various parts of the dermis as discolouring stains of brown, grey, yellow, flesh colour, or rosy red, according to the species and specific pigment. So that in running over a collection we sometimes notice specimens where this morphological process has been arrested in one or other of its stages, and find grasshoppers and leaf-crickets that present portions of their elytra, thorax, or legs that yet retain their primitive hue, while the rest of their body is liveried with the humid brands of vegetable decay. Yet we cannot thereby infer that the species is prematurely decrepid; for in the large voracious Wart Biter, a parrot-like transition after emergence serves to develop an invisible pattern of brown spots, which renders the individuals to human optics considerably more conspicuous and suited to recognise and reproduce their kind. Before passing on it may be likewise observed that structural changes have been reputed

features in climatic variation, and it is stated the common Mole
Cricket is not only smaller in Southern Europe, but that its fore
or digging legs are differently toothed.

In conclusion, the discrimination of species and variety is most
feasible in butterflies and moths, where the specimens take decided
diversity of form in all their stages. On proceeding to consider
the beetles, we find their obscurer transformations less marked and
less studied, and in such cases general similarity, with observation on habit and pairing of the sexes, becomes the sole specific
test. The same remarks apply as forcibly to those classes where
little differentiation is marked during development, as indeed to
the remaining orders inclusively, where there is very little to fix
what is species and what induced variety. Not that there can
be any doubt but that the laws of variation are in operation here
as elsewhere, and in rapacious dragon-flies pattern deviations
of the war-paint occur quite in consonance. Thus we have
a Scottish variety of the Four Spotted Libellula, and a local sort
with the extremities of the wing blackish. In Northern Italy
two others of these large flat-bodied species vary, one being our
common large Blue Dragon Fly, *L. depressa ;* and from Etruria
Rossi mentions a variety of the female of the equally common
Demoiselle, *Caleopteryx Virgo,* that has black body and black wings.
Then *Caleopteryx Vesta*, Charp., of our Epping Forest is likewise
considered a variety where the reddish wings of the male have
not acquired their bluish adult hue, though it also differs in
habit, being found far from water. As bright dragon-flies and
flies mostly acquire their body colours by the sun's action on the
pigment of the skin or dermis, the blue and red males on first
emerging assuming the sober tints of the females, on which by
transition their proper vice is branded, blue exudation being
preceded by feminine yellow or brown, red and carmine by
feminine yellow, orange, or olive ; it is quite palpable an arrest-
ing of this process must evoke parti-coloured varieties, and such
sorts are now and then met with at large. These few observations I extract from the Baron de Selys Longschamp's work on
" European Dragon Flies ; " and on turning to the other group
of Neuroptera containing the May Flies, I find a note by Mr.
M'Lachlan stating that those specimens of the Common Lace
Wing which have hibernated may be known by their reddish

colour—a circumstance that may be doubtless regarded a record of the influx of cold.

The proof of the law of adaptation of form to circumstances has hitherto mainly rested on the conclusions of the general zoologist and geologist, and on such phenomena as we may witness in the prime wings of the table chicken and splint bones of the favourite mare. For who may now-a-days doubt but that the first-named muscular delicacy is aught but the result of prolonged domesticity, terrestrial habit, and inaction of the limb, or that the little spilikins in the second instance are but the rudiments of a five-fingered foot? We know that the early Tertiary horses that roamed the Northern Hemisphere had these five toes, and we can now, thanks to a series of fossil remains lately discovered, trace each of them in process of time dwindling to a splint bone, and eventually disappearing. So when we turn to insects, how endless are the vestiges of parts we behold, and how endless the adaptation, although only in a few cases, indeed, can the laws evoking them be truly indicated! Here we witness various perfection of metamorphosis; tassel fore and hind legs in butterflies and moths, sometimes allotted to a sex or characterising a group; rudimentary wings, elytra, semi-elytra, and hemi-elytra; rudimentary eyes, auditory organs, musical organs, modified segments. The whole being bears the appearance of having passed and re-passed through changing moulds, in harmony with ever-varying geographical features and configuration. For is it not in beetles that live darkling that the eyes are abridged, and in cavern and ant-hill dwellers that they are absent? Have we not shown winged insects become apterous or gain ampler pinions by distribution; and do not they everywhere accommodate themselves to the varied circumstances of life in which they are, so to speak, placed by Providence?

These phenomena of variation in the insect world, again, appear to be presided over by certain laws of natural harmony, termed Natural Selection. It is said this proneness to produce capricious local or temporary variety, observable in species or genera when arranged systematically in a cabinet drawer, is due to an inherited tendency to produce sports, which as they from time to time mimic the surrounding earthy tints, the summer's glow and purple shade, or lurid and brassy reflections of the fresh sea, the

moss or lichen, green or faded leaf, dropping of birds, and that
general tone of reproductive beauty which varies in fervour as the
climatic zones of the globe, acquire for them photographic concealment in their haunts, and here they hide from their enemies,
a juncture at which Nature now and again steps in, exerting a
gradual influence, selecting and perpetuating such races even to
the elimination of an original type.   Other insects, again,
which retain in a locality their conspicuous colours, are either
distasteful to the greater part of the birds that there resort by
possessing weapons of defence or obnoxious secretion, or are those
Nature has shielded by giving protective resemblance to such
species or to surrounding inanimate objects.   Neotropical butterflies thus mimic the acrid kinds (*Heliconidæ*), and many indigenous flies are scarcely distinguishable from stinging bees on
the umbels of flowers, while moths, beetles, and bugs resemble
them in form and actions.   It is, then, evident, were change
induced in natural surroundings, soils, light, vegetation, &c.,
that thus afford protection to groups of insects, a certain portion
would be destroyed, and a new era of protective selection induced.

      Messrs. Buckler, Hellens, Porrit, and others have recently given
considerable attention to the breeding of our indigenous Lepidoptera, and a perusal of their life histories I think will allow us to
gather, firstly, that a change of appearance in the larval state is
not in direct relation to variation of the imago, since the typical
caterpillar often produces a variety and the larval aberration a
typical butterfly or moth ; and, secondly, that certain laws regulate variation in caterpillars.   For example, the differentiation is
generally effected by chromatic change through a scale of yellow,
green, grey, pink, and brown, or by obliteration of the longitudinal
and transverse markings.   Thus the larvæ of the Holly Blue vary
from yellowish-green to green and black, or the extreme segment
is sometimes pinkish ; those of the Spurge Hawk from bronze-
green to blackish bronze ; those of the Death's Head from green
to olive brown ; those of *Noctua umbrosa* from yellowish to
greyish brown ; those of the geometer *Ephyra punctaria* from
green to pinkish brown ; or those of *E. pendularia* from pinkish
purple to bright green ; those of the Yellow Shell Moth from
greenish grey to pinkish grey and pinkish brown.   The " Wave "
Acidalia degeneria again has larvæ blackish brown, marked with

rust colour, and the varieties take one or the other tint. The larvæ of *Philbalapteryx lignata* are yellowish green, more or less suffused with pink, and those of the small *Pyrausta puniccalis* often want the darker stripes and lines. Sometimes, however, it would appear that variation in the caterpillar state coincides with aberration in the imago, since *Merope*, an Alpine variety of our Marsh Fritillary butterfly, according to Boisduval, has a special larval form with yellow dorsal spots, resembling that of *Cynthia*.

Caterpillars and other phytophagous larvæ, as well as amorphous and exposed pupæ, have been noticed to vary according to their aliments and the season, as also, chameleon-like, to assume the colour of the objects on which they rest ; and in the case of the two grey moths *Acronycta tridens* and *psi*, we have a reputed instance of dissimilar hump-backed caterpillars, the one orange red and the other yellowish black, producing perfect insects that can be discriminated by no specific characters, and in the case of the Chalk Hill Blues very similar larvæ produce distinct *imagines*. As regards variation of pupæ when exposed, the chrysalis of the Swallow Tail (*Papilio podalirius*) is in Italy reddish in autumn, green or brownish in the spring, and that of the indigenous *Machaon* varies in colour with the object on which it is placed. The pupa of a South African butterfly (*Anthrocharis Keiskamma*), according to Mr. T. P. M. Weale, varies from greenish-white to bright green, ochreous, and brown, according to the substance to which it is attached, being lightest on a blue surface exposed to light, and darkest on brown leaves in obscurity. The pupa of another species on a white surface became so pale-coloured as to be almost translucent, with the marking on the ventral aspect of the abdomen almost obliterated. The larvæ of the *Acræa Erebia* of Hewitson confined in a dark box produced darker pupæ. The caterpillar of our Peppered Geometer is likewise stated to be yellow when it feeds on birch, yellow or brown when it feeds on elm ; that of the Death's Head Hawk has two varieties. In certain cases larvæ have also been found to vary indifferently as regards their armature of spines and hair, if not as regards their actual forms. So that when, on the other hand, we discover a majority of leaf-feeding larvæ and pupæ protected by mimicry to the leaves, twigs, and stems of plants on which they feed, and that those feeding exposed are distasteful to insectivorous creatures,

we are led to conclude this harmony has been the work of time.
With regard to the chemical aspect of the problem, Mr. Meldola
has stated that the colouring matters of some plants can be found
by the spectroscope in the tissue of caterpillars which feed upon
them in an unaltered condition, and the viscera of many is so
transparent that the undigested food in the stomach confers on
the individual its proper tint, as the yolk often does to an egg.
Thus far, then, change in diet produces change in the larva.
Passing to the markings, it has been frequently remarked by
Sir John Lubbock and some other naturalists, among them Mr. A.
G. Butler, that the longitudinal and diagonal lines of caterpillars
resemble the veins of grasses and the divaricating ribs in the
leaves of trees, enhanced by an illusion of warm shadow-streaks.
Dr. Weisman, commenting on this circumstance in the Elephant
Hawk Moth, where the young caterpillars retain their longitudinal lines only until the third moult, when they are replaced by
diagonal ones and eye-spots, infers that the young larva represents an old form, and that the species in the lapse of ages has
gone through the stage each individual now completes in a few
weeks. This, should the fundamental fact be universal, would
harmonise with the geological opinion that the monocotyledons
preceded the dicotyledons. The Smaller Elephant, again, leaves
the egg with a subdorsal line the larger species does not acquire
until the first moult, and is thus possibly a more recent form.

Having in former chapters traced, however imperfectly, causes
which tend to collect and distribute forms of insect life; having
seen how the stimuli of love and rivalry inspire music and prompt
dances, and how these conspire, with other agencies such as a
common food, to collect certain kinds into flocks, we will now
proceed to examine how such swarms are circulated and disseminated by exterior causes, which, acting mostly in opposition to
volition, subject the species to influences operating such organic
modification as we have just been noticing. For while on the
one hand with insects the tendency to distribution has indirect
relation to their locomotive powers, and is greatest in those which
are winged, or, again, species are introduced with their food or by
animal agency, on the other hand a constant means of dispersion
over the earth's surface must be recognised in the circulation of
aërial and aqueous currents, the development and buoyancy of the

insect tracheal system here conferring increased capability, whether as regards traversing the air or resisting immersion. Migration is more or less checked, on the other hand, by natural barriers, such as mountain-chains and watery expanses, and the beech-line and Alpine snow become the wreck-chart of many species. The effect of migration is change in the habitat, food plant, or habit of the species.

Pyrameis cardui and its varieties sunning on the thistle bank is a cosmopolitan feature in terrestrial scenery we cannot fly from ; it meets us like a friend far from our native land on every gravelly waste, where the gardens of coral islets, deep in the dark Pacific, are overhung by the bread-fruit, or where the dusty sands of Africa and lone savannas of America are imprinted by the hoof of the antelope and buffalo, where the jungle of Bengal echoes to the roar of the tiger, or where the Ceylon elephant crushes through the cane, on ancient lands where the epiornis roved and the emu wanders. Go where you will, there persistently sits the ubiquitous Painted Lady on its heap of shingle or flower-head, just as the Chinese, in their country of gardens, depict it on the rice-tree's pith, from where the eternal snows scarcely melt beneath the spring-tide, to where the equator kindles its glowing heats.

The history of this marvellous distribution cannot remain wholly a sealed scroll to the geologist when we take into consideration that the insect in its wonderful migrations manifestly affords the thread with which to retrace and unravel the problem, while they, on the other hand, render equally patent the reason of its present uncertainty of appearance in various localities. Thus at the present day the migrations of the Painted Lady take a fixed direction in the Northern Hemisphere from the Tropic towards the Pole. " It was," says Colonel Drummond Hay, " as far as I can recollect, in the early part of the summer of 1842, while stationed in Vido, a small island in the harbour of Corfu, that an extraordinary flight of the Painted Lady butterfly took place. The first part of the column reached the island about nine o'clock in the morning, and continued steadily to advance in rolling masses of many thousands for upwards of three hours. Though the density of the column was at no time very great, yet it appeared to extend in breadth as far as one

could see, having the appearance of black drifting snow, if I
may so call it. By one o'clock the flight had completely passed;
the wind at the time was blowing fresh from the south-east.
In the afternoon, on sailing up the channel of Corfu, the traces
of the passage of the flight were very evident from the quantities of dead butterflies which floated on the surface of the
water; and for days afterwards they were to be seen drifting
into the various bays in the island of Corfu. I did not hear
whether this flight had been observed on the Continent, but
as they appeared to be taking the direction of the coast of Italy
they would, in all probability, strike the land in the vicinity of
Otranto."

Onward through Italy proceed these insect-chains. Thus the
late Professor Bonelli, of Turin,* one year in the beginning of
the present century, observed a similar flight of the same species
of butterfly towards the end of March. "Their flight was from
south to *north*, and their numbers were so immense that at night
the flowers were literally covered with them. As the spring
advanced their numbers diminished, though even in June a few
still continued." Then we come to Madame de Meuran Wolff's
account. "She had gone on this or a subsequent summer to
Grandson, and was enjoying the prospect of the lake of Neuf-
chatel and charming Swiss mountains. It was the beginning
of June, we believe, when this lady observed with surprise an
immense flight of butterflies traversing her little garden with
great rapidity. They were all of the species called Belle Dame,
and were flying close together in the same direction from *south*
to *north*, and were so little afraid when any one approached
that they never turned to the right or to the left. The flight
continued for two hours without interruption, and the column
was about ten or fifteen feet broad. They did not stop to alight
on flowers, but flew onwards, low and equally." Their line of
migration still strikes northward. On the 20th of June, 1879,
the following notice appeared in the *Daily Telegraph :*—" Still
bent on showing itself extraordinary, eccentric, and unparalleled,
this year of floods, earthquakes, eruptions, and general crises is
determined to be remarkable also for entomological manifesta-

---

* J. Rennie, "Insect Miscel.," p. 266.

tions. One of its latest marvels has been a visitation of locusts upon a scale of amazing magnitude, with which Southern Russia and the Caucasus have been recently plagued. These locusts, it appears, have positively encountered the troops of his Imperial Majesty the Czar in the field, fought with, and decisively defeated them. A strong detachment of infantry, under the command of Major Lazoff, was on the march from Goatschkai to Elisabetpol for the purpose of reinforcing the column destined to operate against the Tekinz tribes, when it was assailed by a flight of locusts thirty-five versts square, which had alighted in such masses upon the ground that the soldiers could not wade through them, and still filled the air, swarming upon the men in countless myriads. In the hope of dispersing them by noise, Major Lazoff gave his men the command to fire in the air, and kept up a series of volleys for nearly half an hour; but the locusts paid no attention whatsoever to the rapid discharge of Russian musketry directed against them, and the detachment was forced to retreat for the second time to Dzigamskoi. A less unpleasant but equally remarkable visitation is reported from Switzerland, Baden, and the Rhine districts, where, on the 7th instant, incalculable flights of butterflies passed over the country from westwards to eastwards. At St. Gatten and Gossau one flight occupied the entire afternoon in traversing the district, the whole population turning out in amazement to watch its fluttering progress. This gathering appears to be the same which was observed a day later in the Canton of Zurich, at Wetzikon, and in the Thurgau at Dussnang. Another astounding company of butterflies passed over Karlsruhe and Buhl last Sunday week, flying from north to south, and again in the opposite direction on the following Tuesday. All these consisted exclusively of the so-called thistle eater (*Vanessa cardui*), a dark-brown butterfly with reddish-yellow spots. They appear to have crossed the Rhine near Rheinweiler, and to have caused great consternation among the Rhenish farmers. It may be remarked the season had been proverbially wet, and the prevailing winds during the period of migration were south and westerly." Another paper, quoting an account given by Dr. Von Krauss, in the *Verein für Vaterländische Naturkunde*, of this migration, says :—" In three previous years—1741, 1826, and 1857—similar gigantic flights of the same species were observed in Turin and other places. According to the report presented by Professor Eimer, of the University of Tübingen, on the swarms of the present year, the first notice of these massive hosts of butterflies was made at Turin on June 1st; next in Switzerland, from June 2nd to 9th; then in Alsace, France, and Spain, from June 5th to 10th; and latest in Würtemberg, from June 11th to 21st. As a result of his anatomical observations, Dr. Eimer found that eighteen out of every nineteen of these troops were females who were all laden with eggs, and he believes that the object of the great emigration was the discovery of a home in which to deposit them. They seem in the first instance to have been wafted to us from Africa, the swarm traversing Algiers in a north-easterly direction as early as from the 15th to the 20th of April, and reaching Valencia and Italy in about ten or eleven days."

Next to *Pyrameis cardui*, *Danais Archippus* has a wide distribution. It occurs in the New World from Canada to Bolivia, and has spread over some of the islands of the Pacific to Queensland and New Guinea. According to Mr. Riley, these butterflies hibernate through the winter, and then often appear in immense swarms. In September, 1868, accounts were received of their sudden appearance in different parts of the city of Madison, Wisconsin, and at Manteno, Ill.; whilst on the 19th of that month at St. Joseph, Mo., millions of them were seen filling the air to the height of three or four hundred feet for several hours, flying from north to south. And again in the spring of 1870 a remarkable swarm was seen at Manhattan about the middle of April,[*] which, as reported by a resident of that place, came rapidly with a strong wind from the north-west and filled the atmosphere all around for more than an hour, sometimes so as to eclipse the light. The handsome black and green Swallow-tailed, *Urania leilus* and *Marius*, which from their larval forms are sometimes affiliated to the large-bodied moths, migrate periodically from Orizaba in Mexico along the foot of the Cordillera Mountains to Rio Grande in Texas. On the wing alone during the prime of the morning hours their passage occupies from three to four weeks, and then at the expiration of from five to six the cohorts retrograde, much deci-

[*] W. L. Distant, *Trans. Ent. Soc.*, 1877, p. 93.

mated, and the females no longer gravid. Another account of this phenomenon says they fly from sea to sea. The slim South American *Heliconiæ* have likewise individuals that swarm and migrate, for the *H. narcea*, Mr. Moseley informs us, covered the decks of the *Challenger* when making Bahia in calm weather.

That the intolerable heat of the sun, exerting its tendency in amassing insects into flocks by kindling the torch of love and rivalry on vasty breeding-grounds, is the prelude to these migrations, is attested by travellers. A few white and blue butterflies during July weather collect in this country at muddy spots, and from some Alpine rivulet one may startle quite a bevy of thirsty beings. But in tropical climates open spots by a river-side become habitual diurnal resorts for vindictive males, who, flecked in colour and innumerable in species, lend animation to the watery whispers. Such spots form a natural parade-ground for manifestations of rivalry, such as Mr. Wallace has noticed displayed by the butterflies of Borneo, and form also a point of departure for migration, which, regarded in this light, appears intrinsically a modification of the aërial dance; indeed, male Lepidoptera are often noticed following one another when neither migrating nor hurtling in the air—a fact often noticed, and that has been put on record by the acute Edward Newman. "As the waters retreated from the beach," writes Mr. Bates at Obydos, on his voyage up the Amazons, "vast numbers of sulphur-yellow and orange-coloured butterflies congregated on the moist sand. The greater portion of them belonged to the genus *Callidryas*. They assembled in densely packed masses, sometimes two or three yards in circumference, their wings all held in an upright position, so that the beach looked as though variegated with beds of crocuses. These yellow butterflies seem to be migratory insects, and have large powers of dissemination. During the last two days of our voyage (9th and 10th October) the great numbers constantly passing over the river attracted the attention of every one on board. They all crossed in one direction, namely, from *north* to *south*,[*] and the processions

---

[*] "Naturalist on the Amazons," p. 131. In Brazil, March, 1803, an immense flight of butterflies of white and yellow colour continued for many days. They were observed proceeding in a direction from north-west to south-east. (Lindley, "Voyage to Brazil;" or *Royal Military Chronicle*, March, 1815, p. 452.

U

were uninterrupted from an early hour in the morning until
sunset.   All the individuals which resort to the margin of sandy
beaches are of the male sex."

Sir Emerson Tennant confirms the migratory nature of this
genus in his work on Ceylon.   "Butterflies of large size and
gorgeous colouring flutter over the endless expanse of flowers,
and at times the extraordinary sight presents itself of flights of
these delicate creatures, generally of a white or pale yellow hue,
apparently miles in breadth, and of such prodigious extension
as to occupy hours, and even days, uninterruptedly in their
passage."   The butterflies seen in these wonderful migrations
in Ceylon are mostly two kinds of *Callidryas*, with straggling
individuals of the genus *Euplœa*.   Their passage takes place in
April and May, generally in a north-easterly direction.   The
testy, jealous nature of insects is probably one source of the
heterogeneous nature the migratory flocks present.   This we
may see in the Painted Lady, who, when sunning, arises to buffet
and commingle with everything that approaches, and darts even
after an obnoxious bird.   Darwin encountered one very motley
migrating swarm at sea off the north of Patagonia.   "One evening," * he says, "when we were about ten miles from the Bay
of San Blas, vast numbers of butterflies, in bands or flocks of
countless myriads, extended as far as the eye could range.   Even
by the aid of a glass it was not possible to see a space free from
butterflies.   The seamen cried out ' it was snowing butterflies,'
and such in fact was the appearance.   More species than one
were present, but the main part belonged to a kind very similar
to, but not identical with, the common English Clouded Yellow.
Some moths and bees accompanied the butterflies, and a fine
Calosoma flew on board.   The day had been fine and calm, and
the one previous to it equally so, with light and variable airs.
Before sunset a strong breeze sprung up from the north."

As the migration of butterfly flocks appears to originate when
the air is light, and the direction to be ruled by the coming
breeze, and since Lepidoptera that do not thus band from their
large wing-expanse are also influenced by the prevailing winds,

* Darwin, " Voyages of the *Adventure* and *Beagle*," p. 185.   See also
Zoologist, 1846.   In Ceylon the migrations start from the east coast on the
setting in of the north-east monsoon, and travel north to Calcutta.

we are led to surmise an insensible movement of this order
within the trades towards the equator, and a corresponding
movement beyond their action towards the poles. Local emigration on more confined aërial currents also occurs, and to this
source should be due the appearance on these shores of
unimported Camberwell Beauties and rare Sphinx Moths,
and those files of Small and Large Whites bred in large
measure from sea-shore cruciferæ on the sand dunes sometimes observed crossing the Channel. A flight of the Small
White, recorded in Coleman's " British Butterflies " (in the form
of an extract from the *Canterbury Journal* for 1846), reached
England about twelve o'clock at noon. During the sea-passage
the weather was calm and sunny, with scarce a puff of wind
stirring ; but an hour or so after they had touched *terra firma*
it came on to blow great guns from the south-west, the direction
whence the insects came. Towards the middle of July, 1872, a
swarm of the Large White enveloped a boat at the mouth of the
Weser. Many of the insects were to be seen posing themselves with erect wings on the watery surface, others lying flat
on it as if dead, but flew away rapidly if disturbed. They were
accompanied by Dragon-flies (*Æschna*), that preyed upon them,
and by small flies and ichneumons. A similar occurrence in a
neighbouring locality was observed in July, 1868. Among the
most regular of our supposed Channel emigrants must be
reckoned the Common Clouded Yellows and Painted Lady
Butterflies. The first has nearly established itself in the south-
eastern counties, although in the opinion of local collectors it
sometimes seems to miss a season. The inference is that it
comes across the sea from the south-east. It arrives about the
end of May in a more or less faded condition, and from these
emigrants the butterfly is reproduced in about seven weeks.
The most remarkable English migrations on record occurred in
the years 1826, 1859, and 1877, when Clouded Yellows became
nearly as common as white butterflies all over England, and
species turned up in the south of Scotland. In the years 1842,
1857, and 1868, Pale Clouded Yellows were especially plentiful,
and in 1842 they extended as far as Derbyshire and Yorkshire.
In August, 1872, the Camberwell Beauty appeared in nearly every
English county, being most frequent in the east of the island.

Another actual or obsolete line of migration of the Scottish
moth and butterfly fauna southwards may be found in the
elevated mountain chains of the Lake District and Wales; but
few, however, in their distribution pass into the latter district,
which has, again, many species peculiar to it. Mr. Birchall some
time back remarked the similarity of the northern lepidopterous
fauna of Ireland to that of Scotland. Here we have not alone
the Large Heath Butterfly (*C. Davus*) in the Donegal Mountains, but many of the more northern British Moths occur, such
as *Acronycta menyanthidis*, *Epunda nigra*, *Plusia interrogationis*,
Phibalapteryx lapidata, and *Peronea maccana*; and although
present flux and migration have not been noticed to or from
Caledonia, they are more than probable; since Fair Head in Ireland is only fifteen miles distant from the red rocks of Cantyre,
and Belfast Bay only twenty miles from Port Patrick; a less
water interval in fact than that presented by the boisterous
Straits of Dover. As regards Europe, the Mediterranean basin is
more or less regularly traversed in April and May by swarms
of the Cabbage Butterfly and Painted Lady, which on arriving
in France oviposite. These are considered as the *avant-coureurs*
of the quails. Night-fliers uncertain in appearance, and which do
not regularly breed in Europe, as the Hawk Moths, *Chœrocampa
nerii* and *celerio*, or *Deilephila lineata*, similarly arrive, sometimes
in broad day. Since the insects of Eastern Siberia are considered
to have special affinity with those of Western America, Behring
Strait may be regarded as another hypothetical line of distribution; and many European butterflies straggle from the Alps
by the elevated Asian chains into Northern India.

Insects of other orders, with wings more rigid or weaker,
as a rule float on the gale. Prominent among such are the
locusts or larger Orthoptera, whose disastrous name we, from
our childhood, associate with seasons of southern drought, whirl-
winds, hailstorms, earthquakes, plagues, famine, volcanic eruptions, and other slumbering powers that shook the foundations
of the globe in its earliest ages, or of society in its earliest
history. These species, although by their strong fore-wings
and distinct discoidal cell endowed with some powers of flight,
nevertheless do not appear capable of " warping on the eastern
wind," but rather allow themselves to be swept along on the

blast, and in this way they will travel great distances. One was captured off Cape Blanco, 370 oceanic miles from land, and another I myself secured scudding on the trade wind 200 miles by ship's reckonings from Cape Palmas, on the same coast of Africa. It is thus they become dependent on the winds for distribution, and to the same cause they owe an uncertain appearance in Europe, or stragglers some autumns will waft northward to our islands. Their terrestrial aggregation is due to the workings of love and jealousy in their music, or coloration as is probable. Their flocks often consist of more than one species. Borne from their wintering-ground on arid plains or barren mountains in seasons of dryness and sirocco, they alight on the fragrant, well-watered lands of warmer climates, and there deposit their ova. The larvæ hatch when spring is yet glowing in tender green, and then everywhere the canker-worm is seen, in corn-lands, in orchards, in vineyards. These, treading in the wake of the spring caterpillars, form into bands, and advance, mowing and devastating with their jaws, like the tearing harrow, crackling flame, or serried ranks of war, till the first two stages of their voracious life are flown, and the winged state approaches. Then, leaving their nympha cases on plant and bush, they again flock together, animated by their newly-acquired music, till of a sudden, rising like disembodied spirits in a murmuring cloud that casts disastrous twilight, and yielding to the impulse of the first airs, they waft onward for leagues over sea and land, where'er they alight changing the summer's landscape to a bleak and leafless image of northern winter, amid the execrations, fires, and noises of the terrified inhabitants. The assemblage then pairs. Nor is the plague stayed before they leave their drowned carcases, dropped from the zephyrs, in putrefying heaps along the tangle on the shore or margin of lakes and rivers. So wrote a Syrian prophet, poet, and early naturalist of the Migratory Locust, when rousing his depraved countrymen to repel an Assyrian invasion ; and thus modern travellers repeat the history of every such winged serpent they meet with in Asia, Africa, or America. With respect to the course which the Western Asiatic locusts pursue, Hasselquist has observed the flocks migrate in a direct meridian line from south to north, passing from the deserts of Arabia, which form the great cradle of them, to Palestine, Syria, Caramania, Natolia, Bithynia, Constantinople, and Poland; they never appear to turn to the east or to the west. In Northern India they are stated to migrate east, and in Southern Africa southward.* In 1650 a cloud of locusts was seen to enter Russia in three different places, and they afterwards spread themselves over Poland and Lithuania. From Africa they frequently invade Spain, Italy, and Austria, and trend northward. Both locusts and butterflies where they increase inordinately become the food of primitive races; and at Missouri certain cultivated gentlemen have revived this dainty dish of ancient kings and prophets, in an annual feast on spikes of the American locust, *Caloplerus spretus*, with wild honey from the Rocky Mountains. As to occult science and astronomical influences on the revolution of our swift little planet, it would be hard to state what might not be learnt from locust invasions. If we direct our attention to Europe in 591 of our present era, we find they invaded Italy; and in 1478 they were a pest in the Venetian territory; in 1542 they invaded Silesia; in 1556, Maddaloni; in 1613, Marseilles; in 1650, Russia, spreading to Poland; in 1693, Thrace, penetrating to Germany; in 1713, Silesia. In the years intervening between 1747 and 1750 their phalanxes again swept over Central Europe, reaching our own shores; and lastly, in 1780, they appeared in Transylvania, having previously multiplied at Morocco and ravaged at the Cape. From these years of wonder we extract the dark numbers 64, 14, 57, 37, 43, 20, 34, 33—differences whose evident fluctuation might try even an expert algebraist in concocting a series. Yet the more recent ravages of the North American locust, which furnish us with the periods 29, 7, 3, 2, 3, 2, if they continue as they have set in, with recurring twos and threes, may not be so summarily dealt with. In 1818 they first appeared in Red River, and devastated until 1820; and subsequently the years 1857, 1864, 1867, 1869, 1872, 1874, mark successive inroads of these grasshoppers, causing an agricultural gloom doubly heightened by the ravages of the canker-worms, disclosed in the following springs. Did they accommodate themselves to such numbers, one is tempted to ask, on the

* Kirby and Spence, p. 131.

cuneiform tablets of an Assurbanipal and Nebuchadnezzar, and
how does this affect our handy calendars and storm-drums?
There is one common defect observable in accounts given of locust
ravages, namely, an habitual disregard of the scientific name of the
noxious species observed. Thus, for instance, locusts reported to
have invaded Gibraltar and districts adjoining, on the 28th of
November, 1876, from specimens forwarded to the Entomological
Society turned out to be a *Decticus* (*D. albifrons*, Fab.) and not
locusts at all.

The morass with reeds and water-lilies, where dragon-flies
of lineage old emerge from the amber wave, becomes the parade-
ground of migration, from whence species rise in clouds and
wing over the country to form new settlements. In Prussia the
villagers in Anhalt were alarmed one clear summer afternoon,
about 4 p.m., by a passing procession of the common flat-bodied
Libuella depressa, that looked like locusts, and obscured the sun.
This well-watered district appears a rendezvous for these
rapacious insects, and in our fen districts the smaller vividly-
coloured species of *Agrion* have been similarly seen proceeding inland from the sea in numbers sufficient to cast a moving shadow
over the fields of Suffolk. At the end of June, 1867, the West of
Piedmont was invaded by a great number of *Anax Mediterraneus*,
De Selys, apparently brought by the wind from Africa. They
were noticed by Professor Craveri, at Bra, on the 18th of July, and
at Cuneo the 28th of the same month. On the 8th of August
they appeared at Turin and Venaria. Another flight of dragon-
flies arrived in the South of France in 1837, which, having
coupled, returned united to the sea. Many dragon-flies have
acquired a very wide distribution,* and these are " notorious wanderers."† Other migratory flocks of insects may be regarded as
due to an inordinate increase of their food, rather than to the
phenomena of love and rivalry. To this circumstance must be
attributed the autumnal showers of Lady-birds, Aphides, and
blight-devouring Hover-flies; the destroyers and their prey so
frequently wind-driven from hop-fields and cultivated lands. An

---

* *Pantala flavescens*, Fab., or *Tholymis tillarga*, Fab., for instance: or the
European *Libellula quadrimaculata*.

† Kirby and Spence, Lettr. XVI.; M'Lachlan, *Ent. Mon. Mag.*, 1873,
p. 273.

organ of scent is, I conceive, a means of terrestrial migration
with certain beetles, and this would likewise account for the
phenomena of gregarious and processionary caterpillars.

We have evidence, then, of the existence of constant laws
operating to maintain insect species as known to naturalists from
external characters of shape and colour, and can show they retain
such character as long as the laws remain in force. But we likewise witness disturbing causes distributing the species and producing superficial change ; and as it is evident internal structure
is dependent on external form, any variation in the latter must
affect the former. Thus when an insect is dwarfed, this can take
place, as in worker-bees, only at the expense of the reproductive
or other organs ; or when it becomes wingless from climatic
causes, the alar muscles by disuse tend to disappear and change
the shape of the body.

Dealing with geological chronology, the phenomena of generic
and specific variation should also be applicable in explanation of
certain plants and insects of constant character, being discovered
confined to various geological soils within the radius of their
distribution, or to favourite haunts postulating more than simple
dispersion from a centre. And the pale blue of butterflies frequenting limestone and chalky downs need evoke no interference
in the law of albinism if the honeyed cowslips and downy oxlips,
over whose leaves they flutter, are, as reputed by Linnæus and
Professor Henslow, specifically identical with the shadow-seeking
primrose, and may be raised from the same root.* So likewise the
local feature of melanism may be regarded as not only manufacturing annual varieties, but as pervading the black, brown, and
drab tribes of the Alpine, Arctic, and woodland faunas, and may
give a reason for their dark trait of beauty.

But in order that a scientifically defined species should become
diverse from, or identical with, another so as to cross freely, it is
necessary variation should not alone exist as it can be shown to
do in shape, including the form of the anal appendages, in colour
and organs of sound, but also in odours that allow of recognition
and prevent an undue mingling of kind. And when, divesting
ourselves of the inaccurate conventionalities of language, we are

* "British Flora," p. 105.

led to observe and ponder how truly contrary to all natural law and possible definition the proposition of two living things being wholly alike has ever been, we can but allow the rosy charm and dropping perfume to be things alike organically chemic and plastic in their nature. Nor does the classifier find his ideal *species* nearly so averse to uniting with their far-off affinities as might be supposed. Mules are freely obtained by German dealers and by those that possess the floral fancy, from the ruddy Poplar and Eyed Hawks of our parterres;* and even the swift-darting Vollucella Flies of the damp woodland, if we can credit a communication once made by Mons. Lapelletier de Saint Fargeau to the Académie des Sciences, unite very well together. In natural conditions diverse species of other and very various orders have been rarely found united; as in Italy a male of the common Two Spot Lady-bird with the variable *Coccinella dispar;* our bramble-frequenting Fritillary, *Adippe*, in Switzerland with its congener *Niobe;* and the leaf-rolling *Tortrix ribeana*, so common in dusty July, with its blunt-winged congener *cerasana*. Species also flirt and pair with their colour varieties, detecting the strangers doubtless by odour or music, and with the common Clouded Yellow and its variety *Helice* this appears a source of a race of sub-hybrids. Hermaphroditism, where the two halves of the insect on either side of the median line represent the opposite sexes, is frequent, and is found in all orders; or again, the two halves may represent type and variety. But such insects are lawless and monstrous.

Given climate and surroundings are the existing ministers of change that rule the filmy insect fabric; geology will reveal the past evidence of their endless operation in the oscillations of cosmogenic temperature, and in the relics that indicate the shifting of their life's drama from area to area. Let us suppose one of the great changes of climate recorded on the rocks to supervene, with sunken land areas and piercing cold, over the Palæarctic. Now, as the circles of insect distribution gradually move southward, the more northern varieties, such as the blue of Arthur's Seat, would establish themselves in the south of our

---

* *Deilephila vespertilio*, likewise, will cross with *D. euphorbiæ* and *D. hippophaes;* and of the Bombycina, *Saturnia pyri* with *S. spini* and *S. pavonia.*—Dr. Staudinger's Catalogue.

island, and the short or chill season would eliminate summer
forms of the Geometrical Moths from the brushwood. Or, on
the contrary, were the now submerged lands raised, and the
circles of distribution wheeled northward, we should meet the
orange variety of the Brimstone Butterfly flitting along our lanes,
brighter colours would begin to paint our larger-winged moths
among the shady lanes and on the moorland, and generally species
would appear in a greater number of annual broods, and exotic
species commence to take possession of the land.

Now, if we take into account the change of surrounding soils
and vegetation to which the foreigners would be exposed, it is
evident the little-understood Providence of protection by mimicry
would be disturbed, and the introduced species must vary in
harmony or suffer decimation ; or should the change in the subordinate and sustaining floral realm be regarded, larger or smaller
species would be produced, or colours, and probably secretions,
generally change. It is thus the selection of species will gain in
importance when viewed with regard to its application to unravel
enigmas that present themselves in the palæontology or history
of the European flora and fauna. And when we consider the
operation of these vicissitudes in soil, vegetation, and climate,
the incessant struggle for existence or domination, the migrations
and adaptations to which organic life must have been subjected
by the incessant interchange of sea and land during the formation
of the super-granitic crust of our globe, we are thus led to the
conception of existing kinds being climatic or protected forms of
others pre-existing, or in other words, the functions of a variable
environment and the exponents of atavism. Regarded in this
light, our present life is no more to be considered new, but as
having come to us rather the product of a world-old manufactory,
that has stamped with its die on plastic shapes many a curious
scutcheon and by-gone crest of glory, wherein the learned decipher
mechanically but genera and species, and where the schoolboy
sees but delight.

Neither are these views brain-spun, nor do we at all tread the
threshold of a vision. Fluctuations quite unhypothetical continue
in our day to be manifested by the changing seasons, and these
engrave the record of their visitations on the life in districts and
localities. English summers come to us unprecedentedly wet

or dry, and portions of the year vary ; and such evoke invariably the prognostications of insect prodigies, and cause marvellous tales to become current among entomological coteries. A wet season, they will tell you, has a tendency to delay and decrease the number of annual broods, to multiply certain caterpillars, and to produce sports in our fauna. The year 1879, a golden number of the Rocky Mountain Locust, proved proverbially disastrous and rainy in Europe; that brought out a many-pictured plethora of Drinker Moths, Gammas, and Magpie Moths all over the country, while a majority of insects scarcely put in an appearance. It also sent us the Painted Lady or Thistle Butterfly, with other of the African insect population, winging northward on the early sirocco, and pushed a pugnacious horde of locusts into the southern realms of the Czar. A cold and late spring in 1872 brought on a warm autumn that disseminated Camberwell Beauties and Queen of Spain Fritillaries out over the land, and the scarce pink-spotted moth, *Deiopea pulchella*, of Oriental affinities, just previously turned up in the September fallows. And so farther back again, 1860 was a year peculiarly devoid of summer weather, and then our rare Hawk Moths migrated freely to this island during the spring and autumn months, while irregularities in the appearance of insects were everywhere the subject of comment.

As far as would appear, a warm season dawns on us with like potents as the colder cycle; insects, however, are then more plentiful, and the species it fosters are generally different. Such a year was 1857, reported by the Registrar-General as quite unprecedented for brilliant weather and heat. The two Clouded Yellow Butterflies became plentiful on the coasts of Sussex and Kent, the commoner sort extending to the south-west corner of Scotland. Purple Emperors and Convolvulus Hawks were taken freely in this country. Another such year was 1868, unprecedented for the intensity and duration of its summer heat. And then Dr. Knaggs writes in the *Entomologists' Annual*, "The good old butterfly days appear to be returning "—Bath Whites, Camberwell Beauties, and Queens of Spain having put in a much stronger appearance than usual, while as for Pale Clouded Yellows, I suppose they have been never more abundant. The Convolvulus Hawk was also unusually abundant, and so were other rare Hawk Moths.

In the current cycle one last natural selection has begun to
operate of which man is the delegated agent, and the demesnes
of oak and heather, home of the Druid and Saxon, are fast
ceding their primitive mysteries before the steady march of
building, agriculture, horticulture, floriculture, and domestication of species; handmaidens of an era of civilisation, the
creative wonders of whose potent wand can scarcely compensate
an entomologist for the loss of his breezy heath and sylvan
shade where the Fritillaries once sunned, or the lonely classic

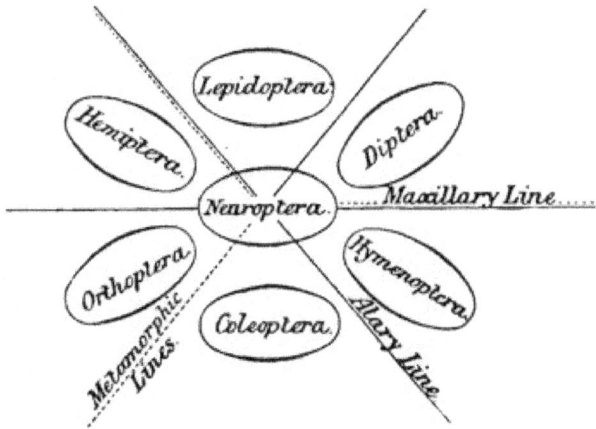

fen-land where the Large Copper Butterfly, *Chrysophanes Hip-
pothoë*, variety *Dispar*, once flew.

Edward Newman, in imitation of MacLeay, has left us an
ingenious circular plan (see above fig.) uniting the metamorphic,
alar, and maxillary systems of authors, that may also teach us something of the insensible descent of insects, and suggest a family tree.

The substance of a last contribution on the subject of metamorphosis by this lamented author is interesting. I abstract it
in substance from the *Magazine of Entomology* which he so long
conducted. Amorpha, including flies, butterflies, and moths, has
a chrysalis provided with neither mouth nor locomotive organs,
which appears to act in the capacity of a mould in the formation of the imago. The penultimate state of flies is different
in different families; in some it is an oblong smooth object,

and in others it approaches that of certain moths. Necromorpha, in which the pupa is provided with mouth, and organs of locomotion detached from the trunk throughout their length, but so swathed and enveloped in separate cases that it can employ neither; this group includes the bees and beetles. Isomorpha, which in all states is active and voracious, of similar form to the imago except as regards wings, including bugs and grasshoppers. Lastly, the Heteroptera, which, from its earliest situation, possesses characters of all the rest as well as some peculiar to itself. It includes the Stegoptera, which have a necromorphous pupa, and the dragon-flies, which have an isomorphous pupa.

It was held by Cuvier and Lamarck that the development of the nervous system corresponds with the degrees of intelligence in the same way as that of the circulatory system does; and that in successively reviewing the different families every organ simplifies by degrees, loses energy, and finishes by disappearing and confounding itself with the mass. Here, then, it would be natural to seek, as in Mollusca, characters in the organs of sensation, and the investigations undertaken to determine the existence of auditory structures detailed in this volume may serve to indicate a maximum of specific volition in certain groups of European insects and a rank in genera and class, although the results signally fail to elevate any one order in entirety, or to arrange them in linear precedence as one is accustomed to regard them in modern classification. Next to the auditory the ocular organ may be looked to for structural characters, and the investigations commenced by Johannes Müller, and carried on by others, may shed some further light when taken in conjunction with actual observation. So also in regard to the senses of smell and touch.

The presence of auditory organs and well-developed eyes, then, places the genera of grasshoppers and crickets representing the Orthoptera Saltatoria first in the list. These would be followed by Homoptera, represented by the genera of Cicadæ, which have the auditory organs greatly developed in the males, but with whom sight appears less potent. Next to these appear to rank the moths of the Noctuina, Bombycina, and a few of the Geometrina, which have complex auditory organs,

and may be taken to represent the Nocturni of the Lepidoptera.
The Diurni do not appear to have any great power of hearing,
but they all possess excellent optic organs and sharp sight.
After the Orthoptera, Hemiptera, Homoptera, and Lepidoptera
would follow the Coleoptera, which certainly give evidence of
possessing an auditory apparatus in certain groups—Lamellicornia, Longicornia, and Malacodermata, represented by the
Death Watch Beetles. But here in the Lamellicorns the visual
organs are imperfect. On the other hand, the species of
Hymenoptera, Neuroptera, and Diptera have the auditory sense,
if present, less potent, but sight, touch, and smell are generally
manifest. In bees and flies the sense of smell is especially
keen, and dragon-flies have enormously developed eyes. These
deductions are in harmony with an exposition of Newman's
table from left to right, thus:—

1. Orthoptera.		1. Hemiptera.
2. Coleoptera.	2*. Neuroptera.	2. Lepidoptera.
3. Hymenoptera.		3. Diptera.

Insects, there is a presumption, originated in Neuroptera.

A TABLE OF REMARKABLE CYCLES OF INSECT MULTIPLICATION
AND MIGRATION.

*Anni Mirabiles marked thus on a provisional quinary method (*), and
apparent dislocations denoted by a (†).*

A.D. 591. An infinite army of locusts grievously ravage part of Italy.

1478. Locusts invade the Venetian territory and produce a famine.

1556. A swarm of butterflies rain blood in Germany.

*1574. A number of chafers fall into the Severn, and stop the wheels of the
water-mills ?

1608. Blood prodigy at Aix.

1613. Locusts visit the South of France.

1650. A cloud of locusts enter Russia in three different places, and from
thence pass into Poland and Lithuania.

1665. *Tomicus typographus*, the "Turk," multiplies, and ravages the Hartz
Forest.

1688. Chafers strip the trees in the county of Galway.

1710. The Scarce Swallow-tail Butterfly taken in England prior to this date.

1730. The Gold-tail Moth and Gipsy multiply, and lay waste the foliage
in France for three years.

1731. Wet season. A swarm of the scarce chafer *Melolontha Fullo*, L., appears at Mark Brandenburg.

1735. Bath White Butterfly captured in England prior to this date.

1740. Cold spring, dry summer. The moth *Charæas graminis* multiplies, and ravages in Sweden for three years.

1741. A migration of Painted Lady Butterflies observed at Turin.

*1747. A legion of locusts invade Wallachia, Moldavia, Transylvania, Hungary, and Poland. *C. graminis* again destroys the grass in Sweden. Both pests continue until '50; and in '48 the locusts reach England, but soon perish.

1753. Locusts devastate in Portugal.

*1757. The Turk attacks the trees in the Hartz Forest.

*1759. Caterpillar destroys the grass of the high sheep farms in Tweeddale. '68, wet season.

1769. The "Turk" attacks the trees in the Hartz Forest.

1776. Thames frozen. Hessian Fly begins to destroy the wheat in Long Island.

1778. Morocco devastated by locusts for three years. *P. podalirius*, the Scarce Swallow-tail Butterfly, taken in England (?)

1780. Locusts invade Transylvania. A grub noxious in France. '81, hot dry summer.

*1782. The Gold-tail Moth multiplies in the vicinity of the metropolis; rewards offered for collecting the caterpillars. Nigger Saw Fly (*Athalia*) attacks turnip crops for two years.

1783. Intense heat in summer, atmospheric and volcanic phenomena in Europe. The ravages of the "Turk" beetle in the Hartz Forest attain a climax.

*1784. Locusts ravage at the Cape, and being driven into the sea, cease their depredations for the next ten years.

1785. Autumn rainy. Cockchafers multiply near Blois in France. Aphides and ants numerous. *Tortrix viridana* strips oak woods in England.

*1787. Cold season. American blight naturalised.

1788. Hessian Fly destructive in America.

1790. The "Turk" appears in the Hartz Forest. Gilbert White notices an abundance of cockroaches and crickets.

1795. The Scarce Swallow-tail captured in England previous to this date.

1802. *C. graminis* (?) infests the high sheep farms in Tweeddale. Beetle grub destroys wheat.

1803. Two specimens of the Bath White Butterfly captured in June. *P. podalirius* during May.

*1804. The Clouded Yellow Butterfly (*C. Edusa*) abundant in England.

1805. A cloud of locusts in Southern France.

*1807. Larvæ of Scarce Swallow-tail Butterfly in England.

1808. The Clouded Yellow Butterfly abundant in England.

1810. *P. podalirius*, Aug. 24th. Black-veined White Butterfly destructive round London.

1811. The Clouded Yellow Butterfly abundant.

1813. *Zabrus gibbus* ravages the wheat in Germany. Wire Worm destructive in England.

1815 to 1818. Cold and wet.

1818. Bath White Butterfly taken in July; *Podalirius*, July 9th; Painted

Lady Butterflies abundant after wet spring. *D. lineata* plentiful in Switzerland, near Berne. The Swallow-tail Butterfly (*P. Machaon*) disappears from Glanville's Wootton.

*1819. Bath White and Queen of Spain Butterflies (*Vanessa Antiopa*) in England. *A. Niobe* (?), and *D. nerii*, the Oleander Hawk Moth, near Southampton.

1821. Pale Clouded Yellow Butterfly (*Colias Hyale*) common in England. *C. graminis* swarms at Meldon Park.

*1822. Caterpillar of *P. podalirius* taken in England. *M. Artemis* abundant at Glanville's Wootton; not seen there since '15.

1823. *Sphinx lineata* at Sunderland. Cockchafers abundant near London.

*1824. Locusts visit the South of France this and also the succeeding year. *C. graminis* destroys the grass on Skiddaw.

1825. Painted Lady Butterflies and Orange Tips (*A. Cardamines*) abundant. Gen. Chrysomela plentiful.

†1826. Mr. Dale's Annus Mirabilis. Clouded Yellow Butterflies (*C. Edusa*) disseminate themselves over England; Hyale also common. A migration of Painted Lady Butterflies at Turin. Caterpillars of the Lackey Ermine Moth and *H. padellus* abundant on the hedges. Black-veined White Butterfly (*A. Crataegi*) plentiful in Hampshire. Lady-birds. Bath White Butterfly taken Aug. 6th, another in '27.

*1827. The Green Tortrix destroys the oak-leaves in Kent. Death's Heads and Privet Hawk Moths. Black Adippe, Sutton Park.

1828. A cloud of Painted Lady Butterflies on the banks of the Lake of Neufchâtel, in Switzerland, during July, passing from N.E. to S.W. Unfavourable year in England. Pale Clouded Yellow Butterflies common. Cockchafers plentiful seven years since. *R. solstitialis* appear in numbers. *V. Hunteri* captured in South Wales. Stinging Fly, *Stom. calcitrans*, abundant.

*1829. Spring cold and wet. Larvae of *P. podalirius* taken.

1830. Warm spring. Pale Clouded Yellows plentiful in England; caterpillars of *H. padella* in millions on the hedges. *D. hippophaes* plentiful on the banks of the Arve. The Comma (*V. C. album*) disappeared from Dover twelve or thirteen years back.

1831. A flock of cockchafers startle the horses of the diligence proceeding from Gournay to Gisors, in France. Clouded Yellow Butterflies plentiful in England. *V. C. album* taken at Dover in November, and Antiopa in August. *D. lineata* in July. Dark varieties of *M. Artemis* taken; this butterfly has now disappeared from Willesden.

*1832. Mild winter, late spring. A migration of locusts at Marseilles, and of White Butterflies at Paris. Lineata taken in July, the caterpillar of Nerii in August, in Devon. Queens of Spain (*Pieris Daplidice*) and Camberwell Beauties in England, the latter extending to Northumberland. Several *D. Galii* taken, one at York. Fifteen or sixteen years since *V. C. album* was in profusion at Epping.

1833. The Clouded Yellows, Edusa and Hyale, plentiful in England. *V. C. album* taken fourteen or sixteen years ago near Coventry, many years ago at Norfolk. Gamma Moths and butterflies plentiful in Jersey.

*1834. A migration of locusts in Southern France. *D. Galii* plentiful near Berne. The Oleander Hawk taken in England.

1835. The Oleander Hawk with its congener Celerio occurs all over France and Germany. Both Clouded Yellow Butterflies common in England. The

Migratory Locust taken in Ireland. Several *D. Galii* bred from caterpillars. A few Bath White Butterflies in England.

†1836. A migrating column of the Painted Lady Butterfly observed in the canton of Vaud, Switzerland. *Argynnis aphrodite*, a North American butterfly, taken in Warwickshire.

*1837. A migration of Dragon-flies in Southern France in September that darkened the air and loaded the plants. Camberwell Beauty taken near London.

*1839. Clouded Yellows common, caterpillars plentiful.

1840. Extraordinary number of leaf-crickets at Saint-Geniz le Bas, in the South of France, at the end of the spring. Caterpillars ravage at Odessa.

1841. Locusts appear during the month of May in Spain in quantities; said to have migrated from Africa.

*1842. A vast flight of White Butterflies (*Pieris Brassicæ*) from the Continent to the coast about Dover, and of the Painted Lady (*Vanessa cardui*) at Corfu, in the direction of Otranto. The Pale Clouded Yellow Butterfly (*Colias Hyale*) abundant, and spreads over England as far as Derbyshire and Yorkshire.

*1844. The Clouded Yellow plentiful in England.

1845. Locusts become a plague in Algeria.

†1846. "Remarkable Year." Mild winter. The Queen of Spain and Camberwell Beauty butterflies abound; all the rare Hawk Moths taken in numbers in this country during the autumn. A flock of White Butterflies crosses the Channel; and a cloud of locusts passes over Banff, Moray, and the Zetland Islands in August. *L. trifolii* at the Land's End. *C. Fraxini* taken. General migration.

*1847. A few of the insects of the previous year turn up in places. Pale Clouded Yellow Butterfly common.

1848. The Clouded Yellow Butterfly captured in Scotland, otherwise rare.

*1849. A few Pale Clouded Yellows. Dia's Fritillary said to have been taken near Birmingham.

*1852. The Clouded Yellow (*Colias Edusa*) extends to Scotland. The Oleander Hawk taken at Brighton. The Convolvulus Hawk Moth common.

1856. Hair Streak Butterflies and other Lepidoptera common.

*1857. Hot summer. The two Clouded Yellow Butterflies plentiful. Hyale on the coast of Sussex and Kent, Edusa reaching the south-west of Scotland. Camberwell Beauties and Convolvulus Hawks. The Rocky Mountain Locust devastates in America. Oleander Hawk taken at Brighton. The Large Copper Butterfly has disappeared from our fens previous to this date.

1858. The Black-veined White Butterfly multiplies in England. Camberwell Beauty taken in the autumn. Edusa in the North of England.

*1859. The Clouded Yellow Butterfly multiplies in England. Bath White (*P. Daplidice*) common.

1860. Hawk Moths. *Deilephila Livornica* in the spring and *Chærocampa celerio* in the autumn. A few Convolvulus Hawks and *O. lunaris*. The caterpillar of the Oleander Hawk found. The variety Cleopatra of the Brimstone Butterfly taken near Rotherham.

*1862. Edusa extends to Scotland. Nerii taken. Several Celerio and Livornica.

1863. A few of the migrating Hawk Moths.

*1864. Rocky Mountain Locusts devastate the Western States of America. Two Queen of Spain Butterflies (*A. Lathonia*) and two Celerio Hawks.

V

1865. Death's Head Moths, Humming Bird Hawks, and Gamma Moths abundant. A few Camberwell Beauties and Queen of Spain Fritillaries. *C. celerio* frequent, and a few other rare Hawk Moths. Edusa and Hyale both common.

1866. Locusts become a plague in Algeria. *E. Cassiope* at a lower elevation in Perthshire than usual.

*1867. One or two specimens of Lineata and Celerio and an Oleander Hawk taken in England. At the end of July a migrating cloud of Dragon-flies, *Anax Mediterraneus*, seen in the East of Piedmont. Rocky Mountain Locusts descend on the United States.

†1868. Hot summer. A swarm of White Butterflies noticed in the Channel at the end of July. Pale Clouded Yellow Butterfly and Queen of Spain Fritillary appear, the former abounding all over England. The Camberwell Beauty likewise taken. The scarce blue *Polyommatus Arion* turned up in new localities. The rare Hawk Moths all captured, *C. lineata* perhaps most freely. In North America, *Danais Archippus* migrates in September.

*1869. White and other butterflies scarce. *Deiopeia pulchella* in October. Rocky Mountain Locusts descend on the western plains of America. Painted Lady Butterfly common in Italy. Numerous melanic and pale varieties taken in this country.

1870. The scarce Hawk Moth, *D. Livornica*, appears all over the United Kingdom. *C. Edusa* common in Italy. *Danais Archippus* migrates in North America about the middle of April.

1871. The spring cold. Then caterpillars of the Lackey (*C. Neustria*) and of the Brown (?) Tail abound in certain districts. *Deiopeia pulchella*, of Eastern affinities, becomes common in Italy and England. Gamma Moths common in Italy. *Chionobas Aello*, a butterfly of Arctic affinity, appears in the Eastern Alps.

*1872. Cold summer. Then appear the Bath Whites, Queens of Spain, and Camberwell Beauties, the last unusually plentiful and extending to Scotland. A flock of Large White Butterflies seen at the mouth of the Weser. The Rocky Mountain Locusts descend on the Western States of America. *S. pinastri* and *Argynnis Dia* in England.

1873. The Camberwell Beauty appears to recede south, a few spring specimens being taken in the north, but all records of its occurrence in the autumn coming from the south of England. Entomologists occupied with discussing the probability of *Argynnis Ilia* and *Niobe* being taken in this country. *C. Aello* on the Alps. *Ophiodes lunaris* taken in England.

*1874. Wet June. Inundation of the Garonne. Rocky Mountain Locusts invade the Western States of America. *D. pulchella* at Folkestone.

1875. Convolvulus Hawk Moth common. Edusa and Hyale abundant. *C. Aello* taken on the Alps.

1876. Locusts multiply in Abyssinia and in South America, where they combine with hail-storms in committing havoc. In November the Leaf-cricket, *Decticus albifrons*, multiplies in Southern Spain. A few locusts turn up in Yorkshire during the autumn, and three specimens of *Danais Archippus* taken in the south of England, supposed to have come from America. Several *D. pulchella* and *S. pinastri* taken at Ipswich.

*1877. Clouded Yellow Butterflies become abundant, and extend to Scotland. *V. Antiopa, S. convolvuli* and *pinastri* taken in England. Yellow variety of *Zyg. filipendulæ* at Cambridge. Caterpillars ravage the island of Ascension. *P. Hunteri* taken in England this and the previous year.

1878. *A. Cratœgi* and *C. cardui* common in Italy. *Chionobas Aello* taken on the Alps. *S. pinastri* in England. The Large Copper taken at Cromer in August.

*1879. Wet season. Many insects do not appear; others are retarded. A migration of Painted Lady Butterflies in April from Algeria through Spain and Italy to Central France and Germany. Gamma Moths swarm in Europe. A cloud of small beetles (*Galeruca capreæ*) in Argyleshire. *C. Hera* in the Isle of Wight. *S. pinastri.* Varieties noticed. Various ravages of grubs and larvæ.

NOTE.—As regards Continental areas, the butterflies *P. Podalirius, V. Xanthomelas, Melitæa maturna, P. Maera* and *Achine* have deserted the environs of Berlin; and the Hawk Moths, *Livornica, Celerio,* and *Nerii,* only go thither when migrating. (Jul. Pfützner; Syst. Verzeich.)

This Table is mainly compiled from papers by the Rev. J. Brec. Mr. Dale, and Messrs. H. T. Stainton, Dr. Knaggs, and others, contained in the *Ann. and Mag. Nat. Hist. ; Entomologist ; Zoologist ; Entomologists' Annual ; Ent. Mon. Mag.,* and some other periodicals.

LIST OF SUBSCRIBERS.

	No. of Copies.
BANSALL, W. H., 4, Woodfield Terrace, Harrow Road	1
BARRETT, C. G., Pembroke, Wales	1
BARRINGTON BROWNE, Mrs. W., Highfield, Guildford	1
BOND, FRED., F.Z.S., 5, Fairfield Avenue, Staines	1
BURGEAS, ED., Boston Nat. Hist. Society, Boston, Mass., U.S.A. ...	1
CALLANDER, Colonel, Bareilly, India.	2
CALLANDER, Miss, Dartington Hall, Devonshire	1
CAMBRIDGE ENT. Soc., The, Jesus College, Cambridge	1
CAMPBELL, Captain, H.M.I.N., Oak Side, Box Grove Road, Guildford	1
CHAMPERNOWNE, A.F.G.S., Dartington Hall, Devonshire	2
CHOLMONDELY, The Marquis, Cholmondely Castle, Cheshire	3
DICKINSON, F. H., F.G.S., Kingweston, Somerset	1
DISTANT, W. L., 1, Selston Villas, Derwent Grove, East Dulwich	1
DONALD, Miss E., Finnart, Greenock, N.B.	6
DUNLOP, Mrs., 25, Randolph Crescent, Maida Hill	1
DUNNING, F.L.S., F.Z.S., 12, Old Square, Lincoln's Inn, W.C. ...	1
FLETCHER, J. E., Happy Land, Worcester	1
FLETCHER, W. H. B., Bersted Lodge, Bognor, Sussex ...	1
FRY, Mrs. A., Thornhill House, Dulwich Wood Park, Norwood ...	1
GALBRAITH, Dr., Southfield, Stirling, N.B.	1
GAMAN, Rev., Warsash, Hampshire	1
GIBB, Miss H. D., 226, St. Vincent Street, Glasgow ...	3
GIBB, Mons. A., 30, Allée de Garonne, Toulouse	1
HAGGERSTON ENT. Soc., The, 10, Brownlow Street, Dalston	1
HEPBURN, Major A. Buchan, 26, Punjáb N.I., B.S.C., Rawulpindi, Punjáb, India	1
HUTCHESON, Miss, Orchard, Renfrew, Scotland	2
KIDD, H. W., Godalming, Surrey	1
LIGARD, A., 2, Woodstock Road, Oxford	1
LOCKYER, B., 27, King Street, Covent Garden, W.C.	1
MATHEW, E. W., Wern, Stoke next Guildford, Surrey	3
MORLAND, H., Cranford, Hounslow	1

	No. of Copies.
PETRIE, W., 8, Crescent Road, Bromley, London, S.E. ...	6
PIGGOTT, RICHARD, Reigate, Surrey	1
REYNOLDS, Miss, Blenheim Lodge, Guildford	1
ROUTLEDGE, THOS., F.L.S., Claxheugh, Sunderland...	1
SCUDDER, Dr. SAM., Harvard College Library, Camb., Mass., U.S.A. ...	1
SHEPHARD, E., 21, Albert Terrace, Clapham Road, S.W.	1
STANDON, R. S., Holmwood Lodge, Surbiton	1
SWINTON, A. CAMPBELL, Kimmerghame, Dunse, N.B.	3
SWINTON, Colonel, Milton House, Woolston, Southampton	1
SWINTON, Miss, Warsash, Hampshire	1
TAIT, P. M., F.S.S., F.R.G.S., Mansion House Buildings, E.C.	2
TAYLOR, J. E., F.L.S., F.G.S., Stanley House, Burlington Road, Ipswich	1
WEMYSS, Gen., Highfield, Guildford	1
WEMYSS, Miss, Highfield, Guildford	1
WEMYSS, Miss B., Highfield, Guildford	1
WESTERN, E., 8, Craven Hill, London, W.	1
WOODS, E., Netherton, Waterden Road, Guildford	1

CASSELL, PETTER, GALPIN & CO., BELLE SAUVAGE WORKS, LONDON, E.C.

www.ingramcontent.com/pod-product-compliance
Lightning Source LLC
Chambersburg PA
CBHW021403210326
41599CB00011B/990

9 783337 088187